THE QUADRATIC FORMULA

The roots of the quadratic equation $ax^2 + bx + c = 0$ are given by

$$x = \frac{-b \pm \sqrt{b^2 - 4ac}}{2a}$$

DETERMINANTS

$$\begin{vmatrix} a_1 & b_1 \\ a_2 & b_2 \end{vmatrix} = a_1b_2 - a_2b_1$$

Cramer's rule: The solution to the system

$$\begin{cases} a_1x + b_1y = c_1 \\ a_2x + b_2y = c_2 \end{cases} \text{ is } x = \frac{\begin{vmatrix} c_1 & b_1 \\ c_2 & b_2 \end{vmatrix}}{\begin{vmatrix} a_1 & b_1 \\ a_2 & b_2 \end{vmatrix}}; \quad y = \frac{\begin{vmatrix} a_1 & c_1 \\ a_2 & c_2 \end{vmatrix}}{\begin{vmatrix} a_1 & b_1 \\ a_2 & b_2 \end{vmatrix}}$$

LAWS OF LOGARITHMS (where $\log_b x = y$ means $x = b^y$)

$$\log_b(M \cdot N) = \log_b M + \log_b N, \quad \log_b\left(\frac{M}{N}\right) = \log_b M - \log_b N$$

$$\log_b(M^x) = x \cdot \log_b M,$$

ANALYTIC GEOMETRY

Straight line: slope $= m = \dfrac{y_2 - y_1}{x_2 - x_1}$; $\quad y = mx + b$; $\quad y - y_1 = m(x - x_1)$

Circle: $\quad x^2 + y^2 = r^2$; $\quad (x - h)^2 + (y - k)^2 = r^2$

Parabola:
- *Vertical axis:* $\quad x^2 = 4ay$; $\quad (x - h)^2 = 4a(y - k)$
- *Horizontal axis:* $\quad y^2 = 4ax$; $\quad (y - k)^2 = 4a(x - h)$

Ellipse:
- *Vert. major axis:* $\quad \dfrac{x^2}{b^2} + \dfrac{y^2}{a^2} = 1$; $\quad \dfrac{(x - h)^2}{b^2} + \dfrac{(y - k)^2}{a^2} = 1$
- *Hor. major axis:* $\quad \dfrac{x^2}{a^2} + \dfrac{y^2}{b^2} = 1$; $\quad \dfrac{(x - h)^2}{a^2} + \dfrac{(y - k)^2}{b^2} = 1$

Hyperbola:
- *Vert. transverse axis:* $\quad \dfrac{y^2}{a^2} - \dfrac{x^2}{b^2} = 1$; $\quad \dfrac{(y - k)^2}{a^2} - \dfrac{(x - h)^2}{b^2} = 1$
- *Hor. transverse axis:* $\quad \dfrac{x^2}{a^2} - \dfrac{y^2}{b^2} = 1$; $\quad \dfrac{(x - h)^2}{a^2} - \dfrac{(y - k)^2}{b^2} = 1$

Ordinary Differential Equations
with Applications

SECOND EDITION

Ordinary Differential Equations

with Applications

SECOND EDITION

Bernard J. Rice
Jerry D. Strange
University of Dayton

Brooks/Cole Publishing Company
Pacific Grove, California

Brooks/Cole Publishing Company
A Division of Wadsworth, Inc.

© 1989, 1986 by Wadsworth, Inc., Belmont, California 94002. All rights reserved. No part of this book may be reproduced, stored in a retrieval system, or transcribed, in any form or by any means—electronic, mechanical, photocopying, recording, or otherwise—without the prior written permission of the publisher, Brooks/Cole Publishing Company, Pacific Grove, California 93950, a division of Wadsworth, Inc.

Printed in the United States of America

10 9 8 7 6 5 4 3 2 1

Library of Congress Cataloging-in-Publication Data
Rice, Bernard J.
    Ordinary differential equations with applications / Bernard J.
Rice, Jerry D. Strange. —2nd ed.
        p.   cm.
    Includes index.
    ISBN 0-534-09906-8
    1. Differential equations.   I. Strange, Jerry D.   II. Title.
QA372.R55 1988
515.3′52—dc19   88-21725
                                                                   CIP

Sponsoring Editors: *Sue Ewing, Jeremy Hayhurst*
Editorial Assistants: *Heidi Wieland, Virge Kelmser*
Production Editor: *Phyllis Larimore*
Manuscript Editors: *Charles Cox, Phyllis Larimore*
Interior and Cover Design: *Kelly Shoemaker*
Cover Photo: *Al Satterwhite, courtesy of The Image Bank West*
Art Coordinator: *Sue C. Howard*
Typesetting: *Weimer Typesetting Co., Inc.*
Cover Printing: *The Lehigh Press, Inc.*
Printing and Binding: *The Maple-Vail Book Manufacturing Group, Binghamton*

Preface

This text is designed for a one-term course in ordinary differential equations that emphasizes solution techniques and applications. Our aim is to provide a text that is understandable for beginning students, particularly those interested in applications. *Ordinary Differential Equations with Applications*, Second Edition, has the following specific strengths:

- Selected topics from calculus, such as integration by parts and partial fractions, are reviewed in the text as required.

- Integration techniques required to solve the exercises are kept relatively simple so the student can concentrate on the solution process.

- Applications are distributed somewhat uniformly throughout the text so that most equation categories have an identifiable application area.

- The applications can be understood by anyone with an elementary physics background.

- Systems of differential equations are solved both with and without linear algebra.

- The answers to all exercises are provided.

Chapters 1 through 8 of this edition of *Ordinary Differential Equations with Applications* are, with minor exceptions, the same as in the first edition. Additional exercises have been inserted into some exercise sets and summaries of important procedures have been added, but the basic presentation is intact. The major change in the book, and the motivation for this edition, is the inclusion in Chapter 9 of four new sections on the use of matrix methods to solve systems of differential equations.

A unique feature of our text is that we show two alternative approaches to using the method of undetermined coefficients in the solution of linear, nonhomogeneous differential equations with constant coefficients. These two approaches are presented in Sections 5.2 and 5.2*; the first uses the method of differential families, the second that of polynomial operator annihilators. Each of these sections stands independently and only one of them needs to be discussed, with the remainder of the book flowing easily from either section.

As with most textbooks, there are more topics included than can be covered in one term. The book can be used for a variety of courses by choosing those topics that meet individual course needs. For instance, not all of the applications need to be discussed. The Laplace transform is presented in Chapters 6 and 7. Chapter 7 can be deleted in its entirety by those who wish to limit this coverage. Many other options of this type are included.

We have highlighted numerical techniques by placing them early in the book—in Chapter 3—and we provide computer programs, in the form of flow diagrams, for the appropriate numerical techniques. This chapter stands as a separate unit and can be covered at any convenient time.

We wish to thank the many users of the first edition who offered suggestions for improvement and our colleagues at the University of Dayton for their encouragement. In particular, we would like to thank Carroll Schleppi for preparing a solutions manual, helping us improve the exercise sets, and correcting the errors in the answers.

We also wish to thank those who reviewed the manuscript for the second edition: Philip Crooke, Vanderbilt University; David Horowitz, Golden West College; Samih Obaid, San Jose State University; Monty Strauss, Texas Tech University; and John Willemain, Western New England College.

Special thanks to our editors Jeremy Hayhurst and Sue Ewing for their assistance in completing this project. Thanks also to Phyllis Larimore and the production staff at Brooks/Cole for getting the book through the production process so successfully.

Bernard J. Rice
Jerry D. Strange

Contents

4 Homogeneous Linear Differential Equations 127

5 Nonhomogeneous Linear Differential Equations 169

6 The Laplace Transform 219

7 More Laplace Transforms 251

Ordinary Differential Equations

with Applications

SECOND EDITION

1

Differential Equations

What does the suspension system of an automobile have in common with an electric circuit? How long will it take a polluted lake to return to its natural state once man-made pollution is stopped? What is the shape of a power line hanging between two utility poles? Does the quantity of fuel burned by a rocket affect its velocity? The answers to these questions are obtained by using the techniques described in this book. If you are interested in the answers to these questions, you should find the study of differential equations both useful and rewarding.

Equations that contain derivatives or differentials of one or more functions are called **differential equations.** The study of differential equations originated in the investigation of laws that govern the physical world and were first solved by Sir Isaac Newton (1642–1727), who referred to them as "fluxional" equations. The term "differential" equation was introduced by Gottfried Leibniz (1646–1716) who, along with Newton, is credited with inventing the calculus. Many of the techniques for solving differential equations were known to mathematicians of the seventeenth century, but it was not until the nineteenth century that Augustin-Louis Cauchy (1789–1857) developed a general theory for differential equations that was independent of physical phenomena.

The three aspects of the study of differential equations—theory, methodology, and application—are treated in this book with the emphasis on methodology and application. The purpose of this chapter is to set the stage for the specific methods and applications that follow.

1–1   SOME TERMINOLOGY

As indicated above, a differential equation is an equation that contains a derivative or a differential of one or more functions. The following are some examples of differential equations:

1. $\dfrac{dy}{dx} = e^{2x}$

2. $dy = (y^2 + x)\, dx$

3. $y'' + 4y = (x^2 + 1)^3$

4. $y'' - (y')^3 + y = \sin x$

5. $x^2\, dy + 3y\, dx = 25\, dx$

6. $yy' + x = 3$

7. $\dfrac{\partial u}{\partial x} + \dfrac{\partial u}{\partial y} = 0$

8. $\dfrac{\partial^2 u}{\partial t^2} + \dfrac{\partial^2 u}{\partial x^2} = t + x$

Classification of Differential Equations

Differential equations are classified in a number of ways. If the unknown function in the equation depends upon only one independent variable, the equation is called an **ordinary** differential equation. Equations 1 through 6 are examples of ordinary differential equations in which y represents the unknown function (or dependent variable) and x represents the independent variable. If the unknown function depends upon two or more independent variables, the equation is called a **partial** differential equation. Equations 7 and 8 are examples of partial differential equations. In this book we shall restrict our discussion to ordinary differential equations.

Differential equations are further classified according to the order of the highest derivative appearing in the equation.

DEFINITION

> The **order** of a differential equation is the order of the highest derivative appearing in the equation.

COMMENT: Notice that Equations 3, 4, and 8 are second-order differential equations and the others are first-order.

The general form of a first-order differential equation is

$$F(x, y, y') = 0$$

of a second-order differential equation is

$$F(x, y, y', y'') = 0$$

and of an nth-order differential equation is

$$F(x, y, y', \ldots, y^{(n)}) = 0 \qquad \left(\text{NOTE: } y^{(n)} = \frac{d^n y}{dx^n}. \right)$$

where F is a real function of the arguments x, y, y', $\cdots$, $y^{(n)}$. An important subclass of this general form is given in the next definition.

DEFINITION

> An nth-order ordinary differential equation is said to be **linear in y** if it can be written in the form
>
> $$a_n(x)y^{(n)} + a_{n-1}(x)y^{(n-1)} + \cdots + a_1(x)y' + a_0(x)y = f(x)$$
>
> where $a_0, a_1, \ldots, a_n$ and f are functions on some interval of x, and $a_n(x) \neq 0$ on that interval. The functions $a_k(x)$ are called the **coefficient functions.**

The definition implies that an ordinary differential equation is linear if the following conditions are met:

- The unknown function and its derivatives algebraically occur to the first degree only.

- There are no products involving either the unknown function and its derivatives or two or more derivatives.

- There are no transcendental functions involving the unknown function or any of its derivatives.

A differential equation that is not linear is said to be **nonlinear.**

Example 1

(a) The following ordinary differential equations are linear:

$$y'' - 3y' + 3y = x^3$$
$$xy''' + ye^x + 5 = 0$$

Notice that the manner in which the independent variable enters the equation has nothing to do with the linear classification.

(b) The ordinary first-order differential equation

$$(y')^3 + 2y = x$$

is nonlinear because the first derivative of the unknown function is to the third degree.

(c) The first-order equation

$$yy' + 3x = 0$$

is nonlinear because yy' involves the product of the unknown function and its derivative.

(d) The second-order equation

$$y'' + 5y = \cos y$$

is nonlinear because $\cos y$ is a transcendental function of the unknown function.

(e) The first-order differential equation

$$y^2 \, dx + x \, dy = \sin y \, dy$$

is nonlinear in the function y, but it is linear if we reverse the roles of x and y and think of x as an unknown function of the variable y. The given equation may be expressed in the form

$$y^2 \frac{dx}{dy} + x = \sin y$$

which is linear in x.   ■

The Origin and Application of Differential Equations

One of the reasons we study mathematics is that it gives us a precise language for describing many of the physical laws and processes of the real world. For instance, the fact that the product of the pressure P and the corresponding volume V of an ideal gas is constant is represented by the mathematical expression $PV = k$. We say that $PV = k$ is a **mathematical model** of the pressure/volume relationship. To construct a mathematical model of a real-world condition, the important variables must be identified and the relationship between them established. Of particular interest to us at this time is the process of constructing mathematical models that involve derivatives.

Recall from differential calculus that the derivative of a function can alternately be interpreted as the slope of the graph of the function (see Figure 1–1) or as the rate of change of one variable with respect to the other. Thus if y is known to be a function of x, then y' is the slope of the graph of the function or, equivalently, the rate of change of y with respect to x. As you will see, both of these interpretations of y' are used in constructing mathematical models.

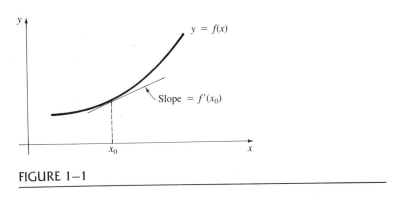

FIGURE 1–1

In this section our purpose is to show a few examples of how differential equations arise in describing real-world situations; we leave the problem of solving differential equations, whatever that may mean, to subsequent chapters.

Growth and Decay Problems

Many physical situations are concerned with either the growth or the decay of a quantity. Experimental data suggests that in some of these problems the rate of change of the quantity with respect to time is directly proportional to the measure of the quantity itself. In these problems we use the derivative to represent the time-rate of change of the quantity. As the next example shows, growth and decay problems are modeled by first-order differential equations.

In most cases a mathematical model is only an approximation of the physical condition being studied. For instance, we know from physics that radioactive elements decay by randomly emitting particles, each of which has a discrete mass. Figure 1–2 shows a graph of the disintegration of the mass of a typical radioactive substance. The discontinuous nature of the graph indicates that for a specified increment of time the mass decreases by discrete amounts; the fact that some jumps are larger than others is intended to reflect the random way in which particles are emitted. The model of radioactive decay developed in Example 2 assumes that the disintegration process is a continuous function of time. Continuous functions are frequently used to represent discrete processes because they tend to simplify the model and ensure that the derivative of the function exists everywhere. Of course, as indicated in Figure 1–2, the continuous nature of the model introduces some error.

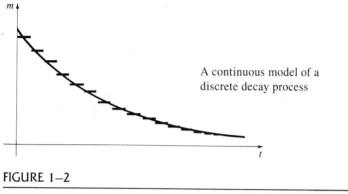

A continuous model of a
discrete decay process

FIGURE 1–2

A typical decay process

Example 2

The rate of decomposition of a radioactive substance is known to be directly proportional to the remaining mass of the substance. Write a differential equation that describes this decay process.

SOLUTION   Let m represent the mass of the radioactive substance at any time t. Then the time-rate at which the substance is decomposing can be represented by dm/dt. We can now describe the fact that the rate of decay of the substance is directly proportional to its remaining mass by

$$\frac{dm}{dt} \propto m$$

Finally, introducing a constant of proportionality k, the desired model is the first-order differential equation

$$\frac{dm}{dt} = km$$

The relationship between the unconverted mass and elapsed time is found by solving this differential equation.   ■

Supply and Demand

Suppose that a company is about to market a new product and wishes to develop a model to describe the behavior of the price of the product. A common economic model for this purpose assumes that the time-rate of change of the price of a product is directly proportional to the difference in the demand and the supply of the product. Conceptually, the model assumes that if the demand exceeds the supply, the price will go up; and if the supply exceeds the demand, the price will go down.

Example 3

Write an economic model for the price of a product under the assumption that the time-rate of change of price is directly proportional to the difference in the demand and the supply of the product.

SOLUTION   We let P represent the price of the product at any time t. Then if D is the demand for the product and S is the supply, the desired model is

$$\frac{dP}{dt} = k(D - S)$$

The relation for P is found by solving this first-order differential equation.   ∎

The Differential Equation of a Family of Curves

A relation between x and y can be represented by a graph in the xy-plane. For instance, the relation $x^2 + y^2 = 4$ describes a circle with its center at the origin and radius 2. More generally, the relation $x^2 + y^2 = c^2$, where c is an unspecified constant called a **parameter** of the relation, describes the **family** of all circles centered at the origin. See Figure 1–3. One member of the family is obtained for each value of c.

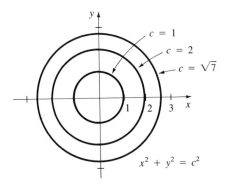

FIGURE 1–3

A one-parameter family of curves such as $x^2 + y^2 = c^2$ may also be represented by a first-order differential equation. If the differential equation is free of the parameter, it is called the **differential equation of the family.** To obtain the differential equation of a family, we differentiate the equation of the family in such a way as to eliminate the parameter. In Section 2–2 we will use the differential equation of a given family of curves to construct a family of curves that is everywhere perpendicular to the given family.

Example 4

Find the differential equation of the one-parameter family of circles shown in Figure 1–3.

SOLUTION   The equation of the family is

$$x^2 + y^2 = c^2$$

Implicitly differentiating this relation with respect to x, we get

$$2x + 2yy' = 0$$

This is the differential equation of the given family of circles.   ∎

Example 5

Find the differential equation of the family of curves $y^2 = 4ax$. See Figure 1–4.

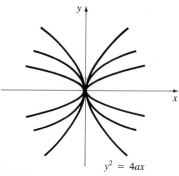

$y^2 = 4ax$

FIGURE 1–4

A family of parabolas: $y^2 = 4ax$

SOLUTION   The expression $y^2 = 4ax$ is differentiated with respect to x to obtain

$$2yy' = 4a$$

The parameter a is eliminated between these two equations by solving for a in the second and substituting into the first. Thus

$$y^2 = (2yy')x$$

After simplification we have the first-order differential equation

$$2xy' - y = 0, y \neq 0$$ ∎

Models Derived from Geometric Considerations

In many cases a differential equation is generated as a result of the geometry of a problem. In situations of this type the differential equation often arises from our interpretation of the derivative as the slope of a curve.

Example 6

A classic problem in differential equations involves a dog pursuing a rabbit in which the pursuit begins at time t_0 when the dog sights the rabbit. To model the situation, we assume that the rabbit runs in a straight line at a constant speed away from the dog and that the dog runs at a constant speed so that its line of sight is always directed at the rabbit. The condition is depicted in Figure 1–5, in which the rabbit runs along the y-axis away from the origin at a rate v_R. The dog is initially at $(b, 0)$ and runs toward the rabbit at a rate v_D. At any time t the coordinates of the rabbit are $(0, y_R)$ and of the dog are (x, y). The object is to find the path that the dog runs in pursuing the rabbit.

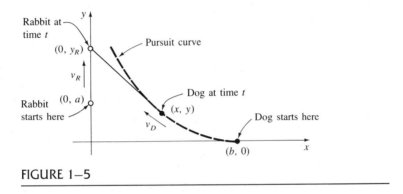

FIGURE 1–5

SOLUTION   Since the dog always runs directly at the rabbit during the pursuit, the slope of the line of sight between the dog and the rabbit at any time t is given by

$$m = \frac{y - y_R}{x - x_R} = \frac{y - y_R}{x}$$

Assuming that the line of sight is tangent to the pursuit curve $y = f(x)$, we know that $m = dy/dx$, and therefore

$$\frac{dy}{dx} = \frac{y - y_R}{x}$$

running after

The solution of this differential equation yields the equation of the pursuit curve. ∎

Example 7

A flexible cable is hung from two points A and B, as shown in Figure 1–6. Assume that the vertical loading (its weight and any external forces) cause the cable to take the shape shown in the figure with its lowest point C. Find the defining differential equation for the shape of the hanging cable.

SOLUTION   Choose the y-axis to pass through the minimum point C on the cable and let P be any other point to the right of C. Assume that the tension in the cable at any point is represented by a vector in the direction of a tangent line to the cable at that point. Further, **assume that the cable is in equilibrium, so the sum of the forces in any direction is zero.** The vertical loading W on this portion of the cable acts through point E, not necessarily the center of the segment. Since C is

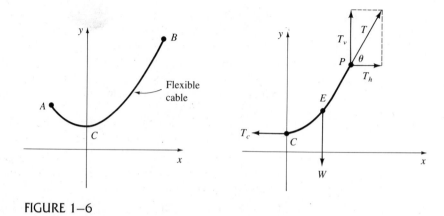

FIGURE 1–6

the minimum point of the cable, the tension T_C at C is horizontal. The tension T at P is in the direction of the tangent line at P. To use the equilibrium conditions that the sum of the forces in any direction is zero, we resolve T into a vertical component $T_v = T \sin \theta$ and a horizontal component $T_h = T \cos \theta$. Thus,

$$T \cos \theta = T_C \qquad (\Sigma \text{ horizontal forces} = 0)$$
$$T \sin \theta = W \qquad (\Sigma \text{ vertical forces} = 0)$$

Dividing the second equation by the first, we get

$$\tan \theta = \frac{W}{T_C}$$

Since $\tan \theta = dy/dx = $ slope of tangent line at P, we have the first-order differential equation

$$\frac{dy}{dx} = \frac{W}{T_C}$$

The differentiation of this equation under the assumption that T_C is constant and W is a function of x yields

$$\frac{d^2y}{dx^2} = \frac{1}{T_C}\frac{dW}{dx}$$

where dW/dx is the weight per unit distance in the horizontal direction. We note that this form is preferable to the first-order equation because dW/dx is usually easier to obtain than $W(x)$. ∎

EXERCISES FOR SECTION 1–1

In Exercises 1–10 classify the given differential equation by order, and tell whether it is linear or nonlinear.

1. $y' + xy = x^2$

2. $y'(y + x) = 3$

3. $y'' + yy' = x$

4. $y'' + x^2y' = x^3$

5. $y \sin y = y''$

6. $y''' - xy' = e^x$

7. $\sin y \, dy = \cos x \, dx$

8. $y' = y'' + x$

9. $y^{(5)} - y = x^2y'$

10. $y' = e^y$

11. If the time-rate of growth of a population is directly proportional to the size of the population, write a differential equation that describes this growth process.

12. When the switch is closed on a capacitive circuit, the time-rate of change of electric current in the circuit is directly proportional to the current itself. Write a differential equation that describes the current in the circuit.

13. The velocity of a particle in a magnetic field is found to be directly proportional to the square root of its displacement. Write a differential equation for this situation in terms of the displacement of the particle and the elapsed time.

14. The intensity of a certain electron beam decreases at a rate that is directly proportional to the square of the intensity. Write a differential equation that describes this situation.

In Exercises 15–24 sketch a few members of the given family and derive the differential equation of the family.

15. Straight lines having a slope of 2.

16. Straight lines with y-intercept equal to the slope.

17. The ellipses with center at the origin, foci on the x-axis, and major axis twice that of the minor axis.

18. The parabolas with vertex at $(0, 0)$ and focus on the x-axis.

19. The hyperbolas $x^2 - y^2 = c$.

20. The cubics $y = cx^3$.

21. The exponentials $y = ce^x$.

22. The logarithms $y = \ln cx$.

23. The circles with center on the x-axis. (HINT: A two-parameter family is described by a second-order differential equation.)

24. The equilateral hyperbolas with foci and center on the y-axis. (HINT: See Exercise 23.)

25. The tangent line to a given curve at any point $P(x, y)$ has an x-intercept equal to one-half of the x-coordinate of P, as shown in Figure 1–7. Write a differential equation that describes this situation.

26. The tangent line of a curve has a slope that is everywhere equal to the ratio x/y. Write a differential equation that describes this situation.

27. A boat is docked at right angles to a long pier. A person holding a rope attached to the bow of the boat starts from A and walks at a constant rate v_m along the pier, as

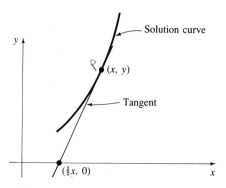

FIGURE 1–7

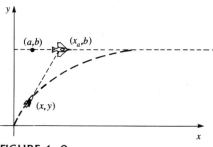

FIGURE 1–8

shown in Figure 1–8. Write a first-order differential equation for the path of the boat as it is pulled to the right.

28. The guidance systems of heat-seeking missiles are able to follow objects that emit heat. Once the guidance system has locked on to the heat source, the missile will move so that it is always pointing toward the source. In Figure 1–9 a missile is initially located at $(0, 0)$ and an airplane is located at (a, b). Assume the airplane moves parallel to the x-axis with a constant speed v_a and that the missile has a constant speed $v_m > v_a$. Write the differential equation of the path of the missile as it moves to intercept the airplane.

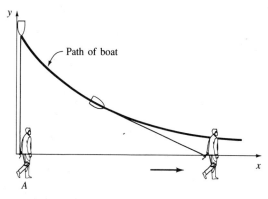

FIGURE 1–9

1–2   SOLUTIONS OF DIFFERENTIAL EQUATIONS

One of the objectives of this book is to show how certain ordinary differential equations are solved. Roughly speaking, to **solve** an nth-order ordinary differential equation

$$F(x, y, y', \cdots, y^{(n)}) = 0$$

means to find a real function that satisfies the given differential equation. We formalize this idea in the following definition.

DEFINITION

> A **solution** of an nth-order differential equation on an interval $a \le x \le b$ is any function possessing all the necessary derivatives, which when substituted for $y, y', y'', \cdots, y^{(n)}$, reduces the differential equation to an identity.

To **verify** that a function is a solution of a given differential equation, we must show that the function satisfies the differential equation. Two procedures for verifying that a function satisfies a given differential equation are illustrated in the following example.

Example 1

(a) Verify that $y = 2 + e^{-x}$ is a solution of $y' + y - 2 = 0$ and indicate any limitations on the solution.
(b) Verify that the equation $x^2 + y^2 = 4$ defines a function that is a solution of $x + yy' = 0$ and indicate any limitations of the solution.

SOLUTION   (a) From $y = 2 + e^{-x}$ we obtain $y' = -e^{-x}$. Substituting y and y' into $y' + y - 2 = 0$, we obtain

$$-e^{-x} + (2 + e^{-x}) - 2 = 0$$

Since the left-hand side is identically zero for all values of x, it follows that the function $y = 2 + e^{-x}$ is a solution of $y' + y - 2 = 0$ on the interval $(-\infty, \infty)$.
(b) If we differentiate $x^2 + y^2 = 4$ with respect to x, we obtain

$$2x + 2yy' = 0$$

which is the given differential equation. Hence $x^2 + y^2 = 4$ satisfies the differential equation. Furthermore, since $y = \pm\sqrt{4 - x^2}$ results from solving $x^2 + y^2 = 4$ for y, there are actually two solution functions defined. Each of the solutions is limited to the interval $[-2, 2]$.   ∎

COMMENT: The two types of solutions in Example 1 are typical of those we encounter in ordinary differential equations. We formalize these two forms of the solution of an nth-order ordinary differential equation in the following definition.

DEFINITION

1. A real function $y = f(x)$ is called an **explicit solution** of the differential equation $F(x, y, y', \cdots, y^{(n)}) = 0$ on $[a, b]$ if

$$F(x, f(x), f'(x), \cdots, f^{(n)}(x)) = 0 \text{ on } [a, b]$$

2. A relation $g(x, y) = 0$ is called an **implicit solution** of the differential equation $F(x, y, y', \cdots, y^{(n)}) = 0$ on $[a, b]$ if $g(x, y) = 0$ defines at least one real function f on $[a, b]$ such that $y = f(x)$ is an explicit solution on this interval.

Example 2

Verify that $y = \sin 2x$ is an explicit solution of $y'' + 4y = 0$ for all real x.

SOLUTION   The first and second derivatives of $y = \sin 2x$ are

$$y' = 2 \cos 2x$$
$$y'' = -4 \sin 2x$$

which, like $\sin 2x$, are defined for all real x. Substituting $\sin 2x$ for y and $-4 \sin 2x$ for y'', the differential equation reduces to the identity

$$(-4 \sin 2x) + 4(\sin 2x) = 0$$

Hence $y = \sin 2x$ is a solution of $y'' + 4y = 0$ for all real x.   ■

Example 3

Verify that the relation $y^2 + x - 4 = 0$ is an implicit solution of $2yy' + 1 = 0$ on the interval $(-\infty, 4)$.

SOLUTION   Differentiating $y^2 + x - 4 = 0$ with respect to x, we obtain

$$2yy' + 1 = 0$$

which is the given differential equation. Hence $y^2 + x - 4 = 0$ is an implicit solution if it defines a real function on $(-\infty, 4)$. Solving this equation for y, we get

$$y = \pm\sqrt{4 - x}$$

Since both $y_1 = \sqrt{4 - x}$ and $y_2 = -\sqrt{4 - x}$ and their derivatives are functions defined for all x in the interval $(-\infty, 4)$, we conclude that the relation $y^2 + x - 4$

$= 0$ is an implicit solution on this interval. Actually, we need only show that $y^2 + x - 4 = 0$ defines one explicit solution on $(-\infty, 4)$ to conclude that it is an implicit solution. ∎

> *CAUTION: A relation $g(x, y) = 0$ can reduce a differential equation to an identity without constituting an implicit solution of the differential equation. For instance, $x^2 + y^2 + 1 = 0$ satisfies $yy' + x = 0$, but it is not an implicit solution since it does not define a real-valued function. This can be seen by solving $x^2 + y^2 + 1 = 0$ to get $y = \pm\sqrt{-1 - x^2}$ which does not give real values. The relation $x^2 + y^2 + 1 = 0$ is called a **formal** solution of $yy' + x = 0$; by that we mean it appears to be a solution of $yy' + x = 0$ since it reduces the differential equation to an identity.*

The determination of which relations actually define implicit solutions can be quite difficult. We will not always bother to prove that a relation can be solved explicitly for y in terms of x to yield some real-valued function. Instead, we will often be satisfied to show that a particular relation is a formal solution with the understanding that the relation makes sense over some interval of real numbers.

Families of Solutions

Differential equations usually have infinitely many solutions. For example, the differential equation $y' - y + 2 = 0$ has infinitely many solutions of the form $y = 2 + ce^x$, where c is a real constant. Such a solution is called a **family of solutions;** one solution is obtained for each value of the parameter c. We label families of solutions by the number of parameters in the family. Thus $y = 2 + ce^x$ is called a **one-parameter family.** Similarly, $y = c_1 + c_2 x$ is called a **two-parameter family.**

Example 4

The family of functions $y = x^2 + c$ is a one-parameter family of solutions of $y' - 2x = 0$. The graphs of some of the members of this family are shown in Figure 1–10. The curves are called **integral curves** of the differential equation.

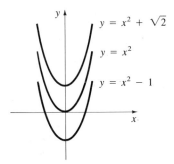

$y = x^2 + \sqrt{2}$

$y = x^2$

$y = x^2 - 1$

FIGURE 1–10

When specifying the number of parameters in a family of solutions, be careful to avoid calling *any* constant a parameter. For example, the function $y = c_1e^{3x} + c_2e^{3x}$ appears to be a two-parameter family, but notice that

$$y = c_1e^{3x} + c_2e^{3x} = (c_1 + c_2)e^{3x}$$

which, by letting $c = c_1 + c_2$, reduces to a one-parameter family. When we speak of an *n*-parameter family, we assume that there are *n* essential constants; that is, there are *n* constants that cannot be reduced by algebraic manipulation to fewer than *n* constants.

DEFINITION

> An *n*-parameter family of solutions of an *n*th-order differential equation is called a **general solution** of the differential equation if all solutions of the differential equation are obtainable from the *n*-parameter family.

For many differential equations the *n*-parameter family and the general solution are the same, but there are exceptions. For example, $y = (x + c)^{-2}$ is a one-parameter solution of $y' = -2y^{3/2}$; but $y = 0$, which is also a solution, is not a member of the one-parameter family. A solution of a differential equation that is not obtainable from an *n*-parameter family of solutions is called a **singular solution.**

Example 5

(a) Show that the one-parameter family $y = cx + \frac{1}{2}c^2$ is a solution of the differential equation $\frac{1}{2}(y')^2 + xy' - y = 0$.
(b) Show that the function $y = -\frac{1}{2}x^2$ is a singular solution of this differential equation.

SOLUTION   (a) To check the validity of $y = cx + \frac{1}{2}c^2$, we note that $y' = c$, and therefore

$$\frac{1}{2}(c)^2 + x(c) - (cx + \frac{1}{2}c^2) = 0$$

reduces to an identity.
(b) The function $y = -\frac{1}{2}x^2$, which is not a member of $y = cx + \frac{1}{2}c^2$, is a solution of the given differential equation since substituting $y = -\frac{1}{2}x^2$ and $y' = -x$ into the equation yields

$$\frac{1}{2}(-x)^2 + x(-x) - \left(-\frac{1}{2}x^2\right) = 0$$

which is an identity. The function $y = -\frac{1}{2}x^2$ is a singular solution of this differential equation since it is not a member of the given one-parameter family. Figure 1–11 shows some of the members of the one-parameter family and also the singular solution. Notice that the members of $y = cx + \frac{1}{2}c^2$ are tangent to the singular solution $y = -\frac{1}{2}x^2$.

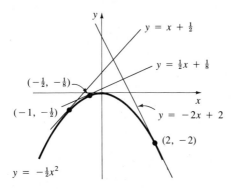

FIGURE 1–11 ∎

COMMENT: Although there is technically a distinction between an n-parameter family of solutions and the set of all solutions to an nth-order differential equation, we choose to use the terms n-parameter family and general solution interchangeably. When a singular solution exists, it will be so noted.

DEFINITION

Any solution of an *n*th-order differential equation obtained from a general solution by assigning values to the *n* parameters is called a **particular solution.**

Example 6

The general solution of $y'' + 9y = 0$ is the two-parameter family

$$y = c_1 \cos 3x + c_2 \sin 3x$$

A particular solution of the differential equation, obtained by letting $c_1 = 2$ and $c_2 = 1$, is

$$y = 2 \cos 3x + \sin 3x$$ ∎

Particular solutions are often determined from a general solution by imposing conditions on the solution function. The next two examples show how conditions on the solution function are used to obtain a particular solution from a general solution.

Example 7

The general solution of $y' + y = 1$ is the one-parameter family

$$y = 1 + ce^{-x}$$

Find the particular solution whose graph passes through the origin.

SOLUTION   The graph of the solution function must pass through the origin, which means that $y = 1 + ce^{-x}$ must be such that $y = 0$ when $x = 0$. Upon substituting $x = 0$ and $y = 0$ into the general solution, we get

$$0 = 1 + c$$

Thus $c = -1$, and the particular solution is

$$y = 1 - e^{-x}$$

Figure 1–12 shows the graph.

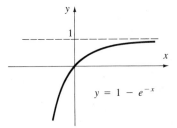

$$y = 1 - e^{-x}$$

FIGURE 1–12   ∎

Example 8

Given that $y = c_1e^{2x} + c_2e^{-x}$ is a general solution of the second-order equation $y'' - y' - 2y = 0$, find the particular solution that satisfies the conditions $y(0) = 2$ and $y'(0) = -1$.

SOLUTION   Here we have conditions given at $x = 0$ on both y and y'. To obtain an expression for y', we differentiate the given solution function. Thus

$$y' = 2c_1e^{2x} - c_2e^{-x}$$

To evaluate c_1 and c_2, we substitute $x = 0$, $y = 2$ and $x = 0$, $y' = -1$ into the appropriate equation. The following equations result from this substitution:

$$2 = c_1 + c_2$$
$$-1 = 2c_1 - c_2$$

Solving these equations for c_1 and c_2, we get $c_1 = \frac{1}{3}$ and $c_2 = \frac{5}{3}$. Thus the particular solution is

$$y = \frac{1}{3}e^{2x} + \frac{5}{3}e^{-x} \qquad\qquad \blacksquare$$

EXERCISES FOR SECTION 1–2

In Exercises 1–10 verify that the expression on the right is a solution of the given differential equation.

DIFFERENTIAL EQUATION	SOLUTION
1. $t\,ds/dt - 2s = 0$	$s = Kt^2$
2. $x\,dy/dx - 2y = -x$	$y = x + cx^2$
3. $V' + V = \cos r - \sin r$	$V = \cos r + ce^{-r}$
4. $y' - 2y = 3e^{2x}$	$y = (3x + c)e^{2x}$
5. $y'' - y' - 2y = 0$	$y = Ae^{-x} + Be^{2x}$
6. $d^2s/dt^2 + 16s = 0$	$s = c_1 \sin 4t + c_2 \cos 4t$
7. $v'' + 9v = 0$	$v = c_1 \cos 3x + c_2 \sin 3x$
8. $d^2s/dt^2 - s = 1$	$s = c_1 e^t + c_2 e^{-t} - 1$
9. $\ddot{x} + 2\dot{x} + x = 0 \left(\text{NOTE: } \dot{x} \equiv \dfrac{dx}{dt}\right)$	$x = (A + Bt)e^{-t}$
10. $y'' + 4y' + 4y = 0$	$y = (c_1 + c_2 x)e^{-2x}$

11. The schematic diagram in Figure 1–13 represents an electric circuit in which a voltage of V volts is applied to a resistance of R ohms and an inductance of L henrys connected in series. When the switch is closed, a current of i amperes will flow in the circuit. Because of the inductance in the circuit the current will vary with time, and it can be shown that a mathematical model for this circuit is the first-order linear differential equation

$$L\frac{di}{dt} + iR = V$$

Verify that the current in the circuit is given by

$$i = \frac{V}{R}(1 - e^{-Rt/L})$$

12. When an object at room temperature is placed in an oven whose temperature is 500° C, the temperature of the object will increase with time, approaching the

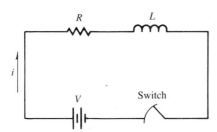

FIGURE 1–13

temperature of the oven. It is known that the temperature Q of the object is related to time through the differential equation

$$\frac{dQ}{dt} = k(Q - 500)$$

Verify that the temperature of the object is given by $Q = 500 + ce^{kt}$, where c and k are constants.

13. Determine those values of m, if any, for which $y = e^{mx}$ is a solution of the differential equation

$$2\frac{d^2y}{dx^2} - 5\frac{dy}{dx} - 3y = 0$$

14. Determine those values of m, if any, for which $y = e^{mx}$ is a solution of the differential equation

$$y''' + 3y'' + 2y' = 0$$

15. Verify that $x(y^2 + 3) = 1$ is an implicit solution of the differential equation $2xyy' + y^2 + 3 = 0$ on the interval $(0, \frac{1}{3})$.

16. Verify that $y^2 - x^2 + x = 0$ is an implicit solution of the differential equation $2yy' + 1 = 2x$ on the interval $(1, \infty)$.

17. Verify that the relation $\ln y + ye^x = c$ formally satisfies the differential equation $(1 + ye^x)y' + y^2e^x = 0$.

18. Verify that the relation $xy + \sin y = K$ formally satisfies the differential equation $(x + \cos y)y' + y = 0$.

In Exercises 19–24 find the particular solution corresponding to the given conditions.

GENERAL SOLUTION	CONDITIONS
19. $y = x^2 + 2x + C$	$y(0) = 3$
20. $y = x + ce^x$	$y(0) = -2$
21. $y = ae^{-x} + be^{2x}$	$y(0) = y'(0) = 1$
22. $s = A \sin 2t + B \cos 2t$	$s(0) = 1, s'(0) = 2$
23. $v = \cos 3t + At + B$	$v(\pi) = 0, v'(\pi) = 1$
24. $y = c_1e^x + c_2e^{-x}$	$y(2) = 1, y'(2) = 3$

1–3 DIFFERENTIAL EQUATIONS OF THE FORM $y' = f(x)$

In the preceding section we showed how to verify a solution of a differential equation; however, we gave no clues as to how to find such a solution. In fact, elementary differential equations of the form $y' = f(x)$ are solved in calculus. In this case the general solution is found to be $y = \int f(x) \, dx + C$. Thus the solution of $y' = f(x)$ is found simply by integrating, although the integration itself might not be simple!

Example 1

Solve the differential equation

$$\frac{dy}{dx} = 3x^2$$

SOLUTION   Integrating $3x^2$, we obtain

$$y = x^3 + C$$

as the general solution to the given differential equation. Figure 1–14 shows some of the integral curves for this differential equation.

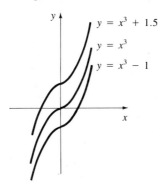

$y = x^3 + 1.5$

$y = x^3$

$y = x^3 - 1$

FIGURE 1–14 ■

Example 2

Solve

$$y' = \frac{3x}{\sqrt{x^2 + 5}}$$

SOLUTION   The solution is given by

$$y = 3 \int x(x^2 + 5)^{-1/2} \, dx = 3\sqrt{x^2 + 5} + C$$

■

Example 3

Solve

$$y' = x \cos 2x$$

SOLUTION   The solution to this equation is given by

$$y = \int x \cos 2x \, dx + C$$

To evaluate $\int x \cos 2x \, dx$, we use the integration-by-parts formula,

$$\int u \, dv = uv - \int v \, du$$

with $u = x$ and $dv = \cos 2x \, dx$. This yields

$$\begin{array}{c|c} u = x & dv = \cos 2x \, dx \\ \hline du = dx & v = \frac{1}{2} \sin 2x \end{array}$$

so that

$$y = \frac{1}{2}x \sin 2x - \frac{1}{2}\int \sin 2x \, dx + C$$

$$= \frac{1}{2}x \sin 2x + \frac{1}{4} \cos 2x + C$$

is the general solution.   ∎

Example 4

Find the general solution of

$$\frac{dy}{dx} = \frac{x + 5}{(x - 3)(x + 1)}$$

SOLUTION   The solution may be written as

$$y = \int \frac{x + 5}{(x - 3)(x + 1)} \, dx + C$$

The integral can be evaluated by decomposing the integrand into a sum of two partial fractions of the form

$$\frac{x + 5}{(x - 3)(x + 1)} = \frac{A}{x - 3} + \frac{B}{x + 1}$$

Multiplying both sides by $(x - 3)(x + 1)$ yields

$$x + 5 = A(x + 1) + B(x - 3)$$

To find A, let $x = 3$, so that

$$8 = 4A \qquad \text{or} \qquad A = 2$$

To find B, let $x = -1$, so that

$$4 = -4B \qquad \text{or} \qquad B = -1$$

The desired partial fractions are then

$$\frac{x + 5}{(x - 3)(x + 1)} = \frac{2}{x - 3} - \frac{1}{x + 1}$$

Thus

$$y = \int \left[\frac{2}{x - 3} - \frac{1}{x + 1} \right] dx + C_1$$

$$y = 2 \ln|x - 3| - \ln|x + 1| + C_1$$

The logarithmic terms suggest that it might be advantageous to express the arbitrary constant in the form $C_1 = \ln C$. Thus we write

$$y = 2 \ln|x - 3| - \ln|x + 1| + \ln C$$

$$= \ln(x - 3)^2 - \ln|x + 1| + \ln C$$

The solution may then be written

$$y = \ln \frac{C(x - 3)^2}{|x + 1|} \qquad \blacksquare$$

*COMMENT: In the next example a differential equation describes a physical situation. Typically, conditions are given at $t = 0$ and are called **initial conditions.***

Example 5

The acceleration of a particle is given by $a = 6t + 10$ cm/sec^2. If the initial velocity of the particle is 3 cm/sec, what is the velocity of the particle at $t = 2$ sec?

SOLUTION   Recall that acceleration is the first derivative of velocity with respect to time. Hence we have

$$\frac{dv}{dt} = 6t + 10, \quad v(0) = 3$$

The velocity is given by

$$v = \int (6t + 10)\, dt = 3t^2 + 10t + c$$

Using $t = 0$, $v = 3$ in this expression, we get $c = 3$, and

$$v = 3t^2 + 10t + 3$$

The velocity at $t = 2$ sec is then

$$v = 3(2)^2 + 10(2) + 3 = 35 \text{ cm/sec} \qquad \blacksquare$$

In some applications we are concerned with phenomena for which the describing differential equation is different over adjoining intervals. To solve this kind of differential equation, we must find solutions to the individual differential equations and then "piece together" the solutions over the adjoining intervals. The piecing together must be done so that the solution function is continuous. More explicitly, if $f(x)$ is the solution function over the interval (a, b) and $g(x)$ is the solution function over (b, c), then we must have that

$$\lim_{x \to b^+} g(x) = \lim_{x \to b^-} f(x)$$

See Figure 1–15.

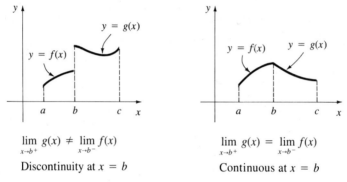

$$\lim_{x \to b^+} g(x) \neq \lim_{x \to b^-} f(x) \qquad\qquad \lim_{x \to b^+} g(x) = \lim_{x \to b^-} f(x)$$

Discontinuity at $x = b$ Continuous at $x = b$

FIGURE 1–15

Example 6

Solve the differential equation which is defined by

$$y' = \begin{cases} 1, & x < 0 \\ x, & 0 \leq x \leq 3 \\ 0, & x > 3 \end{cases}$$

SOLUTION   Note that since this is a first-order differential equation, the general solution will be a one-parameter family. The general solution over the first interval is $y = x + c_1$, over the second interval $y = \frac{1}{2}x^2 + c_2$, and over the third interval $y = c_3$. These three constants *are not completely arbitrary* since any of the solutions must be continuous at $x = 0$ and at $x = 3$. Hence at $x = 0$ we must have $0 + c_1 = 0^2/2 + c_2$, or $c_1 = c_2$. At $x = 3$ we have $(3)^2/2 + c_2 = c_3$. By letting $c_1 = c$ we have the following one-parameter family as the general solution.

$$y = \begin{cases} x + c, & x < 0 \\ \dfrac{1}{2}x^2 + c, & 0 \leq x \leq 3 \\ \dfrac{9}{2} + c, & x > 3 \end{cases}$$

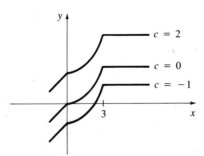

FIGURE 1–16

Figure 1–16 shows the graphs of some of the solutions of this family. Notice that y is everywhere continuous.   ■

EXERCISES FOR SECTION 1–3

In Exercises 1–28 write the differential equation in the form $y' = f(x)$ and solve.

1. $y' - \sqrt{x} = 0$ **2.** $x^2 y' = 1$ **3.** $y' - (x + 2)^{-1} = 0$

4. $S' - e^{2t} = 0$ **5.** $y' \csc 2x = 1$ **6.** $y' \cos \frac{1}{2}x = 3 \sin \frac{1}{2}x$

7. $x^{-1} y' = 2(x^2 + 3)^{3/2}$ **8.** $y' \sqrt{3x^2 + 2} - x = 0$ **9.** $dy = (x^2 + 3)^{-1} dx$

10. $\cos^2 3x \, dy - dx = 0$ **11.** $(x^2 + 4)\dfrac{dy}{dx} = 1$ **12.** $\dfrac{ds}{dt} = \dfrac{t^2 + 2}{t^3 + 6t}$

13. $\dfrac{dy}{dx} - x \cos 3x = 0$ **14.** $e^t \dfrac{ds}{dt} - t = 0$ **15.** $\dfrac{dz}{dt} = t^3 \ln t$

16. $y' = \dfrac{x - 11}{2x^2 + 5x - 3}$ **17.** $y' = \dfrac{x^2}{(x + 1)^3}$ **18.** $y' = \dfrac{9x + 14}{(x - 2)(x^2 + 4)}$

19. $y' = e^x(e^{2x} + 1)^{-1}$ **20.** $y' = e^x(e^x + 1)^{-1}$

21. $(e^x + e^{-x})y' + e^x = e^{-x}$ **22.** $ds/dt = \sin^2 t$

23. $y' = \sin^2 x \cos x$ **24.** $y' = \sin^3 x$ **25.** $v' - x \cos x^2 = \ln 3x$

26. $e^{-3x} y' = x^2$ **27.** $\dfrac{dm}{dt} = \dfrac{t}{t^2 + 2t + 2}$ **28.** $y' = \dfrac{3x + 4}{x^2 + 2}$

In Exercises 29–32 find the general solution (a one-parameter family) and sketch several particular ones.

29. $y' = \begin{cases} \sin x, & x \leq 0 \\ 0, & x > 0 \end{cases}$

30. $y' = \begin{cases} x, & -1 \leq x < 0 \\ x^2, & 0 \leq x \leq 1 \\ x, & 1 < x \leq 2 \\ x^2, & 2 < x \leq 3 \end{cases}$

31. $y' = \begin{cases} e^x, & x < 0 \\ -e^x, & 0 \leq x < 1 \\ 2, & 1 \leq x < \infty \end{cases}$

32. $y' = \begin{cases} \cos x, & x < 0 \\ \sin x, & 0 \leq x < \frac{1}{2}\pi \\ 2, & \frac{1}{2}\pi \leq x < \infty \end{cases}$

33. The graph of a function passes through the point (2, 9). Find the function if its slope is given by $dy/dx = 3x^2$.

34. The velocity of an object is given by $30/\sqrt{6t + 4}$ cm/sec. How far has it traveled after 2 seconds, if its initial displacement is zero?

35. The shear in a beam is given by $S = dM/dx$, where M is the bending moment and x is the distance from one end of the beam. If the shear equation is $S = (3 + 2x)^{-2}$, determine the equation for the bending moment. Assume $M = 10$ when $x = 0$.

36. Electric current is defined as $i = dq/dt$, where q is the electric charge and t is time. If the current in a capacitor is $i = 4e^{-2t}$ amps, find the expression for the electric charge. Assume $q = 0$ when $t = 0$.

37. In Example 7 of Section 1–1 we showed that the shape of a cable of uniform cross section suspended between two points is obtained by solving the second-order differential equation

$$\frac{d^2y}{dx^2} = b\frac{dW}{dx}$$

where b is a constant and dW/dx is weight per unit distance in the horizontal direction. Determine the shape of the cables of the suspension bridge depicted in Figure 1–17 if they support a flat roadway of constant linear density: that is, $dW/dx = k$. Assume the weight of the cables is negligible. The equation can be solved as a first-order differential equation by making the substitution $v = dy/dx$.

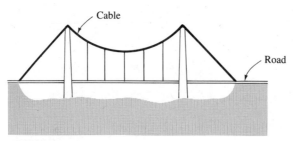
Cable

Road

FIGURE 1–17

1–4   UNIQUENESS OF PARTICULAR SOLUTIONS

A general solution of an nth-order differential equation contains n arbitrary constants. To obtain a particular solution, we specify n conditions on the solution function and its derivatives and thereby hope to solve for the n arbitrary constants. There are two common methods of specifying auxiliary conditions.

DEFINITION

1. If the auxiliary conditions for a given differential equation relate to a single x value, the conditions are called **initial conditions**. The differential equation with its initial conditions is called an **initial-value problem.**

2. If the auxiliary conditions for a given differential equation relate to two or more x values, the conditions are called **boundary conditions** or **boundary values**. The differential equation with its boundary conditions is called a **boundary-value problem.**

Example 1

(a) $y' + y = 3$, $y(0) = 1$ is a first-order initial-value problem.
(b) $y'' + 2y = 0$, $y(1) = 2$, $y'(1) = -3$ is a second-order initial-value problem.
(c) $y'' - y' + y = x^3$, $y(0) = 2$, $y'(1) = -1$ is a second-order boundary-value problem. ■

In the previous section we solved some simple initial-value problems. In this section we look more closely at solutions of both initial-value and boundary-value problems and attempt to give some answers to the following questions:

1. When does a solution exist? That is, can we be sure that an initial-value or a boundary-value problem will have a solution?

2. Is a known solution unique? That is, can we be sure that a solution to an initial-value or a boundary-value problem is the only solution?

The next two examples give some insight into these questions.

Example 2

The general solution of $y'' + y = 0$ can be shown to be

$$y = c_1 \cos x + c_2 \sin x$$

Using this solution, what can we say about the solution to the boundary-value problem $y'' + y = 0$, $y(0) = 0$, $y(\pi) = 2$?

SOLUTION   Using the given boundary conditions in $y = c_1 \cos x + c_2 \sin x$, we get

$$0 = c_1 \cos 0 + c_2 \sin 0$$

and

$$2 = c_1 \cos \pi + c_2 \sin \pi$$

The first equation yields $c_1 = 0$, and the second equation yields $c_1 = -2$. Since c_1 cannot equal both 0 and -2, no solution is possible for this boundary-value problem. See Figure 1–18.   ■

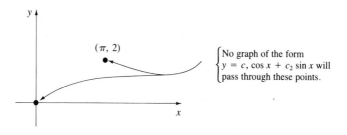

FIGURE 1–18

Example 3

Solve the boundary-value problem

$$y'' + y = 0, \, y(0) = 0, \, y(\pi) = 0$$

SOLUTION   Recall from Example 2 that the general solution of $y'' + y = 0$ is $y = c_1 \cos x + c_2 \sin x$. The boundary values in this case yield

$$0 = c_1 \cos 0 + c_2 \sin 0$$

and

$$0 = c_1 \cos \pi + c_2 \sin \pi$$

Both of these equations lead to the conclusion that $c_1 = 0$. The constant c_2 is not assigned a value and can therefore take on arbitrary values. Thus there are infinitely many solutions represented by

$$y = c_2 \sin x$$

A few of the integral curves on the interval $[0, \pi]$ are shown in Figure 1–19.   ■

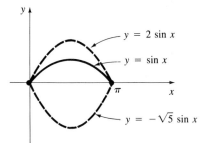

FIGURE 1–19

Examples 2 and 3 show that a boundary-value problem may not have a solution or, if it does, the solution may not be unique. The question of when a boundary-value problem has a unique solution is difficult to answer because there is no simple theory to ensure a unique solution to a boundary-value problem. However, we can specify necessary conditions for which a first-order initial-value problem will have a unique solution.

THEOREM 1–1 *An Existence-and-Uniqueness Theorem for First-Order Initial-Value Problems*

Let f and f_y be continuous functions of x and y in some rectangle R of the xy-plane, and let (x_0, y_0) be a point in that rectangle. Then on some interval I centered at x_0 there is a unique solution $y = \phi(x)$ of the initial-value problem

$$y' = f(x, y), \quad y(x_0) = y_0$$

See Figure 1–20.

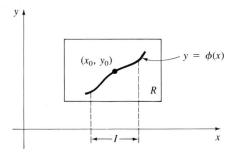

FIGURE 1–20

This existence-and-uniqueness theorem tells us that under the specified conditions a first-order initial-value problem has a unique solution. The conditions that must be met to guarantee a unique solution are

1.   The function f and its partial derivative with respect to y must be continuous in some rectangle R in the xy-plane.

2.   The given point must be in the rectangle R.

The theorem assures us that if these conditions are satisfied, a unique solution $y = \phi(x)$ exists; it does not say that initial-value problems that fail to meet these conditions do not have unique solutions. We will explain the meaning of this existence-and-uniqueness theorem in the following examples.

Example 4

Show that the initial-value problem

$$y' = \frac{y}{x}, \; y(2) = 1$$

satisfies the conditions of the existence-and-uniqueness theorem and therefore has a unique solution.

SOLUTION   To show that the conditions of the theorem are satisfied, we note that

$$f(x, y) = \frac{y}{x} \qquad \text{and} \qquad f_y(x, y) = \frac{1}{x}$$

Both of these functions are continuous except at $x = 0$. Hence f and f_y satisfy the conditions of the theorem in any rectangle R that does not contain any part of the y-axis ($x = 0$). Since the point $(2, 1)$ is not on the y-axis, there is a unique solution. See Figure 1–21. You can verify that $y = \frac{1}{2}x$ is a unique solution.

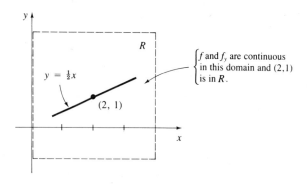

FIGURE 1–21

Example 5

Show that the initial-value problem

$$y' = \frac{y}{x}, \; y(0) = 3$$

does not satisfy the conditions of the existence/uniqueness theorem.

SOLUTION   In this problem neither f nor f_y is continuous at $x = 0$, which means that $(0, 3)$ cannot be included in any rectangle R where f and f_y are continuous. See Figure 1–22. Hence we cannot use Theorem 1–1 to conclude anything about the existence or uniqueness of a solution of this initial-value problem. As an exercise, show that $y = cx$ is the general solution of $y' = y/x$ but that a particular solution cannot be found whose graph passes through the point $(0, 3)$.

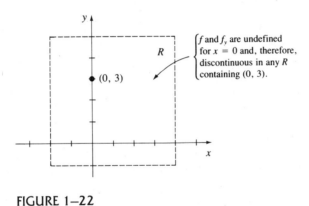

FIGURE 1–22

Example 6

Show that the initial-value problem

$$y' = y^{1/2}, \; y(0) = 0$$

does not satisfy the conditions of the existence/uniqueness theorem.

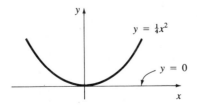

FIGURE 1–23

SOLUTION   In this case $f(x, y) = y^{1/2}$ and $f_y(x, y) = \frac{1}{2} y^{-1/2}$, so that f_y is discontinuous at the point $(0, 0)$. Hence Theorem 1–1 does not guarantee a unique solution. In fact, it is easy to verify that $y = \frac{1}{4}x^2$ and $y = 0$ are both solutions of this problem. Figure 1–23 shows the graphs of both solutions passing through $(0, 0)$. ∎

Example 7

Given the initial-value problem

$$y' = -x, \; y(0) = 2$$

(a) Show that there is a unique solution.
(b) Find the solution.

SOLUTION   (a) Observe that $f(x, y) = -x$ and $f_y(x, y) = 0$ are continuous in every domain in the xy-plane, and $(0, 2)$ lies in some such domain. Thus there is a unique solution function ϕ, defined on some interval about $x_0 = 0$, which satisfies the initial condition $y(0) = 2$. The graph of these initial conditions is shown in Figure 1–24a.
(b) The family of solutions is given by

$$y = \int (-x) \, dx + c = -\frac{1}{2}x^2 + c$$

The condition $y(0) = 2$ yields $c = 2$, and therefore the solution is

$$y = -\frac{1}{2}x^2 + 2$$

The integral curve is shown in Figure 1–24b.

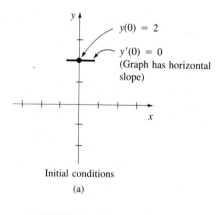

$y(0) = 2$

$y'(0) = 0$
(Graph has horizontal slope)

Initial conditions

(a)

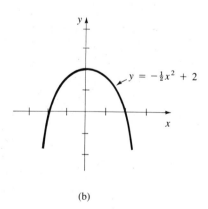

$y = -\frac{1}{2}x^2 + 2$

(b)

FIGURE 1–24   ∎

EXERCISES FOR SECTION 1–4

In Exercises 1–6 determine whether or not the existence-and-uniqueness theorem can be applied to the given initial-value problem. If the theorem does not apply, indicate which condition of the hypothesis is violated.

1. $y' = x^2 + y^2$, $y(0) = 2$

2. $y' = x^2 y$, $y(0) = -1$

3. $xy' + y = 0$, $y(0) = 0$

4. $(x - 2)y' - y = 0$, $y(0) = 2$

5. $y' = y^2/\sqrt{x - 1}$, $y(1) = 2$

6. $y' = y^2/\sqrt{x + 1}$, $y(0) = 0$

Given that $y = c_1 e^{2x} + c_2 e^{-3x}$ is the general solution of the differential equation $y'' + y' - 6y = 0$, solve the boundary-value problems in Exercises 7 and 8, or show that no solution is possible.

7. $y'' + y' - 6y = 0$, $y(0) = 1$, $y(1) = 2e^2 - e^{-3}$

8. $y'' + y' - 6y = 0$, $y(1) = 0$, $y(2) = (e^5 - 1)/e$

Given that $y = c_1 \cos x + c_2 \sin x$ is the general solution of the differential equation $y'' + y = 0$, solve the boundary-value problems in Exercises 9–12, or show that no solution is possible.

9. $y'' + y = 0$, $y(0) = 1$, $y(\pi/2) = 1$

10. $y'' + y = 0$, $y(0) = 2$, $y'(\pi) = 3$

11. $y'' + y = 0$, $y(0) = 0$, $y(\pi) = \pi$

12. $y'' + y = 0$, $y(\pi/2) = 2$, $y(-\pi/2) = 0$

13. Given that $y = c_1 \cos 2x + c_2 \sin 2x$ is the general solution of the differential equation $y'' + 4y = 0$, show that there are infinitely many solutions to the boundary-value problem

$$y'' + 4y = 0, \quad y(\pi/4) = 0, \quad y(5\pi/4) = 0$$

14. Show that $y = c_1 \cosh x + c_2 \sinh x$ and $y = k_1 e^x + k_2 e^{-x}$ are both solutions of the differential equation $y'' - y = 0$.

Given that $y = c_1 \cos kx + c_2 \sin kx$ is the general solution of $y'' + k^2 y = 0$, determine the values of k for which the boundary-value problems in Exercises 15 and 16 have nontrivial solutions, that is, solutions other than $y \equiv 0$.

15. $y'' + k^2 y = 0$, $y(0) = 0$, $y'(\pi) = 1$ **16.** $y'' + k^2 y = 0$, $y(0) = 0$, $y(2) = 0$

In Exercises 17–20 show that there is a unique solution to the given initial-value problem. Find the solution and sketch its graph.

17. $y' = e^{-2x}$, $y(0) = 2$

18. $y' = 1/\sqrt{x}$, $y(4) = 5$

19. $y' = 1/(x^2 + 1)$, $y(1) = 0$

20. $y' = \cos 3x$, $y(\pi) = \frac{1}{3}$

REVIEW EXERCISES FOR CHAPTER 1

1. Classify the following differential equations by order and tell whether each is linear or nonlinear.

(a) $y'' + xy = y$ (b) $y' = 2y^{1/2}$ (c) $dy = (x^{1/2} + y)\, dx$

2. Verify that $y = 2e^{-3x} + 1 + \frac{2}{3}x$ is a solution of $y'' + 3y' - 2 = 0$.

3. Verify that $xy^2 + \text{Arctan } y = 2$ is an implicit solution of

$$(1 + y^2)\, y^2 + 2xyy'\, (1 + y^2) + y' = 0$$

4. Determine values of m for which e^{mx} is a solution of $y'' + 3y' - 5y = 0$. Write the general solution.

In Exercises 5–12 find the general solution of the given equation.

5. $y' = (3x + 2)^{1/2}$ 　　　　　　　**6.** $y' = x + \cot 3x$

7. $dy/dx = 5 + e^{-2x}$ 　　　　　　**8.** $y' = \sin 2x/(1 + \cos 2x)$

9. $ds/dt = t \sec^2 3t$ 　　　　　　**10.** $dy - \text{Arctan } 4x\, dx = 0$

11. $y' = (3x^2 + 2)/(x^3 + 4x)$ 　　　**12.** $ds/dt = 5(t^2 - t - 6)^{-1}$

13. Find the general solution of the first-order equation $y' = f(x)$, where

$$f(x) = \begin{cases} -2, & x < 1 \\ x + 3, & x \geq 1 \end{cases}$$

14. The slope of the graph of a function is everywhere equal to the square of its abscissa. If the graph passes through $(2, 3)$, find the function.

15. Determine whether or not the fundamental existence-and-uniqueness theorem can be applied to

$$y' = xy^{1/2}, \, y(0) = 0$$

16. Determine whether or not the fundamental existence-and-uniqueness theorem can be applied to

$$y' = x + y, \, y(0) = 2$$

17. Use the fundamental existence-and-uniqueness theorem to show that

$$y' = (x + 1)^{-1/2}, \, y(3) = 2$$

has a unique solution.

18. Verify that $y = c_1 \cos 2x + c_2 \sin 2x$ is a solution of $y'' + 4y = 0$. Solve the boundary-value problem $y'' + 4y = 0$, $y(0) = 1$, $y'(\pi/2) = 0$.

2

First-Order
Differential Equations

How do we know that the prehistoric Hopewell Indians lived in the Great Lakes region of our country about 400 B.C.? Archeologists use a technique called carbon dating to estimate the dates of such civilizations from artifacts found at excavation sites. Carbon dating is based on the principles of radioactive decay discovered at the beginning of the century by E. Rutherford. From experimental data Rutherford was able to formulate a simple model to describe radioactive decay based on the assumption that the rate at which atoms disintegrate is proportional to the number of atoms N present in the material. Symbolically, the disintegration process is described by the first-order differential equation

$$\frac{dN}{dt} = -kN$$

where k is a constant of proportionality. In this chapter we develop techniques for solving this differential equation and, consequently, its application in carbon dating.

Recall from Section 1–2 that $y' = f(x, y)$ is a first-order differential equation whose general solution is a one-parameter family. This chapter has two purposes:

- To discuss methods used to find solutions of first-order differential equations.

- To show how differential equations arise in the solution of physical problems.

First-order equations are grouped into various categories or types, each having a clearly outlined technique of solution. Learning the type of equation is an important part of learning the technique of solution.

2–1   SEPARABLE EQUATIONS

Consider a general first-order differential equation that may be written in *derivative form* as

$$y' = f(x, y) \tag{2-1}$$

If $f(x, y)$ is written as

$$f(x, y) = -\frac{M(x, y)}{N(x, y)}$$

then Equation 2–1 has the *differential form*

$$M(x, y)\,dx + N(x, y)\,dy = 0 \tag{2-2}$$

These two general forms are interchangeable. For example, the equations

$$y' = \frac{2 + xy}{x^2 + y^2} \qquad \text{and} \qquad (2 + xy)\,dx - (x^2 + y^2)\,dy = 0$$

mean the same thing.

A differential equation that can be put into the form

$$y' = -\frac{f(x)}{g(y)} \tag{2-3}$$

or equivalently, either

$$g(y)\,dy + f(x)\,dx = 0 \qquad \text{or} \qquad g(y)\,dy = -f(x)\,dx \tag{2-4}$$

is called a **separable** differential equation.

We see that in Equation 2–4 the variables x and y are "separated" and are associated with their respective differentials. The process of associating $g(y)$ with dy and $f(x)$ with dx is called **separating variables.** In most cases the differential equation is not immediately in the form of Equation 2–4 and some algebraic manipulation must be performed actually to separate the variables. Once the variables are separated, the general solution of Equation 2–4 is given in implicit form by

$$\int g(y)\, dy + \int f(x)\, dx = C \tag{2–5}$$

CAUTION: *Before using Equation 2–5, the integrand that accompanies dx must be a function of x only and that with dy a function of y only.*

Example 1

Solve the equation $3(y^2 + 1)\, dx + 2xy\, dy = 0$.

SOLUTION   As written, this equation is not in separated form; however, the variables can be separated by dividing each term by $x(y^2 + 1)$. Thus

$$\frac{3dx}{x} + \frac{2y\, dy}{y^2 + 1} = 0$$

Taking the integral of each term, the solution in implicit form is

$$3 \ln |x| + \ln (y^2 + 1) = C \tag{2–6}$$

The solution obtained in Equation 2–6 may be simplified by using properties of logarithms. Hence

$$\ln |x|^3 + \ln (y^2 + 1) = C$$
$$\ln |x|^3 (y^2 + 1) = C \tag{2–7}$$

is another form of the solution. An additional simplification can be made to Equation 2–7 by writing the arbitrary constant in the form $C = \ln |C_1|$, so that

$$\ln |x|^3 (y^2 + 1) = \ln |C_1|$$

and therefore

$$x^3(y^2 + 1) = C_1 \qquad \blacksquare$$

COMMENTS: *(a) Notice that in the first step of the solution in Example 1 each term of the equation was divided by x. This step assumes that $x \neq 0$.*

(b) The integral of 3/x led to $\ln |x|$. As a general rule we will keep the absolute values only when a logarithm term occurs in the final form of a solution.

(c) The arbitrary constant was written in the form $\ln |C_1|$ to simplify the final solution. Any form of the arbitrary constant may be used; however, we usually choose a form that simplifies the result.

Example 2

Solve the equation $(1 + \cos \theta) \, dr = r \sin \theta \, d\theta$.

SOLUTION   To separate variables, divide by $r(1 + \cos \theta)$. Thus

$$\frac{dr}{r} - \frac{\sin \theta \, d\theta}{1 + \cos \theta} = 0$$

Integrating, we obtain the desired solution:

$$\ln |r| + \ln |1 + \cos \theta| = \ln |C|$$

$$\ln |r(1 + \cos \theta)| = \ln |C|$$

$$r(1 + \cos \theta) = C$$

$$r = \frac{C}{1 + \cos \theta}$$

∎

✕ Example 3

Solve the equation $\dfrac{dy}{dx} = \dfrac{xy}{\sqrt{x^2 - 4}}$.

SOLUTION   Separating variables, we have

$$\frac{dy}{y} = x(x^2 - 4)^{-1/2} \, dx$$

Integrating yields

$$\ln |y| = (x^2 - 4)^{1/2} + c_2$$

$$\ln |y| + \ln |c_1| = (x^2 - 4)^{1/2}$$ (NOTE: $c_2 = -\ln|c_1|$.)

$$\ln |c_1 y| = (x^2 - 4)^{1/2}$$

$$|c_1 y| = \exp(x^2 - 4)^{1/2}$$ (NOTE: $\exp u = e^u$)

$$y = c \exp(x^2 - 4)^{1/2}$$

∎

COMMENT: *The division by y in the process of separating the varia-bles in Example 3 implies that $y \neq 0$. However, it is easy to verify that $y = 0$ is a solution of the given differential equation. Also, we note that $y = 0$ is not a singular solution since it is obtained from the general solution when $c = 0$.*

Example 4

Solve the initial-value problem

$$\frac{dy}{dx} = xy^2 e^x, \quad y(0) = 2$$

SOLUTION   Separating variables yields

$$y^{-2} \, dy = xe^x \, dx$$

Integration of the left-hand side is immediate. We use integration by parts on the right-hand side. Thus

$$\int xe^x \, dx = xe^x - \int e^x \, dx$$

$$= xe^x - e^x$$

$$\begin{array}{c|c} u = x & dv = e^x \, dx \\ \hline du = dx & v = e^x \end{array}$$

The general solution is then

$$-\frac{1}{y} = xe^x - e^x + C$$

Letting $x = 0$ and $y = 2$, we have

$$-\frac{1}{2} = 0 - 1 + C$$

or

$$C = \frac{1}{2}$$

The particular solution satisfying the given condition is then

$$-\frac{1}{y} = xe^x - e^x + \frac{1}{2}$$

or, solving for y, we write

$$y = \frac{2}{2e^x - 2xe^x - 1}$$ ∎

EXERCISES FOR SECTION 2–1

Obtain the family of solutions of each of the given differential equations in Exercises 1–20.

1. $2y \, dx + 2x \, dy = 0$

2. $x^2 \, dy + y^2 \, dx = 0$

3. $xy \, dx + y^2 \, dy = 0$

4. $xy \, dx + (1 + x^2) \, dy = 0$

5. $\dfrac{ds}{dt} = ts^3$

6. $y \, dx + (3x + 2) \, dy = 0$

7. $x(1 - y) \, dx + (x^2 - 2) \, dy = 0$

8. $\cot y \, dx = x \, dy$

9. $dx + (1 + x^2) y \, dy = 0$

10. $\dfrac{dp}{dt} = 2 - p$

11. $xy\,dx + e^x\,dy = 0$

12. $\dfrac{dq}{dt} + \dfrac{t\sin t}{q} = 0$

13. $dx + e^{x+y}\,dy = 0$

14. $v(t + 3)\,dv + (v^2 - 2v - 3)\,dt = 0$

15. $(x^2 - 1)\,dy + x(y^2 + 5y + 4)\,dx = 0$ **16.** $e^x\cos^2 y\,dx - (1 + e^x)\,dy = 0$

17. $\tan x\,dx + y\cos^2 x\,dy = 0$

18. $(x + 1)\,dy = 2xy\,dx$

19. $x^2\dfrac{dy}{dx} = y(1 - x)$

20. $4\,dt = t\sqrt{t^2 - 4}\,ds$

In Exercises 21–26 obtain the particular solution of each differential equation for the given condition.

21. $(y + 2)\,dx + (x - 4)\,dy = 0; x = 2, y = 0$

22. $\dfrac{ds}{dt} = -\dfrac{t}{s}; t = -1, s = 3$

23. $xy\,dx + \sqrt{9 + x^2}\,dy = 0; x = 4, y = 1$

24. $\cos y\,dx + x\sin y\,dy = 0; x = 3, y = \pi/3$

25. $(1 + x^2)\,dy + (1 + y^2)\,dx = 0; x = 1, y = \sqrt{3}$

26. $dy + y\,dx = x^2\,dy; x = 2, y = e$

27. Students sometimes confuse the expression $(y')^2$ with $(y^2)'$. They are not equal in general, of course, but for what functions are the two expressions the same?

28. Find all functions such that the derivative is the cube of the function.

29. Find all functions with derivative two more than the square of the function.

30. Find all functions whose derivative is equal to their reciprocals.

2–2   APPLICATIONS OF SEPARABLE EQUATIONS

Recall that a mathematical expression used to describe a physical condition is called a mathematical model. Many problems that involve a rate of change may be modeled by separable differential equations. In this section we discuss a few of these application areas.

Growth and Decay Problems

Many quantities encountered in science—such as population growth, electric current in a capacitor, and radioactive decay—vary with time in such a way that *the time-rate of change of the quantity is proportional to the quantity itself.* Quantities of this kind are described by the differential equation

$$\frac{dy}{dt} = ky$$

where y is the quantity that is changing and k is a constant of proportionality.

Example 1

The electric circuit shown in Figure 2–1, containing a resistor and capacitor in series, is called a series RC circuit. When the switch is closed, there is a current in the capacitor. The time-rate of change of current i in a capacitor is known to equal $(-1/RC)i$, where $1/RC$ is the **time constant** of the circuit. Find the expression for the current in the capacitor at any time t, if $i = V/R$ when $t = 0$. Assume $t \geq 0$ and $i \geq 0$.

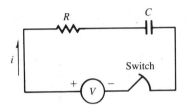

FIGURE 2–1

SOLUTION   The time-rate of change of the current is given by

$$\frac{di}{dt} = -\frac{1}{RC}i$$

Separating variables,

$$\frac{di}{i} = -\frac{1}{RC}dt$$

Hence

$$\ln|i| + \ln|K| = -\frac{t}{RC}$$

$$\ln|iK| = -\frac{t}{RC}$$

$$i = K_1e^{-t/RC} \qquad\qquad \text{(NOTE: } K_1 = 1/K.)$$

Letting $t = 0$ and $i = V/R$ in this equation, we find $K_1 = V/R$ and, consequently,

$$i = \frac{V}{R}e^{-t/RC}$$

is the equation for the current. See Figure 2–2.   ■

A radioactive substance decomposes at a rate proportional to its mass. This rate is called the **decay rate.** If $M(t)$ represents the mass of a substance at any time, then the decay rate dM/dt is proportional to $M(t)$. The **half-life** of a substance is the amount of time for it to decay to one-half of its initial mass.

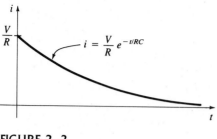

FIGURE 2–2

Example 2

A radioactive isotope has an initial mass of 100 mg, which two years later is 75 mg. Find the expression for the amount of the isotope remaining at any time. What is its half-life?

SOLUTION   Let M be the mass of the isotope remaining after t years, and let $-k$ be the constant of proportionality. Then the rate of decomposition is given by

$$\frac{dM}{dt} = -kM \qquad \text{(NOTE: } \textit{The minus sign indicates the mass is decreasing.)}$$

Separating the variables, integrating, and adding a constant in the form $\ln C$, we find

$$\ln M + \ln C = -kt$$

Simplifying,

$$\ln MC = -kt$$
$$MC = e^{-kt}$$
$$M = C_1 e^{-kt} \qquad (C_1 = 1/C) \tag{2–8}$$

To find the value of C_1, recall that $M = 100$ when $t = 0$. Substituting these values into Equation 2–8, we get $C_1 = 100$ and

$$M = 100e^{-kt} \tag{2–9}$$

The value of k may now be determined by substituting the condition $t = 2, M = 75$ into Equation 2–9. Thus

$$75 = 100e^{-2k}$$

or

$$e^{-2k} = \frac{3}{4}$$

so $\quad -2k = \ln \frac{3}{4}$

$$k = \frac{1}{2} \ln \frac{4}{3} = \frac{1}{2}(0.2877) = 0.1438$$

The mass of the isotope remaining after t years is then given by

$$M = 100e^{-0.14t}$$

The half-life t_h is the time corresponding to $M = 50$ mg. Thus

$$50 = 100e^{-0.14t_h}$$

$$\frac{1}{2} = e^{-0.14t_h}$$

$$t_h = -\frac{1}{0.14} \ln 0.5 = \frac{-0.693}{-0.14} = 4.95 \text{ years} \qquad \blacksquare$$

The next example is concerned with the process of heating and cooling. Experiments show that if an object is immersed in a medium and if the medium is kept at a constant temperature, a good approximation to the temperature of the object can be found by using Newton's law of cooling.

NEWTON'S LAW OF COOLING: The temperature Q of an object will change at a rate proportional to the difference in temperature of the object and the temperature Q_a of the surrounding medium.

Example 3

A thermometer reading $100°$ F is placed in a pan of oil maintained at $10°$ F. What is the temperature of the thermometer when $t = 10$ sec, if its temperature is $60°$ F when $t = 4$ sec?

SOLUTION   The physical situation may be described mathematically by using Newton's law of cooling. If Q is the temperature of the thermometer at any time t, then dQ/dt is the time-rate of change of temperature, and $Q - 10$ is the difference between the temperature of the thermometer and that of the oil. The boundary conditions are $Q = 100$ when $t = 0$ and $Q = 60$ when $t = 4$. Thus the boundary-value problem to be solved is

$$dQ/dt = -k(Q - 10); Q(0) = 100, Q(4) = 60$$

where k is the constant of proportionality (the minus sign is used to indicate that the temperature of the thermometer will decrease). Separating variables yields

$$dQ/(Q - 10) = -k \, dt$$

Integrating, we get

$$\ln |Q - 10| = -kt + \ln |C|$$

which leads to

$$Q = Ce^{-kt} + 10$$

The constant C is evaluated, using $T(0) = 100$, to get

$$100 = C + 10 \qquad \text{or} \qquad C = 90$$

So, we have

$$Q = 90e^{-kt} + 10$$

To solve for k, use the second condition $Q(4) = 60$. Making the substitution

$$60 = 90e^{-4k} + 10$$

or

$$k = \frac{1}{4}\ln\frac{9}{5} = 0.147$$

Therefore the temperature function is

$$Q = 90e^{-0.147t} + 10$$

The temperature variation is shown graphically in Figure 2–3. Notice that the limiting temperature is $10°$ F. Finally, the temperature of the thermometer at $t = 10$ sec is

$$Q(10) = 90e^{-1.47} + 10$$

$$= 90(0.2299) + 10 = 30.7° \text{ F}$$

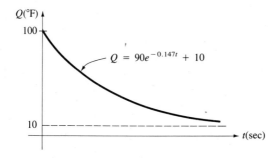

FIGURE 2–3

Orthogonal Families

Recall from Section 1–1 that the differential equation of a one-parameter family of curves is obtained by differentiating the equation of the family in such a way as to eliminate the parameter of the family.

Example 4

Find the differential equation of the family of curves defined by the expression $y = 2 + ce^{-3x}$. See Figure 2–4.

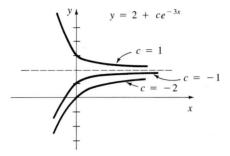

FIGURE 2—4

SOLUTION To find the differential equation of the given family, we differentiate $y = 2 + ce^{-3x}$ with respect to x, to obtain

$$y' = -3ce^{-3x}$$

The parameter c is eliminated between these two equations by solving $y' = -3ce^{-3x}$ for c and substituting this expression into the equation of the family. Since $c = -\frac{1}{3}y'e^{3x}$, we have

$$y = 2 + (-\frac{1}{3}y'e^{3x})e^{-3x}$$

After simplification we write the differential equation of the given family as

$$\frac{1}{3}y' + y = 2 \qquad \blacksquare$$

Two families of curves are said to be **orthogonal** (or perpendicular) if every curve of one family intersects every curve of the other family at right angles. Figure 2–5 shows two curves that are orthogonal at $x = x_0$. Notice that the tangent lines to the two curves are perpendicular.

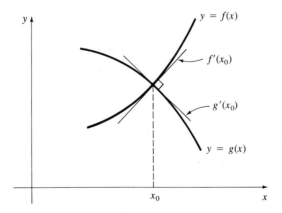

FIGURE 2—5

Orthogonal trajectories

alternating;

We can easily give the equation governing the relationship between two orthogonal families since the *slopes of lines that intersect at right angles are negative reciprocals of each other*. Thus if $y = f(x)$ is one family and $y = g(x)$ is the other, then

$$f'(x) = -\frac{1}{g'(x)}$$

or using y_0 to denote one of the orthogonal families,

$$\frac{dy}{dx} = -\frac{1}{dy_0/dx}$$

The family of curves orthogonal to a given family is said to be the **orthogonal trajectories** to the given family of curves. To find the orthogonal trajectories, we proceed as follows:

1. Find the differential equation of the given family by the method described previously.

2. Replace dy/dx by $-1/(dy_0/dx)$, and y by y_0. This is now the differential equation of the orthogonal trajectories.

3. Solve the differential equation for the orthogonal trajectories.

Example 5

Find the orthogonal trajectories to the family of circles with center at the origin.

SOLUTION   The equation of the family of circles with center at the origin is

$$x^2 + y^2 = r^2$$

Differentiating this equation with respect to x, we get

$$x + yy' = 0$$

This is the differential equation of the family of circles. Replacing y' with $-1/y_0'$, *and* y with y_0, the differential equation of the orthogonal trajectories is

$$x + y_0(-1/y_0') = 0$$

$$x\,dy_0 - y_0\,dx = 0$$

The solution of this equation is found, by separating variables, to be

$$\ln y_0 - \ln x = \ln m \qquad (m = \text{a constant})$$

$$y_0 = mx$$

This is the family of straight lines through the origin. See Figure 2–6.

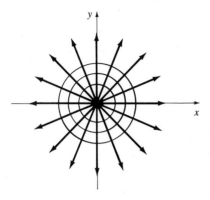

FIGURE 2–6   ■

Example 6

Find the orthogonal trajectories to the family of parabolas with vertex at the origin and foci on the y-axis.

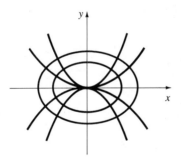

FIGURE 2–7

SOLUTION   The equation of the given family is

$$x^2 = 4ay$$

Differentiating with respect to x,

$$2x = 4ay'$$

Eliminating the parameter a between these two equations, we obtain the differential equation of the family,

$$x^2\, dy/dx - 2xy = 0$$

or

$$dy/dx = 2y/x$$

Consequently, the differential equation of the orthogonal trajectories is

$$dy_0/dx = -x/2y_0$$

Solving by separation of variables, we have

$$\frac{1}{2}x^2 + y_0^2 = C$$

which is the family of ellipses shown in Figure 2–7.   ∎

Models Derived from Geometric Considerations

In many cases a differential equation is generated as a result of the geometry of a problem. The analysis of such problems is an important part of understanding applications of differential equations. The differential equation for the shape of a flexible hanging cable, which was derived in Example 5 of Section 1–1, came from the geometry of the cable. In Example 7 we consider the flow of a liquid through an orifice in terms of the geometry of the container.

Example 7

The cylindrical container shown in Figure 2–8 has a constant cross-sectional area A. The orifice at the base of the container has a constant cross-sectional area B. If the container is filled with water to a height h, water will flow out through the orifice. Show that the rate at which the water level is dropping is given by

$$\frac{dh}{dt} = -\frac{B\sqrt{2gh}}{A}$$

SOLUTION   Let h be the height of the water at time t, and let $h + \Delta h$ be the height at time $t + \Delta t$. Then

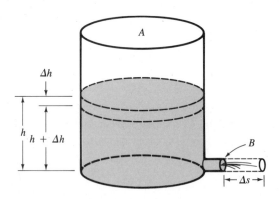

FIGURE 2–8

$$\begin{bmatrix} \text{Volume of water lost} \\ \text{when the level drops} \\ \text{by an amount } \Delta h \end{bmatrix} = \begin{bmatrix} \text{Volume of water} \\ \text{that escapes through} \\ \text{the orifice} \end{bmatrix}$$

The volume of water lost in time Δt is given by $-A\Delta h$, where the minus sign indicates that the volume is a loss. The volume of water flowing through the orifice in time Δt is the volume contained in a cylinder of cross section B and length Δs. Thus we have

$$-A\Delta h = B\Delta s$$

Dividing by Δt and taking the limit as $\Delta t \to 0$, we get

$$-A\frac{dh}{dt} = B\frac{ds}{dt} \tag{2--10}$$

where ds/dt is the velocity of the water through the orifice. The velocity of the water issuing from the orifice decreases as the level of the water decreases and is given by Torricelli's law to be

$$\frac{ds}{dt} = \sqrt{2gh} \tag{2--11}$$

Using this relation in Equation 2–10, we get

$$\frac{dh}{dt} = -\frac{B\sqrt{2gh}}{A} \qquad\blacksquare$$

EXERCISES FOR SECTION 2–2

Growth and Decay Problems

1. The circuit shown in Figure 2–9, consisting of a resistor and an inductor in series, is called a series RL circuit. When the switch is opened, the current in the circuit decays at a rate equal to $-Ri/L$. Find the current in the circuit at $t = 1$. Assume $i = 50$ amp when $t = 0$.

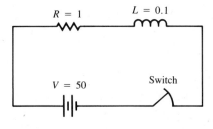

FIGURE 2–9

2. A steel casting at a temperature of $20°$ C is put into an oven that has a temperature of $200°$ C. One minute later the temperature of the casting is $30°$ C. Find the temperature of the casting 5 minutes after it is put into the oven.

3. How long will it take the temperature of the casting in Exercise 2 to reach $190°$ C?

4. A steel ball is heated to a temperature of $80°$ C and at time $t = 0$ is placed in water, maintained at $30°$ C. At $t = 3$ minutes the temperature of the ball if $55°$ C. When is the temperature of the ball reduced to $40°$ C?

5. A certain radioactive substance decays at a rate that is always equal to 25% of the remaining mass. Find the formula for the mass of the substance remaining after t hours if $M = 200$ g when $t = 0$.

6. What is the half-life of a radioactive substance of which 20% disappears in 100 years?

7. After 20 years a quantity of a radioactive substance has decayed to 60 grams, and at the end of 50 years to 40 grams. How many grams were there in the first place?

8. (a) The half-life of carbon 14 (C^{14}) is 5568 years. Show that the formula for the mass at time t is

$$m = m_0 e^{-0.0001245t} \tag{2-12}$$

Carbon dating is one of the tools used by archeologists to estimate the date of origin of organic artifacts is carbon dating, a technique developed by W. F. Libby in the 1950s. Carbon dating is based on the principles of radioactive decay of C^{14} in plants and animals that have died.

   Cosmic rays produce neutrons in the atmosphere that combine with nitrogen to form C^{14}. In living plants and animals, the rate of absorption of C^{14} is balanced by its rate of decay so that the amount of C^{14} is in a state of equilibrium. When the living organism dies, C^{14} atoms decay, but no new C^{14} is absorbed. The decay process at death is described by Equation 2–12. The date when the organism died can then be estimated by measuring the C^{14} level in the specimen and using Equation 2–12 to compute the corresponding time. Carbon dating assumes that the level of C^{14} in living organisms is the same today as it was in the past.

(b) If a piece of charcoal from an archeological excavation is found to contain 15% as much C^{14} as living wood, use Equation 2–12 to estimate the age of the specimen.

9. The acceleration of a solenoid-actuated plunger is found to be proportional to the velocity of the plunger. If the initial velocity of the plunger is 10 cm/sec, and 0.1 sec later it is 100 cm/sec, what is the velocity when $t = 0.2$ sec?

10. The intensity of light I emitted by the phosphor of a television tube decreases at a rate proportional to the intensity I. Find the expression for the intensity of light being emitted at any time t if the initial intensity is $I = 25$ and the intensity at $t = 1$ is $I = 20$.

11. The rate of change of atmospheric pressure P with altitude h is given by $dP/dh = -kP$, where k is a constant of proportionality. If the atmospheric pressure at ground level is 15 psi, and at 10,000 ft it is 10 psi, find the expression for the pressure as a function of altitude. What is the pressure at 3,000 ft?

12. It is found in chemistry that simple chemical conversions take place at a rate proportional to the amount of unconverted substance remaining. If initially there

were 50 g of a substance, and 1 hr later 20 g remained, how long will it take to convert 80% of the substance?

Orthogonal Trajectories

Find the orthogonal trajectories of each family of curves in Exercises 13–22 and sketch a few curves of each family.

13. Straight lines having a slope of 2.

14. Straight lines with y-intercept equal to the slope.

15. The ellipses with center at the origin, foci on the x-axis, and major axis twice that of the minor axis.

16. The parabolas with vertex at $(0, 0)$ and focus on the x-axis.

17. The hyperbolas $x^2 - y^2 = c$.   **18.** The cubics $y = cx^3$.

19. The exponentials $y = ce^x$.   **20.** The logarithms $y = \ln cx$.

21. In the transfer of heat through a body the points through which the temperature is constant are called **isotherms.** The orthogonal trajectories of the isotherms give the direction of heat flow and are, consequently, called **lines of flow.** Find the lines of flow in a long copper wedge if the isotherms are given by $y = kx$.

22. If the equipotential lines of the electric field of a single-phase induction motor are given by $c^2x^2 + y^2 = c^2$, find the equation for the lines of flux for the electric field if they are orthogonal trajectories of the equipotential lines.

Equations Derived from the Geometry of the Problem

23. In Example 7 we showed that the rate at which the water level drops in a tank like that shown in Figure 2–10 is given by

$$\frac{dh}{dt} = -\frac{B\sqrt{2gh}}{A}$$

where A is the cross-sectional area of the tank and B is the cross-sectional area of the orifice. (a) If the initial depth of the water in the tank is h_0, find the function for the depth of the water at time t. (b) How much time does it take to empty the tank completely?

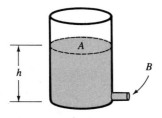

FIGURE 2–10

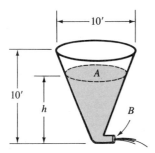

FIGURE 2–11

24. The container in Figure 2–11 is filled with water. (a) Show that the rate at which the water drops is given by $dh/dt = -4B\sqrt{2g}/\pi h^{3/2}$. (HINT: Water lost in time Δt = Change in volume of water in tank.) (b) Solve for h, given that $h(0) = h_0$. (c) How long will it take to empty the tank?

25. The tangent line to a given curve at any point $P(x, y)$ has an x-intercept equal to $\frac{1}{2}x$. If the curve passes through $(1, 2)$, find its equation.

26. The tangent line to a given curve at any point $P(x, y)$ has a slope everywhere equal to y^2. If the curve passes through $(1, 1)$, find its equation.

27. The tangent line to a curve has a slope that is everywhere equal to the ratio x/y. If the curve passes through $(1, 0)$, find its equation.

28. A cylindrical container filled with water is rotated about its axis with a constant angular velocity ω, which causes the water surface to take on the shape of a paraboloid. See Figure 2–12(a). The forces experienced by a particle on the surface are shown in Figure 2–12(b). In addition to the weight $w = mg$, there is a normal reaction r, due to the surrounding particles, and a centrifugal force $f = m\omega^2 x$. Show that the equation of the parabola is

$$y = \frac{\omega^2 x^2}{2g}$$

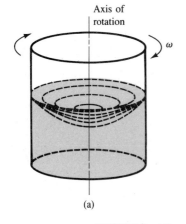

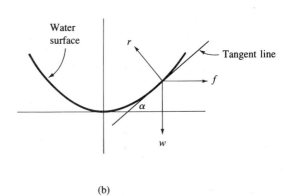

(a)

(b)

FIGURE 2–12

29. Snow began falling at a constant rate sometime during the morning of December 3rd. A snowplow starts out at noon, clearing 2 miles of road the first hour and 1 mile the second. Assume that the snowplow clears snow at a constant rate R. This means that the deeper the snow cover, the slower the truck will move forward. Determine the time that the snow started to fall. See Figure 2–13. The initial steps in the solution are outlined following Figure 2–13.

FIGURE 2–13

To solve this problem, we note that the volume of snow taken into the plow in time Δt is equal to the volume of snow discharged in that same time interval. If the truck moves Δx miles in Δt hours, the volume of snow taken in is given by

$$h \cdot w \cdot \Delta x$$

and the snow blown out is equal to

$$R \cdot \Delta t$$

Therefore since the rate in must equal the rate out, we have

$$hw\Delta x = R\Delta t$$

or

$$hw \frac{\Delta x}{\Delta t} = R$$

In the limit as $\Delta t \rightarrow 0$ we get

$$hw \frac{dx}{dt} = R$$

which is the desired model. An additional observation is that the depth of snow h is given by

$$h = r(t + t_0)$$

where r is the constant snowfall rate, t is the number of hours from noon that the truck has been on the road, and t_0 is the time in hours that it had been snowing prior to noon. The boundary conditions on this problem are

$x(0) = 0$ (Zero miles of road have been plowed at $t = 0$.)

$x(1) = 2$ (Two miles of road have been plowed after one hour.)

$x(2) = 3$ (Three miles of road have been plowed after two hours.)

30. Figure 2–14 shows a diagram of a dog chasing a rabbit. This is similar to the problem described in Example 6 of Section 1–1. The rabbit starts at (0, 0) and runs at a constant speed v_R along the y-axis. The dog starts the chase at (1, 0) and runs at a constant speed v_D so that its line of sight is always directed at the rabbit. If $v_D > v_R$, the dog will catch the rabbit; otherwise the rabbit gets away. Determine the equation of the pursuit curve. An outline of the solution follows Figure 2–14.

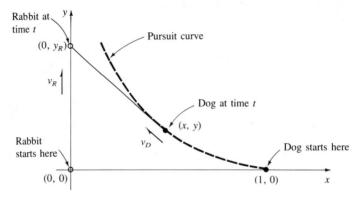

FIGURE 2–14

It was shown previously that the differential equation of the pursuit curve is

$$\frac{dy}{dx} = \frac{y - y_R}{x}$$

The position of the dog at any time $t > 0$ is (x, y), and the y-coordinate of the rabbit at the corresponding time is $y_R = 0 + v_R t = v_R t$, so

$$\frac{dy}{dx} = \frac{y - v_R t}{x}$$

or

$$xy' = y - v_R t$$

Implicitly differentiating this expression with respect to x yields

$$xy'' + y' = y' - v_R(dt/dx)$$

This may be rewritten as

$$\frac{xy''}{v_R} = -\frac{dt}{dx} \tag{2-13}$$

Finally, we note that the speed of the dog can be written

$$v_D = \frac{ds}{dt} = \sqrt{\left(\frac{dx}{dt}\right)^2 + \left(\frac{dy}{dt}\right)^2} = -\sqrt{1 + \left(\frac{dy}{dx}\right)^2}\,\frac{dx}{dt}$$

See Figure 2–15.

Solving this for dt/dx, we have

$$\frac{dt}{dx} = -\frac{1}{v_D}\sqrt{1 + (y')^2}$$

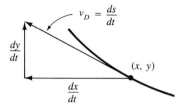

FIGURE 2–15

Substituting this result into Equation 2–13 yields

$$\frac{xy''}{v_R} = \frac{1}{v_D}\sqrt{1 + (y')^2}$$

This equation may now be made separable by using the substitution $w = y'$. The initial conditions are $y(1) = 0$ and $y'(1) = 0$.

31. A power company's transmission line hangs under its own weight between two towers, as shown in Figure 2–16. Determine the equation of the shape of the cable.

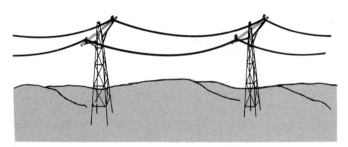

FIGURE 2–16

We showed in Example 7 of Section 1–1 that the differential equation of a hanging cable is

$$\frac{d^2y}{dx^2} = \frac{1}{T_C}\frac{dW}{dx} \tag{2–14}$$

For a cable hanging under its own weight we can obtain an expression for dW/dx by examining a portion of the cable, as shown in Figure 2–17. If the linear density of the cable w is constant, the weight of the segment is $W = ws$. Since s depends on x, it follows that

$$\frac{dW}{dx} = w\frac{ds}{dx} \tag{2–15}$$

FIGURE 2–17

Further, we know from integral calculus that the arc length of the segment of cable is given by

$$s = \int_{x_1}^{x} \sqrt{1 + (y')^2}\, dx$$

Or, differentiating with respect to x,

$$\frac{ds}{dx} = \sqrt{1 + (y')^2} \tag{2-16}$$

Combining the results of Equations 2–15 and 2–16 and substituting into Equation 2–14 yields

$$\frac{d^2y}{dx^2} = \frac{w}{T_C} \sqrt{1 + (y')^2}$$

To find the equation for the shape of the cable, let $p = y'$ and solve the resulting separable equation. For convenience we assume that the low point of the cable is $(0,\, T_C/w)$.

2–3   THE METHOD OF SUBSTITUTION

The differential equation

$$y' = \frac{x + y - 1}{x + y + 1}$$

is a nonseparable differential equation; that is, it cannot be expressed directly in the form

$$f(x)\, dx + g(y)\, dy = 0$$

In this section we discuss several types of nonseparable equations that can be made separable by an appropriate substitution. The substitution to be used depends upon the form of the given differential equation.

Substitutions Suggested by the Equation

Some substitutions are suggested by the occurrence of similar groupings of terms in the differential equation. The following example shows an equation of this type.

Example 1

Solve the differential equation

$$y' = \frac{x + y - 1}{x + y + 1}$$

SOLUTION   The occurrence of the expression $x + y$ in both the numerator and the denominator suggests that we let $u = x + y$ in the equation. From $y = u - x$ we get $y' = u' - 1$. Substituting $u = x + y$ and $y' = u' - 1$ into the differential equation yields

$$u' - 1 = \frac{u - 1}{u + 1}$$

$$u' = \frac{u - 1}{u + 1} + 1$$

Adding the terms on the right-hand side, we get

$$u' = \frac{2u}{u + 1}$$

Separating variables,

$$\frac{u + 1}{u} \, du = 2 \, dx$$

Integrating both sides,

$$u + \ln |u| = 2x + c$$

Finally, since $u = x + y$, the solution in implicit form is

$$y - x + \ln |x + y| = c$$ ■

Homogeneous Equations

Consider the nonseparable differential equation

$$(2x^2 + y^2) \, dx - xy \, dy = 0$$

We rearrange the equation into the form

$$\frac{dy}{dx} = \frac{2x^2 + y^2}{xy} = 2\left(\frac{x}{y}\right) + \left(\frac{y}{x}\right)$$

In this form a substitution for the quotient y/x is suggested. Letting $v = y/x$, it follows that $y = vx$ and $y' = v + v'x$. Substituting into the differential equation,

$$v + x\frac{dv}{dx} = \frac{2}{v} + v$$

from which

$$x\frac{dv}{dx} = \frac{2}{v}$$

The variables v and x can now be separated to give

$$v \, dv = \frac{2}{x} \, dx$$

Integrating both sides gives

$$\frac{1}{2}v^2 = 2 \ln |x| + \ln c_1$$

or

$$v^2 = \ln cx^4 \qquad\qquad \text{(NOTE: } c = c_1^2.)$$

To get the solution in terms of x and y, replace v with y/x. Thus

$$\frac{y^2}{x^2} = \ln cx^4$$

or

$$y^2 = x^2 \ln cx^4$$

At this point the question in your mind should be "Are there other differential equations that can be made separable by the substitution $y = vx$ and, if there are, how can they be identified?" To answer this question, we introduce the concept of a homogeneous function.

DEFINITION

A function $f(x, y)$ is said to be **homogeneous** of degree n in x and y if, for every k,

$$f(kx, ky) = k^n f(x, y)$$

where k is a real parameter.

Example 2

(a) $f(x, y) = x^2 + xy$ is homogeneous of degree 2 since

$$f(kx, ky) = (kx)^2 + (kx)(ky) = k^2(x^2 + xy) = k^2 f(x, y)$$

(b) $f(x, y) = \sqrt{x + y}$ is homogeneous of degree $\frac{1}{2}$ since

$$f(kx, ky) = (kx + ky)^{1/2} = k^{1/2}(x + y)^{1/2}$$

(c) $f(x, y) = e^{x/y}$ is homogeneous of degree zero since

$$f(kx, ky) = e^{kx/ky} = k^0 e^{x/y}$$

(d) $f(x, y) = x^2 + y^2 + 5$ is not homogeneous since

$$f(kx, ky) = k^2 x^2 + k^2 y^2 + 5$$

which *cannot* be expressed as $k^2 f(x, y)$.   ∎

DEFINITION

A first-order differential equation

$$M(x, y) \, dx + N(x, y) \, dy = 0$$

is said to be **homogeneous** if $M(x, y)$ and $N(x, y)$ are both homogeneous functions of degree n.

The next example shows that the differential equation solved earlier in this section by the substitution $y = vx$ is homogeneous.

Example 3

Show that

$$(2x^2 + y^2) \, dx - xy \, dy = 0$$

is a homogeneous equation.

SOLUTION   In this equation we see that $M(x, y) = 2x^2 + y^2$ and $N(x, y) = -xy$. Therefore

$$M(kx, ky) = 2k^2x^2 + k^2y^2 = k^2(2x^2 + y^2) = k^2M(x, y)$$

and

$$N(kx, ky) = -(kx)(ky) = -k^2xy = k^2N(x, y)$$

$M(x, y)$ and $N(x, y)$ are both homogeneous functions of degree 2; hence the given differential equation is homogeneous.   ∎

The following theorem tells us that first-order homogeneous differential equations can be transformed into separable equations.

THEOREM 2–1

If $M(x, y) \, dx + N(x, y) \, dy = 0$ is a homogeneous equation of degree n, then the substitution **y = vx** transforms the given equation into a separable equation in the variables v and x.

PROOF   Write the given homogeneous equation in the form

$$\frac{dy}{dx} = -\frac{M(x, y)}{N(x, y)}$$

Let $y = vx$, so that

$$\frac{d}{dx}(vx) = -\frac{M(x, vx)}{N(x, vx)} = -\frac{x^n M(1, v)}{x^n N(1, v)}$$

Then

$$v + x\frac{dv}{dx} = -\frac{M(1, v)}{N(1, v)}$$

Observing that the expression on the right side is a function of v, we let $g(v) = -M(1, v)/N(1, v)$ and write

$$v + x\frac{dv}{dx} = g(v)$$

Separating the variables,

$$\frac{dx}{x} + \frac{dv}{v - g(v)} = 0$$

Therefore we have shown that, regardless of the form of a homogeneous equation, we can always obtain a separable form by the substitution $y = vx$.

Example 4

Solve the equation

$$xy' = xe^{-y/x} + y, \qquad x > 0$$

SOLUTION   Dividing by x, we have

$$y' = \frac{xe^{-y/x} + y}{x}$$

The numerator and denominator are both homogeneous of degree 1, so we let $y = vx$. Thus $y' = v + xv'$ and the given equation becomes

$$v + xv' = e^{-v} + v$$

$$xv' = e^{-v}$$

Separating variables yields

$$e^v \, dv = dx/x$$

Integrating, we get

$$e^v = \ln x + \ln c$$

or, since $v = y/x$,

$$e^{y/x} = \ln cx$$

Solving for y,

$$y = x \ln |\ln cx| \qquad\qquad\blacksquare$$

COMMENT: By interchanging the roles of x and y in a homogeneous equation, the substitution $x = vy$ can be used to obtain a separable equation. Either substitution will produce a separable equation, but sometimes the substitution $x = vy$ produces a separable equation that is easier to solve than that obtained by the substitution $y = vx$.

Example 5

Use the substitution $x = vy$ to solve the differential equation

$$xy\, dx - (2x^2 + y^2)\, dy = 0$$

SOLUTION   Here $M(x, y) = xy$ and $N(x, y) = -(2x^2 + y^2)$. Both of these functions are homogeneous of degree two, so the equation is homogeneous. We write the given equation in the form

$$\frac{dy}{dx} = \frac{xy}{2x^2 + y^2}$$

and observe that the substitution $y = vx$ produces the separable equation

$$\frac{v^2 + 2}{v(v^2 + 1)}\, dv + \frac{dx}{x} = 0$$

While this equation can be integrated by using partial fraction techniques, the substitution $x = vy$ produces a separable equation that is easier to solve.

To use the substitution $x = vy$, we write the given equation in the form

$$\frac{dx}{dy} = \frac{2x^2 + y^2}{xy}$$

Then, since $x = vy$, $dx/dy = v + y(dv/dy)$, and we get

$$v + y\frac{dv}{dy} = \frac{2(vy)^2 + y^2}{(vy)y} = \frac{2v^2 + 1}{v}$$

or

$$y\frac{dv}{dy} = \frac{2v^2 + 1}{v} - v = \frac{v^2 + 1}{v}$$

Separating variables yields

$$\frac{v\, dv}{v^2 + 1} = \frac{dy}{y}$$

This integrates as

$$\ln(v^2 + 1) = 2\ln|y| + \ln c$$

Using the properties of logarithms, this can be written

$$v^2 + 1 = cy^2$$

Substituting $v = x/y$ yields

$$x^2 + y^2 = cy^4$$ ∎

Linear Substitutions

Consider the equation

$$M(x, y)\, dx + N(x, y)\, dy = 0$$

in which M and N are linear functions of x and y. That is,

$$M(x, y) = a_1 x + b_1 y + c_1$$

and

$$N(x, y) = a_2 x + b_2 y + c_2$$

Equations of this type can be made separable by a substitution; the form of the substitution depends upon the values of the coefficients as follows:

1. If c_1 and c_2 are both zero, the equation is homogeneous, and we may use the substitution $y = vx$.

2. If $a_1 b_2 = a_2 b_1$, then $a_1 x + b_1 y = k(a_2 x + b_2 y)$, and we may use the substitution $u = a_2 x + b_2 y$.

3. If c_1 and c_2 are not both zero and if $a_1 b_2 \neq a_2 b_1$, then the equation can be made homogeneous by the substitution $x = X + h$ and $y = Y + k$, where h and k are constants to be determined.

To show the validity of this substitution, we observe that from $x = X + h$ and $y = Y + k$, we get $dx = dX$ and $dy = dY$. Then substituting into the given equation yields

$$(a_1 X + a_1 h + b_1 Y + b_1 k + c_1)\, dX$$
$$+ (a_2 X + a_2 h + b_2 Y + b_2 k + c_2)\, dY = 0$$

The constant terms in each of these coefficients are equated to 0.

$$a_1 h + b_1 k + c_1 = 0$$

$$a_2 h + b_2 k + c_2 = 0$$

Now taking h and k to be the solution pair of this linear system, the differential equation becomes

$$(a_1 X + b_1 Y)\, dX + (a_2 X + b_2 Y)dY = 0$$

which is homogeneous of degree one in X and Y. We formalize this discussion in the following theorem.

THEOREM 2–2

A differential equation of the form

$$(a_1x + b_1y + c_1) \, dx + (a_2x + b_2y + c_2) \, dy = 0, \ a_1b_2 \neq a_2b_1$$

can be made homogeneous of degree one with the substitutions

$$x = X + h \qquad and \qquad y = Y + k$$

Example 6

Solve the differential equation

$$(y - x - 1) \, dx + (x + y - 1) \, dy = 0$$

by using the substitutions $x = X + h$ and $y = Y + k$.

SOLUTION   By substituting $x = X + h$ and $y = Y + k$, the differential equation becomes

$$(Y + k - X - h - 1) \, dX + (X + h + Y + k - 1) \, dY = 0$$

The values of h and k are calculated to satisfy simultaneously

$$-h + k - 1 = 0 \qquad h + k - 1 = 0$$

The solution to this system is $h = 0$, $k = 1$. Using these values in the given differential equation, we obtain

$$(-X + Y) \, dX + (X + Y) \, dY = 0$$

The coefficients $Y - X$ and $Y + X$ are both homogeneous of degree one, so we use the substitution $Y = vX$ to find the solution. Writing the equation in the form

$$\frac{dY}{dX} = -\frac{Y - X}{Y + X}$$

the substitution $Y = vX$ yields

$$v + Xv' = -\frac{v - 1}{v + 1} \qquad \text{or} \qquad Xv' = -\frac{v^2 + 2v - 1}{v + 1}$$

Separating variables, we get

$$\frac{dX}{X} + \frac{(v + 1) \, dv}{v^2 + 2v - 1} = 0$$

Integrating, the solution is

$$2 \ln |X| + \ln |v^2 + 2v - 1| = \ln c_1$$

which can be written

$$X^2(v^2 + 2v - 1) = c_1$$

Substituting $v = Y/X$ yields

$$Y^2 + 2XY - X^2 = c$$

Letting $X = x - h = x$ and $Y = y - k = y - 1$, the solution of the differential equation in this example is

$$(y - 1)^2 + 2x(y - 1) - x^2 = c$$ ■

EXERCISES FOR SECTION 2–3   1–25 odd, 29, 31, 35

Solve each of the equations in Exercises 1–18. Identify the homogeneous equations.

1. $y' = 2 + \dfrac{y}{x}$ **2.** $x\,dy - (x + y)\,dx = 0$ **3.** $\dfrac{dy}{dx} = \dfrac{y}{x} + \sin\dfrac{y}{x}$

4. $[y + x\tan(y/x)]\,dx - x\,dy = 0$ **5.** $xy\,dx - (x^2 + y^2)\,dy = 0$

6. $(x^2 + y^2)\,dx - xy\,dy = 0$ **7.** $y(x + y)\,dx - x^2\,dy = 0$

8. $xy\,dx - (x^2 + 1)\,dy = 0$ **9.** $xy^2\,dx - \sqrt{x^2 + 4}\,dy = 0$

10. $y' = \dfrac{(x + y)^2}{x^2}$ **11.** $y' = \dfrac{2x - y}{x - 2y}$ **12.** $x^2\dfrac{dy}{dx} = y^2 + 4xy + 2x^2$

13. $x\,dx + (y - 2x)\,dy = 0$ **14.** $y' = \dfrac{x - y}{x + y}$

15. $y' = \dfrac{y^2 - xy - 8x^2}{x^2}$ **16.** $x^2\,dy - (y^2 - xy)\,dx = 0$

17. $x\,dy - (y + \sqrt{x^2 - y^2})\,dx = 0$ **18.** $\dfrac{x}{y}\dfrac{dy}{dx} = \dfrac{x^2 + y^2}{x^2 - y^2}$

In Exercises 19–22 find the indicated particular solution.

19. $xy' = x - y;\ y(2) = -3$ **20.** $\dfrac{dy}{dx} = \dfrac{y - \sqrt{x^2 + y^2}}{x};\ y(3) = 1$

21. $xy^2\,dx - (x^2y - 4y)\,dy = 0;\ y(4) = 2$

22. $(x - ye^{y/x})\,dx + xe^{y/x}\,dy = 0;\ y(1) = 1$

In Exercises 23–28, use a linear substitution to reduce the equation to homogeneous form. Show that each is in the same form as the homogeneous equation in the exercise in parentheses. Solve the given equation by making an appropriate substitution into the solution of the exercise in parentheses.

23. $y' = \dfrac{2x + y + 4}{x - 2}$, (Exercise 1)   **24.** $y' = \dfrac{x + y + 3}{x + 5}$, (Exercise 2)

25. $(x - 2y + 1)\,dy + (y - 2x + 2)\,dx = 0$, (Exercise 11)

26. $(x - 2) \, dx + (y - 2x + 1) \, dy = 0$, (Exercise 13)

27. $(y + x + 3) \, dx - (x + 2) \, dy = 0$, (Exercise 2)

28. $y' = \dfrac{x - y}{x + y + 2}$, (Exercise 14)

In Exercises 29–30 use a linear substitution to reduce the given equation to homogeneous form, and then solve.

29. $(x - 2y + 1) \, dy + (y - 2x + 2) \, dx = 0$

30. $(x + y + 3) \, dx - (x + 2) \, dy = 0$

In Exercises 31–35 use a substitution as suggested by the differential equation.

31. $y' = \dfrac{x + 2y - 1}{x + 2y + 7}$ **32.** $y' = \dfrac{x - y}{x - y + 2}$

33. $(2x + y) \, dx + (2x + y - 1) \, dy = 0$

34. $y' = \dfrac{1 - \sqrt{x + y}}{\sqrt{x + y}}$ **35.** $y' = \dfrac{(2x + y)^2 - 3}{(2x + y) + 2}$

36. Show that if two functions are homogeneous of the same degree in x and y, then their quotient is homogeneous of degree 0.

37. Show that if a function is homogeneous of degree 0 in x and y, then it is a function of y/x only. Use this to show that if f is homogeneous of degree k in x and y, then f can be written as a product of x^k and a function homogeneous of degree 0.

38. Prove Euler's theorem: If f is a homogeneous function of degree k in x and y, then $xf_x + yf_y = kf$.

2–4   EXACT DIFFERENTIAL EQUATIONS

In this section we consider a special kind of nonseparable equation called an exact differential equation. Recall from calculus that the **total differential** of a function of two variables $u(x, y)$ is given by

$$du = \frac{\partial u}{\partial x} \, dx + \frac{\partial u}{\partial y} \, dy \qquad (2\text{–}17)$$

For instance, if $u = x^2 y^3$, then the total differential of u is

$$du = 2xy^3 \, dx + 3x^2 y^2 \, dy$$

As the following example shows, a knowledge of the total differential of a function u is sufficient to determine the function within an arbitrary constant.

Example 1

Suppose the total differential of a function is given by

$$du = 3x(xy - 2) \, dx + (x^3 + 2y) \, dy$$

Find $u(x, y)$.

SOLUTION   Since du is the total differential of u, we know that

$$\frac{\partial u}{\partial x} = 3x^2y - 6x \qquad \text{and} \qquad \frac{\partial u}{\partial y} = x^3 + 2y$$

Integrating the first equation with respect to x, holding y constant, gives

$$u(x, y) = x^3y - 3x^2 + C(y)$$

$C(y)$ stands for an arbitrary function of y since y is considered to be constant for the partial antiderivative. From this function $\partial u/\partial y = x^3 + C'(y)$. Equating this to the given expression for $\partial u/\partial y$, we have

$$x^3 + C'(y) = x^3 + 2y$$
$$C'(y) = 2y$$

This means that $C(y) = y^2 + k$, and therefore

$$u(x, y) = x^3y - 3x^2 + y^2 + k \qquad\qquad \blacksquare$$

The following definition establishes the connection between the concept of a total differential and the subject of first-order ordinary differential equations.

DEFINITION

The first-order differential equation

$$M(x, y)\, dx + N(x, y)\, dy = 0$$

is said to be **exact** if $M(x, y)\, dx + N(x, y)\, dy$ is the total differential of some function $u(x, y)$.

Since an exact equation can be written in the form $du = 0$, its solution in implicit form may be written $u(x, y) = c$. The problem is to find the function $u(x, y)$. Note that **$u(x, y)$ is not the solution but, when equated to c, it defines an implicit solution.**

Example 2

Solve the differential equation

$$3x(xy - 2)\, dx + (x^3 + 2y)\, dy = 0$$

SOLUTION   In the previous example we found a function u such that du is the left-hand side of this equation. Hence the implicit solution is

$$x^3y - 3x^2 + y^2 + k = c$$
$$x^3y - 3x^2 + y^2 = C \qquad (C = c - k) \qquad\qquad \blacksquare$$

COMMENT: Since the constant of integration can be combined with the arbitrary constant of the family of solutions, we usually omit the constant term when integrating C'(x) or C'(y).

The next example shows what happens if we try to apply the technique for exact equations to one which is not exact.

Example 3

Attempt to solve the equation $(x + y)\, dx + (y - x)\, dy = 0$ as if it were an exact equation.

SOLUTION   Assuming (erroneously) that the differential equation is exact, we obtain the two requirements on the function $u(x, y)$:

$$\frac{\partial u}{\partial x} = x + y \qquad \frac{\partial u}{\partial y} = y - x$$

Integrating the first of these with respect to x, we get

$$u(x, y) = \frac{x^2}{2} + xy + C(y)$$

where $C(y)$ is assumed to be a function of y only. Finding $\partial u/\partial y$ from this expression and equating it to the given one, we get

$$x + C'(y) = y - x$$

or

$$C'(y) = y - 2x$$

This contradicts the assumed condition that $C(y)$ is independent of x, and therefore our attempt fails to produce a solution. ∎

Example 3 emphasizes that difficulties can arise if we try to treat an inexact equation as if it were exact. The following theorem provides a simple conclusive test of the exactness of a differential equation.

THEOREM 2–3

Let $M(x, y)$ and $N(x, y)$ be functions with continuous first partial derivatives for $a < x < b$ and $c < y < d$. Then the differential equation

$$M(x, y)\, dx + N(x, y)\, dy = 0$$

is exact if and only if $\partial M/\partial y = \partial N/\partial x$.

PROOF   If $M(x, y) \, dx + N(x, y) \, dy = 0$ is exact, then a function $u(x, y)$ exists such that the left-hand side of the equation is of the form $\dfrac{\partial u}{\partial x} \, dx + \dfrac{\partial u}{\partial y} \, dy$. Thus

$$M = \frac{\partial u}{\partial x} \qquad \text{and} \qquad N = \frac{\partial u}{\partial y}$$

Taking the partial derivative of the first of these with respect to y and the second with respect to x,

$$\frac{\partial M}{\partial y} = \frac{\partial^2 u}{\partial y \, \partial x} \qquad \text{and} \qquad \frac{\partial N}{\partial x} = \frac{\partial^2 u}{\partial x \, \partial y}$$

Since $\partial M / \partial y$ and $\partial N / \partial x$ are continuous, we know from a theorem from elementary calculus that

$$\frac{\partial^2 u}{\partial y \, \partial x} = \frac{\partial^2 u}{\partial x \, \partial y}$$

Therefore if the given equation is exact,

$$\frac{\partial M}{\partial y} = \frac{\partial N}{\partial x}$$

The converse of the theorem is a bit more difficult for it requires the construction of a function $u(x, y)$ from $M(x, y)$ and $N(x, y)$. The procedure parallels that of Example 1. Define $u(x, y)$ by the equation

$$u(x, y) = \int_x M(x, y) \, dx + C(y)$$

where $C(y)$ is to be determined later. (The x under the integral sign means to antidifferentiate with respect to x, holding y constant.) From this expression we compute $\partial u / \partial y$:

$$\frac{\partial u}{\partial y} = \frac{\partial}{\partial y} \int_x M(x, y) \, dx + C'(y)$$

and this we set equal to $N(x, y)$. Therefore $C'(y)$ is determined from

$$C'(y) = N(x, y) - \frac{\partial}{\partial y} \int_x M(x, y) \, dx$$

There will be a solution if $C'(y)$ is independent of x, and this will be the case if the derivative of the right-hand side with respect to x is zero. Carrying out this differentiation,

$$\frac{\partial}{\partial x} \left[N(x, y) - \frac{\partial}{\partial y} \int_x M(x, y) \, dx \right]$$

$$= \frac{\partial N}{\partial x} - \frac{\partial}{\partial x} \frac{\partial}{\partial y} \int_x M(x, y) \, dx = \frac{\partial N}{\partial x} - \frac{\partial M}{\partial y}$$

under the assumption that the partial differentiations with respect to x and y may be interchanged. The last expression is zero because of the assumed hypothesis. This completes the proof.

Example 4

Solve the differential equation

$$(2xy + 3y)\, dx + (4y^3 + x^2 + 3x + 4)\, dy = 0$$

SOLUTION   Since $M = 2xy + 3y$ and $N = 4y^3 + x^2 + 3x + 4$ are continuous and $\partial M/\partial y = 2x + 3 = \partial N/\partial x$, the equation is exact. Therefore the partial derivatives of the desired function $u(x, y)$ are given by

$$\frac{\partial u}{\partial x} = 2xy + 3y \qquad \text{and} \qquad \frac{\partial u}{\partial y} = 4y^3 + x^2 + 3x + 4$$

Integrating the first of these with respect to x yields

$$u(x, y) = x^2 y + 3xy + C(y)$$

From this expression we find $\partial u/\partial y$ and equate it to the known expression for $\partial u/\partial y$. Thus

$$x^2 + 3x + C'(y) = 4y^3 + x^2 + 3x + 4$$

Therefore

$$C'(y) = 4y^3 + 4$$

and

$$C(y) = y^4 + 4y$$

Hence the implicit solution to the differential equation is

$$x^2 y + 3xy + y^4 + 4y = c \qquad\qquad\blacksquare$$

Example 5

Solve the equation

$$(x^2 + y^2)\, dx + 2xy\, dy = 0$$

SOLUTION   The equation is exact since $M = x^2 + y^2$ and $N = 2xy$ are continuous and

$$\frac{\partial M}{\partial y} = \frac{\partial}{\partial y}(x^2 + y^2) = 2y \qquad \text{and} \qquad \frac{\partial N}{\partial x} = \frac{\partial}{\partial x}(2xy) = 2y$$

Let the solution to the given equation be $u(x, y) = c$, where

$$\frac{\partial u}{\partial x} = M = x^2 + y^2 \qquad \text{and} \qquad \frac{\partial u}{\partial y} = N = 2xy$$

Since the second of these is the simplest, integrate $\partial u/\partial y$ with respect to y, holding x constant, to get

$$u = xy^2 + C(x)$$

From this expression we find

$$\frac{\partial u}{\partial x} = y^2 + C'(x)$$

Equate this to the known expression for $\partial u/\partial x$:

$$y^2 + C'(x) = x^2 + y^2$$

Therefore

$$C'(x) = x^2$$

and

$$C(x) = \frac{x^3}{3}$$

Hence

$$u(x, y) = \frac{x^3}{3} + xy^2$$

and the desired solution is

$$\frac{x^3}{3} + xy^2 = C \qquad\qquad \blacksquare$$

Solving Exact Equations by Inspection

Exact differential equations can often be solved by inspection. The procedure requires that we be able to recognize the differentials of products of the elementary functions. For example, we must recognize $y\, dx + x\, dy$ as the differential of xy, $e^y\, dx + xe^y\, dy$ as the differential of xe^y, $2x \sin y\, dx + x^2 \cos y\, dy$ as the differential of $x^2 \sin y$, and so on.

Example 6

Solve the equation

$$3x(xy - 2)\, dx + (x^3 + 2y)\, dy = 0$$

by inspection.

SOLUTION The equation is exact since

$$\frac{\partial}{\partial y}(3x^2y - 6x) = 3x^2 = \frac{\partial}{\partial x}(x^3 + 2y)$$

Now write the given equation in the form

$$3x^2y\,dx - 6x\,dx + x^3\,dy + 2y\,dy = 0$$

Notice that the second and fourth terms are exact differentials as they stand. This is not true for the first term $3x^2y\,dx$ and the third term $x^3\,dy$. However, together their sum suggests the differential of the product of x^3 and y and, in fact, $3x^2y\,dx + x^3\,dy = d(x^3y)$. The given equation may now be written as

$$d(x^3y) - 6x\,dx + 2y\,dy = 0$$

Integrating the terms of this equation, we get

$$x^3y - 3x^2 + y^2 = C \qquad \blacksquare$$

Example 7

Solve the equation

$$(x + \sin y)\,dx + (y^2 + x \cos y)\,dy = 0$$

by inspection.

SOLUTION The equation is exact since

$$\frac{\partial M}{\partial y} = \cos y = \frac{\partial N}{\partial x}$$

Expanding the equation into the form

$$x\,dx + \sin y\,dx + y^2\,dy + x \cos y\,dy = 0$$

leads at once to the grouping

$$x\,dx + y^2\,dy + (\sin y\,dx + x \cos y\,dy) = 0$$

Hence the solution is

$$\frac{1}{2}x^2 + \frac{1}{3}y^3 + x \sin y = C \qquad \blacksquare$$

EXERCISES FOR SECTION 2–4   1–25 odd

Test each equation in Exercises 1–12 for exactness and solve the equation. Those that are not exact should be solved by an appropriate method.

1. $(x + y)\,dx + (x - y)\,dy = 0$ **2.** $y' = -\dfrac{2x + 3y}{3x + 2y}$

3. $(5x + y^2) \, dx + y(2x - y^2) \, dy = 0$ **4.** $(3x^2 + 4y^2) \, dx + 8xy \, dy = 0$

5. $(x^2 + y^2) \, dy + 2xy \, dx = 0$ **6.** $y(2x + 3y) \, dx + x(x + 6y) \, dy = 0$

7. $(\sin x \cos y + x) \, dx + \cos x \sin y \, dy = 0$

8. $(2xe^y + 1) \, dx + x^2 e^y \, dy = 0$

9. $(3x^2 + 6xy - y^2) \, dx + (3x^2 - 2xy + 3y^2) \, dy = 0$

10. $(1 + y^2) \, dx - 2xy \, dy = 0$ **11.** $(ye^x - 2x) \, dx + e^x \, dy = 0$

12. $\left(x + \dfrac{y}{x}\right) dx + (4y + \ln x) \, dy = 0$

Test each equation in Exercises 13–19 for exactness. Solve those that are exact by inspection. Solve the others by appropriate methods.

13. $x(y + x) \, dx + y(x + y) \, dy = 0$ **14.** $(x - y) \, dx + (x + y) \, dy = 0$

15. $(\cos y + 4) \, dx - x \sin y \, dy = 0$

16. $(y^3 + x^2 y - x) \, dy + (x^3 + xy^2 - y) \, dx = 0$

17. $y' = \dfrac{y \cos x}{1 + \sin x}$ **18.** $y' = -\dfrac{x^2(3y^2 + 1)}{y(2x^3 + 1)}$

19. $[y \sec^2 x + \tan x] \, dx + \tan x \, dy = 0$

20. $(ye^{xy} + 2xy) \, dx + (xe^{xy} + x^2) \, dy = 0$

In Exercises 21–24 find the indicated particular solution.

21. $2xy \, dx + (y^2 + x^2) \, dy = 0; \; y(2) = 2$

22. $(x + y) \, dx + (x - y) \, dy = 0; \; y(0) = 4$

23. $\sin y \, dx + [2y + x \cos y] \, dy = 0; \; y(0) = \pi$

24. $2xe^{2y} \, dx + 2[1 + x^2 e^{2y}] \, dy = 0; \; y(2) = 0$

25. Show that a first-order differential equation in which the variables are separated is exact.

2–5   INTEGRATING FACTORS; FIRST-ORDER LINEAR EQUATIONS

Sometimes a first-order equation can be "made" exact by multiplying the equation by an expression called an **integrating factor.** Finding an integrating factor for an equation is equivalent to solving it since an exact equation may be solved by finding two partial antiderivatives or even by inspection.

Example 1

(a) x^{-2} is an integrating factor of $(x^2/y) \, dy + 2x \, dx = 0$ because the resulting equation $(1/y) \, dy + (2/x) \, dx = 0$ is exact.

(b) y is also an integrating factor of $(x^2/y) \, dy + 2x \, dx = 0$.

(c) e^x is an integrating factor of $y' + y = x$. ■

Integrating factors are usually not unique, as the first two parts of Example 1 show. Further, as in the case of separable equations, some integrating factors can be found by inspection. But most often an integrating factor is assumed to be of some *general form*, and then the condition for exactness $(\partial M/\partial y = \partial N/\partial x)$ is applied to obtain the *specific form*.

Example 2

Solve the nonseparable differential equation $y^2 \, dx + (4xy + 1) \, dy = 0$ by first finding an integrating factor which is a function of y.

SOLUTION   Assume an integrating factor exists that has the general form $g(y)$. Then multiplying each term of the given differential equation by $g(y)$, we get

$$M(x, y) = y^2 g(y) \qquad \text{and} \qquad N(x, y) = (4xy + 1)g(y)$$

Applying the condition for exactness, we get

$$2y \, g(y) + y^2 g'(y) = 4y \, g(y)$$

This is a differential equation in y and $g(y)$. The variables may be separated to obtain

$$\frac{g'(y)}{g(y)} = \frac{2}{y}$$

Observing that $\int [g'(y)/g(y)] dy = \ln |g(y)|$ and $\int (2/y) \, dy = 2 \ln |y| = \ln y^2$, we have

$$\ln g(y) = \ln y^2$$

from which we obtain the integrating factor,

$$g(y) = y^2$$

Multiplying the given differential equation by y^2, it becomes the exact equation

$$y^4 \, dx + 4xy^3 \, dy + y^2 \, dy = 0$$
$$d(xy^4) + y^2 \, dy = 0$$

whose solution is, by inspection,

$$xy^4 + \tfrac{1}{3}y^3 = c \qquad\qquad ■$$

First-Order Linear Equations

Recall that a first-order linear differential equation defined on the interval $[a, b]$ has the form

$$a_1(x)y' + a_0(x)y = f(x), \qquad a_1(x) \neq 0, \text{ for } x \text{ in } [a, b] \qquad (2\text{--}18)$$

Dividing each member of this equation by the coefficient $a_1(x)$, we obtain

$$y' + p(x)y = q(x) \qquad (2\text{--}19)$$

which is called the **standard form** of a linear differential equation of the first order.

We now show that an integrating factor can always be found for first-order linear differential equations. Assume that an integrating factor exists *that is a function of x only*, say $v(x)$. Thus since $v(x)$ is an integrating factor of Equation 2–19, it follows that

$$v(x)y' + v(x)p(x)y = v(x)q(x)$$

must be an exact equation. We rewrite this as

$$v(x)\, dy + [v(x)p(x)y - v(x)q(x)]dx = 0$$

and then apply the condition for exactness to get

$$\frac{\partial}{\partial x} v(x) = \frac{\partial}{\partial y} [v(x)p(x)y - v(x)q(x)] \qquad (2\text{--}20)$$

Since the function v is a function of x only, we replace $\partial/\partial x\ [v(x)]$ with $d/dx\ [v(x)]$ on the left-hand side of Equation 2–20. Then performing the indicated partial differentiation on the right-hand side of Equation 2–20, we get

$$\frac{d}{dx} v(x) = v(x)p(x)$$

or

$$\frac{d[v(x)]}{v(x)} = p(x)\, dx$$

Taking the antiderivative of both functions yields

$$\ln v(x) = \int p(x)\, dx$$

The constant of integration is omitted since only one integrating factor is desired. In exponential form we write

$$v(x) = e^{\int p(x)\, dx} \qquad (2\text{--}21)$$

which is the desired integrating factor. Therefore we can solve *any* first-order linear differential equation. This important technique is summarized below.

PROCEDURE

> *Solving First-Order Linear Differential Equations*
>
> 1. Put the equation into standard form $y' + p(x)y = q(x)$. (NOTE: The coefficient of y' must be 1.)
>
> 2. Identify $p(x)$ and compute $v(x) = e^{\int p(x)\,dx}$.
>
> 3. Multiply the standard form of the equation by $v(x)$.
>
> 4. Integrate both sides of the modified equation and solve for y.

COMMENT: Step 3 of the procedure—multiplying the standard equation by $v(x)$—makes the given equation exact. Of particular interest is the fact that the left-hand side of the equation will always be the total differential of the product $y \cdot v(x)$. You can check the validity of an integrating factor by taking the differential of $y \cdot v(x)$ and comparing it to the left-hand side of the exact form of the equation.

Example 3

Solve the linear equation $2y' - 4y = 16e^x$.

SOLUTION   Rearranging the equation in standard form, we get

$$y' - 2y = 8e^x$$

We note that $p(x) = -2$, so an integrating factor is

$$v(x) = e^{\int(-2)\,dx} = e^{-2x}$$

Multiplying the given differential equation (in standard form) by e^{-2x}, we get

$$e^{-2x}y' - 2ye^{-2x} = 8e^{-x}$$

or, in differential form,

$$e^{-2x}\,dy - 2ye^{-2x}\,dx = 8e^{-x}\,dx$$

The left-hand side of this equation is the expression for the differential of the product $y \cdot$ (integrating factor) $= ye^{-2x}$, so we write

$$d(ye^{-2x}) = 8e^{-x}\,dx$$

Integrating both sides yields

$$ye^{-2x} = -8e^{-x} + C$$

or

$$y = -8e^x + Ce^{2x}$$

■

Example 4

Solve $x^2y' + 3xy = \dfrac{1}{x}\cos x$.

SOLUTION   Writing the equation in standard form, we have

$$y' + \frac{3}{x}y = \frac{\cos x}{x^3}$$

In this form we see that $p(x) = 3/x$ and

$$v(x) = e^{\int 3/x\,dx} = e^{3\ln x} = e^{\ln x^3} = x^3$$

(NOTE: Integrating factors of the form $e^{\ln f(x)}$ should be written as $f(x)$, or they will be relatively useless in obtaining solutions.)

Applying the integrating factor x^3 to the standard form of the given equation, we get

$$x^3y' + 3x^2y = \cos x$$

or

$$x^3\,dy + 3x^2y\,dx = \cos x\,dx$$

Then the desired solution is

$$x^3y = \sin x + c$$
$$y = (\sin x + c)x^{-3}$$

■

Bernoulli's Equation

An important nonlinear equation that can be reduced to linear form with an appropriate substitution is Bernoulli's equation.

DEFINITION

An equation of the form

$$\frac{dy}{dx} + P(x)y = Q(x)y^n$$

is called a **Bernoulli differential equation.**

Notice that if $n = 0$ or 1, the Bernoulli equation is linear. If $n \neq 0$ or 1, the following theorem tells us how to transform the Bernoulli equation into a linear equation.

THEOREM 2–4

If $n \neq 0$ or 1, then the Bernoulli equation

$$y' + P(x)y = Q(x)y^n$$

is reduced to a linear equation by the transformation

$$v = y^{1-n}$$

PROOF   First multiply the differential equation by y^{-n} to obtain

$$y^{-n}y' + P(x)y^{1-n} = Q(x)$$

Letting $v = y^{1-n}$, then

$$v' = (1 - n)y^{-n}y' \qquad \text{or} \qquad \frac{y'}{y^n} = \frac{v'}{1 - n}$$

Substituting these values into the given differential equation, we have

$$\frac{v'}{1 - n} + vP(x) = Q(x)$$

which is linear in v.

Example 5

Solve the Bernoulli differential equation

$$y' + \frac{1}{x}y = 3y^3$$

SOLUTION   In this equation $n = 3$, so we multiply the differential equation by y^{-3} to obtain

$$y^{-3}y' + \frac{1}{x}y^{-2} = 3$$

Now let $v = y^{1-n} = y^{-2}$. Then

$$v' = -2y^{-3}y' \qquad \text{or} \qquad \frac{y'}{y^3} = -\frac{v'}{2}$$

Substituting these values into the given differential equation, we get

$$-\frac{v'}{2} + \frac{1}{x}v = 3$$

or, multiplying by -2,

$$v' - \frac{2}{x}v = -6$$

which is the standard form of a linear equation in v. An integrating factor for this equation is

$$e^{-\int (2/x)\, dx} = x^{-2}$$

Multiplying by x^{-2}, we have

$$x^{-2}v' - 2x^{-3}v = -6x^{-2} \qquad \text{(NOTE: The left-hand side is } d(vx^{-2}).)$$

Integrating, the solution is

$$x^{-2}v = 6x^{-1} + c$$

$$v = 6x + cx^2$$

Finally, $v = y^{-2}$, so

$$y^{-2} = 6x + cx^2$$

$$y = \pm \frac{1}{\sqrt{6x + cx^2}}$$ ∎

EXERCISES FOR SECTION 2–5   1- 33 odd

In Exercises 1–18 solve the given differential equation.

1. $dy + 2y\, dx = 4\, dx$

2. $2\, dy + 8xy\, dx = x\, dx$

3. $dy - 4y\, dx = 3\, dx$

4. $\dfrac{di}{dt} + i - 2 = 0$

5. $\dfrac{dy}{dx} + \dfrac{y}{x} = x + 1$

6. $x^2\, dy - 2xy\, dx = dx$

7. $(t + 2q)\, dt - dq = 0$

8. $x(3 + y)\, dx + dy = 0$

9. $(t^2 + 1)\, ds + 2ts\, dt = 3\, dt$

10. $(x^2 - 4)\, dy + xy\, dx = x\, dx$

11. $(x^2 + 1)\, dy - xy\, dx = 2x\, dx$

12. $dy + 2y\, dx = e^{-x}\, dx$

13. $\dfrac{dv}{dr} + v \tan r = \cos r$

14. $dy = (\csc x - y \cot x)\, dx$

15. $dy - y\, dx = 2e^{-x}\, dx$

16. $\dfrac{di}{dt} + i = e^{-t}$

17. $t\dfrac{ds}{dt} + 2s = \dfrac{3t^2 + 2}{t^4 + 2t^2}$

18. $(y \cot x - x)\, dx + dy = 0$

In Exercises 19–22 find the particular solution corresponding to the given boundary conditions.

19. $x\, dy - y\, dx = x^2 e^x\, dx;\ x = 1,\ y = 0$

20. $dy/dx + 2y = 4;\ x = 0,\ y = 1$

21. $dy/dx + y \cot x = 10; x = \frac{1}{3}\pi, y = 0$

22. $dr/d\theta + r \cot \theta = \sec \theta; \theta = \frac{1}{3}\pi, r = 2$

In Exercises 23–30 solve the given differential equation.

23. $y' - y/x = y^3$ **24.** $y' + 2y/x = y^2$

25. $y' + y \tan x = 2y^2$ **26.** $y' - \frac{1}{2}y \tan x = y^3$

27. $ds - s(1 + ts)\, dt = 0$ **28.** $y' = 4x(2y^{-3} - y)$

29. $6x^2\, dy - y(2y^3 + x)\, dx = 0$ **30.** $2t^3\, ds - s(s^2 + 3t^2)\, dt = 0$

In Exercises 31–33 use a substitution to solve the given equation as a linear differential equation.

31. $xe^{-y}y' + e^{-y} = x$ **32.** $y' + xy \ln y - y = 0$ **33.** $xyy' + y^2 = x - 1$

34. Show that if M and N are homogeneous of degree k in x and y, then $1/(xM + yN)$ is an integrating factor of the equation $Mdx + Ndy = 0$. (HINT: Use Euler's theorem, Section 2–3, Exercise 38, and the condition for exactness.)

2–6   APPLICATIONS OF LINEAR EQUATIONS

Dynamics

Dynamics is the study of the motion of objects and the forces that cause motion. In this section we study the motion of an object through a resisting medium. We begin with a basic definition and a fundamental principle.

DEFINITION

> The momentum P of a moving object is the product of its mass m and its velocity v. As an equation we write
>
> $$P = mv$$

NEWTON'S SECOND PRINCIPLE

> *The time-rate of change of the momentum of an object is equal to the applied force F. Thus*
>
> $$F = \frac{d}{dt}(mv)$$

*COMMENT: If the mass of the object is constant, Newton's second
principle is usually stated in the form*

$$F = m\frac{dv}{dt} = ma$$

*where a is the acceleration of the object and m is its mass. This statement
of the second principle is valid only if the mass is constant.*

Newton's second principle, together with a knowledge of the technique of
solving first-order linear equations, enables us to analyze and solve a problem from
dynamics involving a rocket moving through the atmosphere.

Consider a rocket moving in a straight horizontal path, as shown in Figure 2–18.

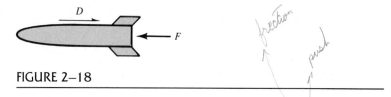

D

F

direction

push

FIGURE 2–18

Assume that the rocket is propelled by a force (called the thrust) F and that as it
moves through a "resisting" medium, it experiences a drag force D; hence the
effective external force on the rocket is $F - D$. From Newton's second principle
we can write

$$\frac{d}{dt}(mv) = F - D \tag{2-22}$$

Since the mass of the rocket changes with time as the propellant is burned
and the velocity is also changing due to the thrust, both m and v are functions of
time, and Equation 2–22 becomes

$$m\frac{dv}{dt} + v\frac{dm}{dt} = F - D \tag{2-23}$$

This equation is linear in v and can be solved if we are given the expressions for
the functions F, D, and m.

The drag force experienced by an object moving through a resisting medium
typically increases as the velocity increases. Consequently, we are encouraged to
express drag force as a function of velocity. A common model for drag force assumes
that it is directly proportional to its velocity—that is,

$$D = kv$$

This model agrees with experience as long as the velocity is not too great.

*COMMENT: The use of Equation 2–22 requires consistent units. In all
of our examples and exercises length is in feet, force in pounds, time in
seconds, and mass in slugs.*

Example 1

Consider a rocket that starts from rest and is propelled by a constant thrust of 2000 lb in a straight horizontal path for 20 seconds. Suppose that, as the propellant is burned, the mass of the rocket varies with time in accordance with $m = 50 - t$ slugs, and that the drag force is numerically equal to three times the velocity. Find the velocity equation of the rocket. What is its velocity when $t = 10$ sec?

SOLUTION   Note that $m = 50 - t$, $F = 2000$, and $D = 3v$. Using these in Equation 2–23,

$$(50 - t)\frac{dv}{dt} + v\frac{d}{dt}(50 - t) = 2000 - 3v$$

or

$$(50 - t)\frac{dv}{dt} + 2v = 2000$$

which, when put into standard form, becomes

$$\frac{dv}{dt} + \frac{2}{50 - t}v = \frac{2000}{50 - t}$$

Here $p = \dfrac{2}{50 - t}$, so the integrating factor is $(50 - t)^{-2}$. Thus

$$d[v(50 - t)^{-2}] = 2000(50 - t)^{-3}\, dt$$

and therefore

$$v(50 - t)^{-2} = 1000(50 - t)^{-2} + C$$

so that

$$v = 1000 + C(50 - t)^2$$

Using the fact that $v = 0$ when $t = 0$, we get $C = -0.4$. Hence the velocity equation is

$$v = 1000 - 0.4(50 - t)^2 \text{ ft/sec}$$

See Figure 2–19.

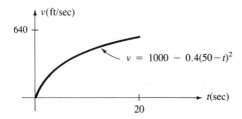

FIGURE 2–19

When $t = 10$ sec,

$$v(10) = 1000 - 0.4(1600) = 360 \text{ ft/sec}$$ ∎

Example 2

A 96-lb testing sled that rides on a horizontal track is propelled by a horizontal force that varies with time according to

$$F(t) = \begin{cases} t \text{ lb,} & 0 \le t < 3 \text{ sec} \\ 0, & t \ge 3 \text{ sec} \end{cases}$$

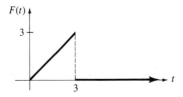

FIGURE 2–20

If the drag force is equal to the velocity of the sled, what is the velocity equation? Assume $v(0) = 0$. See Figure 2–20.

SOLUTION   The mass of an object is obtained from its weight by

$$m = \frac{W}{g}$$

where W is the weight and g is the acceleration due to the force of gravity. This formula gives mass in slugs if weight is in pounds and g is in ft/sec². (Throughout this book we use $g = 32$ ft/sec².)

In this problem the mass of the sled is $m = 96/32 = 3$ slugs. The drag force experienced by the sled is equal to the velocity, so the drag force is given by $D = v$ lb. Finally, since the thrust is defined by two rules, the velocity must be determined separately for $0 \le t < 3$ sec and $t \ge 3$ sec.

For $0 \le t < 3$ we must solve the initial-value problem

$$3\frac{dv}{dt} = t - v, \qquad v(0) = 0$$

Rearranging this equation, we get

$$\frac{dv}{dt} + \frac{v}{3} = \frac{t}{3}$$

The integrating factor is $e^{(1/3)\,dt} = e^{t/3}$. Using this integrating factor, the equation becomes

$$d(ve^{t/3}) = \frac{t}{3} e^{t/3} \, dt$$

Integration by parts on the right-hand side yields the solution

$$ve^{t/3} = te^{t/3} - 3e^{t/3} + c$$

$$v = t - 3 + ce^{-t/3}$$

Since $v = 0$ when $t = 0$, the arbitrary constant is $c = 3$. Thus

$$v = t - 3(1 - e^{-t/3}) \text{ ft/sec}$$

For $t \geq 3$ sec, $F(t) = 0$, so we must solve the initial-value problem

$$3\frac{dv}{dt} = -v, \, v(3) = 3e^{-1}$$

where $v(3)$ is obtained from the solution in the preceding interval. By separating variables, we get

$$\frac{dv}{v} = -\frac{1}{3} \, dt$$

$$v = c_1 e^{-t/3}$$

Substituting $t = 3$, $v = 3e^{-1}$, we find that $c_1 = 3$. Thus

$$v = 3e^{-t/3} \text{ ft/sec}$$

Combining the above results, the velocity function may be written

$$v = \begin{cases} t - 3(1 - e^{-t/3}) \text{ ft/sec,} & 0 \leq t < 3 \text{ sec} \\ 3e^{-t/3} \text{ ft/sec,} & t \geq 3 \text{ sec} \end{cases}$$

See Figure 2–21.

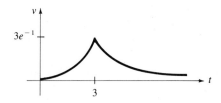

FIGURE 2–21   ∎

Mixture Problems

Another problem that is modeled with a linear differential equation involves the process of mixing liquids that contain dissolved substances such as salt or dye. A typical situation involves a tank filled with a liquid containing a **concentration** (the amount of the dissolved substance per unit volume of the liquid) of some substance, a liquid flowing into the tank with a different concentration of the same substance, and the simultaneous release of the thoroughly mixed solution from a valve at the bottom of the tank. The problem is to determine the amount Q of the dissolved substance in the tank at any time t. Assuming the two liquids are thoroughly mixed, we can express the rate at which the substance is accumulating in the tank by

grow into a mass.

$$\frac{dQ}{dt} = \textbf{(Rate of substance in)} - \textbf{(Rate of substance out)} \qquad (2\text{–}24)$$

The initial-value problem must satisfy the condition $Q(t_0) = Q_0$.

Example 3

A tank with a capacity of 300 gal initially contains 100 gal of pure water. A salt solution containing 3 lb of salt per gal is allowed to run into the tank at a rate of 8 gal/min, and the mixture is then removed at a rate of 6 gal/min, as shown in Figure 2–22. Find the expression for the number of pounds of salt in the tank at any time t.

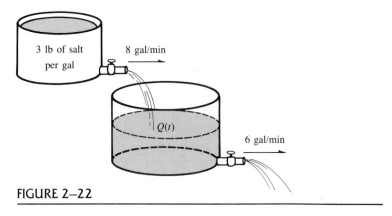

FIGURE 2–22

SOLUTION   Since the water is initially pure, $Q(0) = 0$. The rate at which salt is being added to the tank is

(3 lb/gal)(8 gal/min) = 24 lb/min

The rate at which brine (salt solution) is entering the tank is greater than the rate at which the mixture is being removed, so at any time t the amount of solution in the tank is increasing at the rate of

8 gal/min − 6 gal/min = 2 gal/min

Therefore the number of gallons of solution in the tank at any time t is given by $n = 100 + 2t$, and the concentration of salt in solution in the tank is expressed by

$$\frac{Q}{100 + 2t}$$

The rate at which salt is being removed from the tank is then

$$\left(\frac{Q}{100 + 2t} \text{ lb/gal}\right) (6 \text{ gal/min}) = \frac{6Q}{100 + 2t} \text{ lb/min}$$

Combining the previous results in Equation 2–24, we write the initial-value problem

$$\frac{dQ}{dt} = 24 - \frac{6Q}{100 + 2t}, \ Q(0) = 0$$

The linear equation

$$\frac{dQ}{dt} + \frac{3Q}{50 + t} = 24$$

has an integrating factor

$$v = e^{\int 3/(50+t) \, dt} = e^{3 \ln (50+t)} = (50 + t)^3$$

This integrating factor yields

$$Q(50 + t)^3 = \int 24(50 + t)^3 \, dt$$
$$= 6(50 + t)^4 + c$$

Solving for Q, we get

$$Q = 6(50 + t) + c(50 + t)^{-3}$$

The initial condition, $Q(0) = 0$, yields $c = -300(50^3)$. Hence the desired equation is

$$Q = 6(50 + t) - 300 \left(\frac{50}{50 + t}\right)^3 \text{ lb} \qquad \blacksquare$$

Electric Circuits

In elementary physics you learn that a constant voltage V applied across a conductor produces a current i that is directly proportional to the constant applied voltage. This relationship, known as **Ohm's law,** is written

$$V = iR$$

where R is a constant called the **resistance** of the circuit. Circuits that involve constant voltage and resistance are easy to analyze by using Ohm's law and do not require the solution of differential equations. However, if the circuit contains an inductance, then the current changes with time, and Ohm's law does not completely describe the circuit. Because of the inductance there is a voltage induced in the circuit defined by

$$v_L = L\frac{di}{dt}$$

where L is a constant of proportionality called the **inductance** of the circuit.

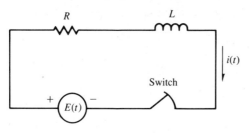

FIGURE 2–23

A circuit including a resistance and an inductance is shown in Figure 2–23. The components are said to be in a **series** arrangement. A circuit of this type obeys Kirchhoff's second law: **"At any instant the algebraic sum of the voltage drops around a circuit is zero."** Thus in the RL series circuit shown in Figure 2–23 we have

$$v_L + v_R = E(t) \tag{2-25}$$

or

$$L\frac{di}{dt} + Ri = E(t) \tag{2-26}$$

where $E(t)$ is the applied, or driving, voltage. The response, or current $i(t)$, is then the solution of a first-order linear differential equation.

> COMMENT: Consistent electrical units are voltage in volts, resistance in ohms, current in amperes, and inductance in henrys.

Example 4

Find the current in a series RL circuit in which the resistance, inductance, and voltage are constant. Assume the initial current is zero.

SOLUTION   Replacing $E(t)$ with V in Equation 2–26, we have

$$L\frac{di}{dt} + Ri = V$$

which can be written in standard form as

$$\frac{di}{dt} + \frac{R}{L}i = \frac{V}{L}$$

An integrating factor for this equation is $e^{Rt/L}$, which, when applied to the standard form, yields

$$d(ie^{Rt/L}) = \frac{V}{L}e^{Rt/L}\,dt$$

Integrating, we get

$$ie^{Rt/L} = \frac{V}{R}e^{Rt/L} + C$$

Or, solving for i,

$$i = \frac{V}{R} + Ce^{-Rt/L}$$

The condition $i(0) = 0$ yields $C = -V/R$, and therefore the current is

$$i = \frac{V}{R}(1 - e^{-Rt/L}) \tag{2-27}$$

The graph of this function is shown in Figure 2–24.

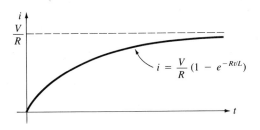

FIGURE 2–24

COMMENT: *Equation 2–27 is a very famous formula for the response of a series RL circuit to a constant voltage. Notice that the equation is valid only for circuits in which R, L, and V are constants. Can you tell where, in the derivation of this formula, it was necessary to use this fact?*

Example 5

A 6-volt battery is applied in series with a resistance of 1 ohm and an inductance in henrys that varies with time according to

$$L = \begin{cases} 2 - t \text{ henry,} & 0 \le t < 2 \text{ sec} \\ 0, & t \ge 2 \text{ sec} \end{cases}$$

Find the expression for the current in the circuit. Assume $i(0) = 0$. (NOTE: Do not use Equation 2–27 to solve this problem. Why not?)

SOLUTION   On the interval $0 \le t < 2$ the inductance is $2 - t$ henrys, so

$$(2 - t)\frac{di}{dt} + i = 6$$

or

$$\frac{di}{dt} + \frac{1}{2 - t}i = \frac{6}{2 - t}$$

Using $p = 1/(2 - t)$, an integrating factor is

$$v = e^{-\ln(2 - t)} = (2 - t)^{-1}$$

Multiplying the integrating factor through the standard equation, we get

$$d[i(2 - t)^{-1}] = 6(2 - t)^{-2}\, dt$$

Integrating,

$$i(2 - t)^{-1} = 6(2 - t)^{-1} + C$$

$$i = 6 + C(2 - t)$$

The value of C is found by using $i(0) = 0$ in this equation. Thus

$$0 = 6 + 2C \qquad \text{or} \qquad C = -3$$

So the equation for the current for $0 \le t < 2$ is

$$i = 6 - 3(2 - t) = 3t \text{ amp}$$

For $t \ge 2$ the inductance is equal to zero, so

$$0\frac{di}{dt} + i = 6$$

which means $i = 6$ amp for $t \ge 2$. Summarizing what we know about the current in this circuit, we see that

$$i = \begin{cases} 3t \text{ amp,} & 0 \le t < 2 \text{ sec} \\ 6 \text{ amp,} & t \ge 2 \text{ sec} \end{cases}$$

The current is shown in Figure 2–25.

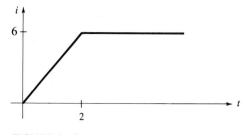

FIGURE 2–25   ■

Example 6

Find the response of a series RL circuit to a unit square wave. See Figure 2–26(a).
Assume the initial current is zero.

SOLUTION   For the interval $0 \le t \le 1$ Equation 2–27 may be used. Substituting
$V = 1$ in Equation 2–27 yields

$$i = \frac{1}{R}(1 - e^{-Rt/L}), \ 0 \le t \le 1$$

For $t \ge 1$ Equation 2–27 does not apply because the initial current for this interval
is not zero. In this case we write the initial-value problem

$$L\frac{di}{dt} + Ri = 0, \ i(1) = \frac{1}{R}(1 - e^{-R/L})$$

The differential equation is solved by separation of variables to yield

$$i = c_1 e^{-Rt/L}$$

Applying the initial condition for the interval $t \ge 1$,

$$\frac{1}{R}(1 - e^{-R/L}) = c_1 e^{-R/L}$$

from which

$$c_1 = \frac{1}{R}(e^{R/L} - 1)$$

Therefore the current for $t \ge 1$ is

$$i = \frac{1}{R}(e^{R/L} - 1)e^{-Rt/L}$$

$$= \frac{1}{R}(e^{-R(t-1)/L} - e^{-Rt/L})$$

See Figure 2–26(b).

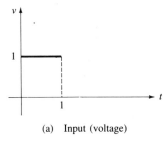
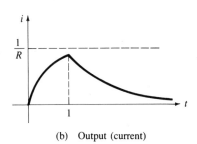

(a) Input (voltage)   (b) Output (current)

FIGURE 2–26

EXERCISES FOR SECTION 2–6

Dynamics

1. Find the equation for the velocity of an object having a weight of 64 lb if the resisting force is $D = 2v$ lb and the thrust is $F = 10t$ lb. Assume $v(0) = 0$.

2. A block weighing 32 lb is moved along a horizontal plane by a force of $e^{-0.01t}$ lb. If friction produces a retarding force equal to $0.01v$ lb, describe the motion of the block. Assume the block starts from rest.

3. A rocket starting from rest is moved in a straight-line path by a thrust of 1000 lb. During the first three seconds of propulsion the mass of the rocket decreases according to $m = 25 - t$ slugs, and the drag force is equal to $1.5v$ lb. Determine the velocity equation of the rocket.

4. Draw the graph of the velocity equation found in Exercise 3.

5. A constant thrust of 2000 lb is generated by the motor of a rocket during the first five seconds of flight. During this interval the mass of the rocket varies with time according to $m = 25 - 4t$ slugs, and the drag is equal to $8v$ lb. Assuming the initial velocity of the rocket is 100 ft/sec, find its velocity at the end of the burn period.

6. A 1600-lb rocket sled is accelerated from rest by a thrust that varies with time according to

$$F = \begin{cases} 1000t \text{ lb}, & 0 \le t \le 2 \text{ sec} \\ 2000 \text{ lb}, & t > 2 \text{ sec} \end{cases}$$

Determine the velocity equation for the sled if the mass remains constant and the drag is equal to ten times its velocity.

7. A 160-lb parachutist, including equipment, free-falling toward the earth, experiences a drag force (in pounds) equal to one-half her velocity in ft/sec. When the parachute opens, the drag is equal to five-eighths of the square of her velocity. (a) Find the equation for her velocity before the parachute opens. (b) Assuming she falls for 30 seconds before her chute opens, find her velocity equation after it opens.

8. Assume the inclined plane shown in Figure 2–27 is frictionless and that the only force acting on the 25-lb block is gravity. Assume $x = 5$ ft and $v = 2$ ft/sec when

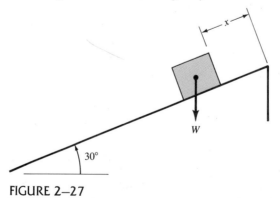

30°

FIGURE 2–27

$t = 0$. (a) Write the differential equation describing the motion of the block. (b) Determine the equation for the velocity of the block. (c) Determine the equation for the displacement of the block. (d) Determine how long it will take the block to move 20 ft down the incline.

An object that moves over a rough surface encounters a resistance (called **friction**) due to the roughness of the surface. The friction experienced by an object is shown in physics to be a force given by $f = \mu N$, where μ is a constant called the **coefficient of friction** and N is the normal force exerted by the plane on the object.

9. A 64-lb block is released from the top of a plane inclined at $30°$ to the horizontal. The coefficient of friction of the block is 0.25. (a) Determine the equation for the velocity of the block. (b) Determine the equation for the displacement of the block. (c) Calculate the displacement and velocity of the block 5 sec after it is released.

10. A 64-lb block is released from the top of a plane inclined at $30°$ to the horizontal. As the block slides down the plane, its coefficient of friction is 0.25, and it experiences a drag force due to air resistance equal to one-half its velocity in ft/sec. (a) Determine the equation for the velocity of the block. (b) Determine the equation for the displacement of the block. (c) Calculate the displacement and velocity of the block 5 sec after it is released.

Mixtures

11. A tank contains a brine solution in which 5 lb of salt is dissolved in 10 gal of water. Brine containing 3 lb of salt per gal flows into the tank at 2 gal/min, and the well-stirred mixture flows out at the same rate. (a) Determine the equation for the amount of salt in the tank at any time $t > 0$. (b) How much salt is present after a long time?

12. One hundred gallons of a 25% acid solution is contained in a tank. A 40% acid solution is allowed to enter the tank at a rate of 10 gal/hr, and the well-stirred solution is removed from the tank at the same rate. Determine the equation for the amount of acid in the tank at any time $t > 0$.

13. A tank contains a brine solution in which 10 lb of salt is dissolved in 50 gal of water. Brine containing 2 lb of salt per gal flows into the tank at the rate of 5 gal/min, and the well-stirred solution flows out at a rate of 3 gal/min. (a) Determine the equation for the amount of salt in the tank at any time $t > 0$. (b) How much salt is in the tank after 5 min?

14. A 250-gal tank contains 100 gal of pure water. Brine containing 4 lb of salt per gallon flows into the tank at 5 gal/hr. If the well-stirred mixture flows out at 3 gal/hr, find the concentration of salt in the tank at the instant it is filled to the top.

15. Two tanks each contain 100 gal of pure water. A solution containing 3 lb/gal of a dye flows into Tank 1 at 5 gal/min. The well-stirred solution flows out of Tank 1 into Tank 2 at the same rate. Assuming the solution in Tank 2 is well-stirred and that this solution flows out of Tank 2 at 5 gal/min, determine the amount of dye in Tank 2 for any $t > 0$.

16. A 600-gal brine tank is to be cleaned by piping in pure water at 1 gal/min and allowing the well-stirred solution to flow out at the rate of 2 gal/min. (a) If the tank initially contains 1500 lb of salt, how much salt is left in the tank after 1 hour? (b) After 9 hr and 59 min?

17. Water pollution is a major problem in the world. Rivers and lakes become polluted with various waste products and, left unchecked, can kill the marine life. If the source of pollution is stopped, the rivers and lakes will clean themselves by the natural process of replacing the polluted water with clean water. The rate of change of pollution in a lake is a mixture problem and can be modeled by using Equation 2–24. In this model we assume water entering the lake is perfectly mixed so that the pollutants are uniformly distributed, that the volume of the lake is constant, and that the pollutants are removed from the lake only by outflow.

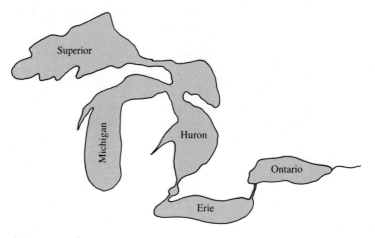

FIGURE 2–28

If P is the pollution concentration in a lake and V is its volume, then the total amount of pollutant is $Q = VP$. Let r be the rate of water both in and out of the lake and P_i be the pollutant concentration of the water coming into the lake. Then by Equation 2–24 we get

$$\frac{d}{dt}(VP) = (P_i - P)r$$

Since the volume is constant, this equation can be written

$$\frac{dP}{dt} + \frac{r}{V}P = \frac{P_i r}{V}$$

This is a linear equation with integrating factor $e^{\int r/V \, dt} = e^{rt/V}$. Thus

$$\frac{d}{dt}(e^{rt/V}P) = \frac{r}{V}e^{rt/V}P_i$$

Integrating and evaluating at $t = 0$ yields

$$e^{rt/V}P = \frac{r}{V}\int_0^t e^{rt/V}P_i \, dt + P(0) \tag{2-28}$$

The fastest possible cleanup will occur if the value of the pollutant concentration P_i of the incoming water is zero. In this case Equation 2–28 becomes

$$e^{rt/V}P = P(0)$$

Solving this for t, we get

$$t = \frac{V}{r} \ln \frac{P(0)}{P}$$

as the time in years for the pollution level of a given lake to drop to a given percentage of its present value.

(a) Verify that $e^{rt/V} P = P(0)$ yields

$$t = \frac{V}{r} \ln \frac{P(0)}{P}$$

(b) Given that $V/r = 2.9$ for Lake Erie, determine how long it will take to reduce the pollution level to 5% of its present value.

18. If P_i is a nonzero constant in Equation 2–28, determine an expression for $P(t)$.

Electrical Circuits

19. A series RL circuit has a resistance of 20 ohms, an inductance of 0.2 henry, and an impressed voltage of 12 volts. Find the current equation if the initial current is zero.

20. Find the current in a series RL circuit with $R = 2$ ohms, $L = 1$ henry, and $E = e^{-t}$ volt. Assume the initial current is zero.

21. Find the current in a series RL circuit with $R = 10$ ohms and $L = 5$ henrys, if the applied voltage varies with time according to

$$E(t) = \begin{cases} 2t \text{ volts}, & 0 \leq t \leq 2 \text{ sec} \\ 4 \text{ volts}, & t > 2 \text{ sec} \end{cases}$$

Assume $i(0) = 0$.

22. Sketch the graphs of voltage and current for the circuit in Exercise 21.

23. The inductance in a series RL circuit is given by

$$L(t) = \begin{cases} 1 + t \text{ henrys}, & 0 \leq t \leq 1 \text{ sec} \\ 2 \text{ henrys}, & t > 1 \text{ sec} \end{cases}$$

If $R = 2$ ohms, $E = 6$ volts, and $i(0) = 0$, find the equation for the current in the circuit. Sketch the graph of the current equation.

24. The initial current in a series RL circuit is zero. Find the current in the circuit when $t = 2$ sec if $R = 2$ ohms, $E = 10$ volts, and $L = 4 - t$ henrys.

25. A **capacitor** is a device for storing electric charge. The amount of charge that a capacitor will store is called its **capacitance.** When a constant voltage of V volts is applied to a resistance of R ohms and a capacitance of C farads in series, the rate at which the capacitor is charged is given by

$$R \frac{dq}{dt} + \frac{q}{C} = V$$

where q is the electric charge measured in coulombs on the left plate in Figure 2–29. Find the equation for the charge transferred, if the capacitor is initially discharged.

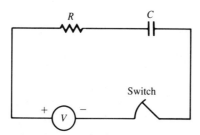

FIGURE 2–29

26. Using the results of Exercise 25 and the fact that $i = dq/dt$, find the equation for the current in the capacitor.

27. Referring to Exercise 25, show that for large values of t the charge on the plates of the capacitor is given by $q = CV$.

28. Consider a series RL circuit in which the voltage is given by $E \sin \omega t$ and $i(0) = 0$. Show that the current in the circuit is described by

$$i = EZ^{-2}(R \sin \omega t - \omega L \cos \omega t + \omega L e^{-Rt/L})$$

where $Z^2 = R^2 + \omega^2 L^2$.

29. Find the current in a series RL circuit with input voltage given by the following functions:
    (a) $V = \sin t$.
    (b) $V = t$.
    Assume R and L are constant and $i(0) = 0$.

30. Find the current in a series RL circuit with $R = 1$ ohm, $L = 1$ henry, and input voltage as shown in Figure 2–30. Assume the initial current is zero.

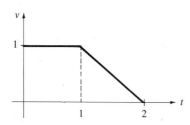

FIGURE 2–30

31. Consider a series RL circuit with $R = 10$ ohms, $V = 20$ volts, and

$$L = \begin{cases} 5 - t \text{ henrys}, & 0 < t \le 5 \text{ sec} \\ 0, & t > 5 \text{ sec} \end{cases}$$

Find the current in the circuit if the initial current is zero.

32. Repeat Exercise 31 with the assumption that the initial current is 2 amp.

33. Find the current in the series *RL* circuit of Exercise 30 if the input voltage
is given by the graph in Figure 2–31
and the initial current is 1 amp.

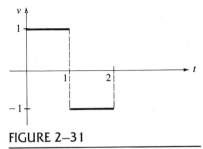

FIGURE 2–31

*A periodic input voltage does not ensure a periodic current in the circuit. Exercises
34 and 35 show two cases in which the periodicity of the current depends on the
initial current.*

34. If the input voltage in Exercise 33 is extended periodically, what should the initial
current be to ensure a periodic current?

35. Find the current if the input
voltage to the series *RL* circuit is the periodic function
shown in Figure 2–32. What
should the initial current be
in order to ensure a periodic
current?

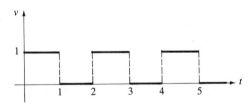

FIGURE 2–32

SUMMARY OF FIRST-ORDER TECHNIQUES

When confronted with a first-order differential equation to solve, begin by identifying its type so you will know what procedure to use. One approach is to ask the
following four questions.

1. Do the variables separate? Carefully examine the coefficients of dx and
dy to determine if factors can be determined so that factors of x and y
can be separated. If so, the differential equation can be solved with two
integrations.

2. Is there an obvious substitution that can be made to simplify the equation? If the coefficients are homogeneous in x and y and of the same
degree, use the substitution $u = y/x$ to reduce the equation to one in
which the variables will separate.

3. Is the equation exact? Remember, when testing for exactness, first put
the equation into the form $Mdx + Ndy = 0$. If the condition for exactness $M_y = N_x$ is satisfied, the differential equation can probably be
solved by inspection.

4. Is the equation linear? If so, remember to first put the equation into the form $y' + p(x)y = q(x)$ before determining the integrating factor.

Note that some equations may be solved by more than one procedure, and it is not possible to give a general rule as to which will be the "easiest." Suffice it to say that if we can find *one* technique for solving an equation we feel quite fortunate.

REVIEW EXERCISES FOR CHAPTER 2

Solve the following differential equations.

1. $(x^2 + 2) \, dy + 3xy \, dx = 0$

2. $y' = 2xy, \, y(0) = 1$

3. $ds/dt = 2(1 - s), \, s(0) = 2$

4. $y' = x(4 - y^2)^{1/2}$

5. $y' = xy/(x^2 + x - 6)$

6. $t \, dt - s \sec t \, ds = 0$

7. $\dfrac{dy}{dx} = \dfrac{x + y + 2}{x + y}$

8. $\dfrac{dy}{dx} = \dfrac{\sqrt{x + y}}{\sqrt{x + y} + 2}$

9. $e^{2x} \, dy + x \cot 3y \, dx = 0$

10. $x(3 - y) \, dy - y^2(x + 1) \, dx = 0$

11. $y' = \dfrac{x + y}{x - 4y}$

12. $(x + y) \, dx + (x - 2) \, dy = 0$

13. $y' = \dfrac{x + y + 2}{x - 4y - 3}$ (HINT: Use a linear substitution and the result of Exercise 11.)

14. $(x + y + 1) \, dx + (x - 2) \, dy = 0$ (HINT: Use a linear substitution and the result of Exercise 12.)

15. $\dfrac{dy}{dx} = \dfrac{y^2}{x^2 + xy}$

16. $xy \, dy + (x^2 + 4y^2) \, dx = 0$

17. $2xy \, dy + (3x^2 + y^2) \, dx = 0$

18. $y' = \dfrac{2x - y}{x + 2y}$

19. $\dfrac{dv}{dt} = \dfrac{-v \sec^2 t}{3v^2 + \tan t}$

20. $y^2 \, dx + (e^y + 2xy) \, dy = 0$

21. $xy' = y + 2xy^2$

22. $y' - y \cot x + y^2 = 0$

23. $dp/dt - 2p = 4$

24. $dy/dx = 3y + e^x$

25. $x \, dy + 2y \, dx = x^2 \, dx$

26. $(2x + 1) \, dy - 3y \, dx = 10 \, dx$

27. $\cos t \, dv = (v \sin t + 10) \, dt$

28. $xy' + 2y = \dfrac{3}{x^3 + 4x}$

Applications

29. The growth of the size of a population increases at a rate that is proportional to the population size. Determine the doubling period of a population if it increases 70% in 30 years. The doubling period is the time required for the population to double in size.

30. The acceleration of a charged particle is found to be directly proportional to its velocity in m/sec and inversely proportional to the square root of the elapsed time in seconds. If $v(0) = 1.5$ m/sec and $v(4) = 3.2$ m/sec, determine how long it takes for the particle to reach a velocity of 10 m/sec.

31. Find the equation of the graph in Figure 2–33 if

$$\tan \theta = \frac{y - 2}{2x}$$

and the curve passes through (4, 6).

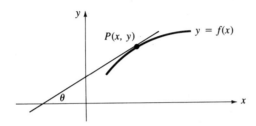

FIGURE 2–33

32. Determine the orthogonal trajectories of the family $x - 4y = c$. Draw a few members of each family.

33. Determine the orthogonal trajectories of the family $y^2 = cx^3$. Draw a few members of each family.

34. A rocket motor generates 10,000 lb of thrust for a period of 5 seconds. During this interval the mass of the rocket varies with time according to $m = 30 - 2t$ slugs, and it experiences a drag force equal to $2.5v$ lb, where v is the velocity of the rocket at any time t. Find the velocity equation of the rocket. Determine the velocity of the rocket at the end of the burn period. Assume $v(0) = 0$.

35. Calculate the velocity of the rocket in Figure 2–34 assuming its mass remains constant and the following values apply:

$$F(t) = \begin{cases} 90e^{0.5t} \text{ lb}, & 0 \le t < 5 \text{ sec} \\ 0, & t \ge 5 \text{ sec} \end{cases}$$

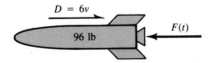

FIGURE 2–34

If its initial velocity is 20 ft/sec determine the velocity equation for the rocket.

36. A 200-gallon tank contains 80 gallons of pure water. A salt solution containing 2 lb of salt per gallon flows into the tank at 6 gallon/min and, simultaneously, the stirred mixture flows out at 4 gallons/min, as shown in Figure 2–35. (a) Determine the equation for the amount of salt in the tank as a function of time. (b) Determine how much salt is in the tank at the end of 5 minutes.

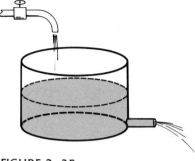

FIGURE 2–35

37. Find the equation for the current in the circuit of Figure 2–36 if the voltage is

$$V = \begin{cases} 3t \text{ volts}, & 0 \le t < 0.5 \text{ sec} \\ 1.5 \text{ volts}, & 0.5 \le t \le 1 \text{ sec} \end{cases}$$

Assume $i(0) = 0$.

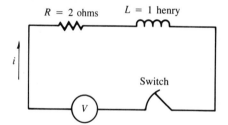

FIGURE 2–36

3

Approximate Methods

The differential equation

$$\frac{dy}{dx} + y^2 = e^{x^2}$$

cannot be solved exactly by writing the solution as a linear combination of elementary functions such as x^n, $\sin bx$, or e^{ax}. In this chapter we introduce methods for estimating solutions to such equations. The numerical methods described are easily adapted to computer programming.

Many first-order differential equations such as $y' = xy^2 + e^{x+y}$ cannot be solved by the methods of Chapter 2. In those cases we turn to techniques that yield approximate solutions. Approximation techniques are often very general in their application, but they have the shortcoming that they do not give the solution in terms of the elementary functions. In this chapter we limit our discussion to the following approximation techniques:

- **Graphical** methods in which the solution is approximated by a curve in the plane.

- **Numerical** methods in which the solution values are approximated by a table of numbers.

- **Taylor series** methods in which the solution function is approximated by the first two or three terms of the Taylor series.

- **Iterative** methods in which the solution is approximated by a sequence of functions.

3–1   DIRECTION FIELDS

In this section we look at the basic geometric ideas underlying what it means for a curve to be a solution of a differential equation. Consider the equation

$$y' = f(x, y)$$

This equation specifies the value of the slope of any solution curve at any point in the plane. Therefore at each point (x_0, y_0) we attach a third number $f(x_0, y_0)$ representing the slope of the solution curve at that point. We visualize these numbers by locating the point (x_0, y_0) in the plane and drawing a short line with a slope of $f(x_0, y_0)$ through the point. The line is called a **lineal element.**

Example 1

Draw lineal elements at $(0, 0), (1, 2), (1, -1), (-1, 0)$ for the differential equation

$$y' = 2x + y$$

SOLUTION   The slope of the lineal element at $(0, 0)$ is $2(0) + 0 = 0$; at $(1, 2)$ it is $2(1) + 2 = 4$; at $(1, -1)$ it is $2(1) + (-1) = 1$; and at $(-1, 0)$ it is $2(-1) + 0 = -2$. These elements are shown in Figure 3–1.   ∎

A set of lineal elements for a differential equation is called a **direction field** for the differential equation. The process of sketching a directon field can be tedious if it is not done in a systematic way. The usual approach is to identify and draw several curves in the direction field along which the slope of the field is constant. Toward this end, we make the following definition:

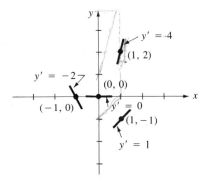

FIGURE 3–1

DEFINITION

A curve along which the first-order differential equation

$$y' = f(x, y)$$

has a constant value c is called an **isocline** of the differential equation. The isoclines for a differential equation are obtained by letting $f(x, y) = c$ for various values of c.

Example 2

Draw several isoclines for the direction field associated with the differential equation $y' = 2x + y$. Show the lineal elements on each isocline.

SOLUTION   The isoclines are given by

$$2x + y = c$$

The members of this family for $c = -2, -1, 0, 1, 2, 3$ are shown in Figure 3–2. Notice that along the isocline $2x + y = 0$, the lineal elements all have slope 0; along $2x + y = -1$, the lineal elements all have slope -1; and so forth.   ∎

Isoclines are *not* the integral curves of the differential equation; isoclines are curves that are used to construct the integral curves. The integral curve must cross each isocline with the same slope as the lineal elements on that isocline. To draw a particular curve, begin at the point (x_0, y_0) and continuously modify the shape of the curve so that it has the same inclination as the lineal elements as it crosses each isocline.

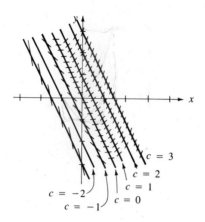

$c = 3$
$c = 2$
$c = 1$
$c = -2$
$c = 0$
$c = -1$

FIGURE 3–2

Example 3

Given the differential equation $y' = 2x + y$, sketch the integral curve through $(2, 0)$.

SOLUTION   See Figure 3–3.

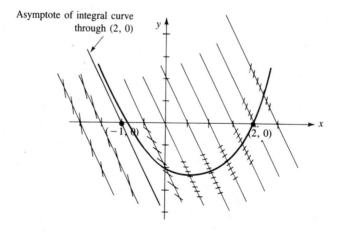

Asymptote of integral curve
through (2, 0)

$(-1, 0)$

$(2, 0)$

FIGURE 3–3

COMMENT: *The method of direction fields is very general and applies to all first-order differential equations and, although it gives a good qualitative idea of the nature of an integral curve, it is relatively inaccurate.*

Example 4

Use the method of direction fields to find the integral curve for the initial-value problem

$$y' = x^2 + y^2, y(1) = 0$$

SOLUTION   The isoclines are given by $x^2 + y^2 = c$, which is the family of concentric circles shown in Figure 3–4. Lineal elements are drawn on each circle with the appropriate slope c. The particular solution is approximated by the integral curve through $(1, 0)$.

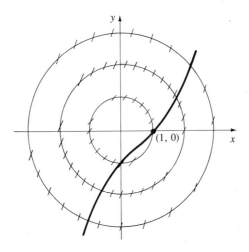

FIGURE 3—4   ■

COMMENT: Although isoclines are not in general integral curves of a differential equation, it is possible for an isocline to double as an integral curve.

Example 5

Find those isoclines that are also integral curves of

$$y' = 2x + y$$

SOLUTION   The isoclines are the straight lines $2x + y = c$, or $y = c - 2x$. Differentiating, we obtain $y' = -2$. Hence the isocline $2x + y = -2$ is also an integral curve. See Figure 3–3 for the direction field of $y' = 2x + y$.   ■

EXERCISES FOR SECTION 3–1   1–17 odd

In Exercises 1–6 sketch the lineal elements for the given differential equation at the following points: $(0, 0)$, $(1, 0)$, $(2, 1)$, $(-1, 2)$, $(2, -1)$.

1. $y' = x$ **2.** $y' = y$

3. $y' = x^2 - y$ **4.** $y' = y^2 - x$

5. $y' = y^2 + 1$ **6.** $y' = 2 - x^2$

In Exercises 7–12 sketch the family of isoclines for the given differential equation and determine if any isocline is also an integral curve.

7. $y' = y$ **8.** $y' = x$

9. $y' = x^2 + y$ **10.** $y' = x/y$

11. $y' = (x + y)/(2x + y)$ **12.** $y' = (x - y)/y$

In Exercises 13–22 sketch the direction field associated with each initial-value problem and draw the integral curve through the given point.

13. $y' = y$, $(0, 1)$ **14.** $y' = x + 1$, $(0, 1)$

15. $y' = 1/x$, $(1, 0)$ **16.** $y' = 2/y$, $(2, 0)$

17. $xy' - y = 0$, $(2, 2)$ **18.** $y' = x + y$, $(0, 1)$

19. $y' = y - 2x$, $(0, 1)$ **20.** $y' = x^2 + y$, $(0, 1)$

21. $y' - xy = 0$, $(0, 1)$ **22.** $x + 2yy' = 0$, $(1, 1)$

3–2   STRAIGHT-LINE APPROXIMATIONS

One of the most elementary techniques for approximating solutions of first-order initial-value problems assumes that the integral curve can be approximated by a

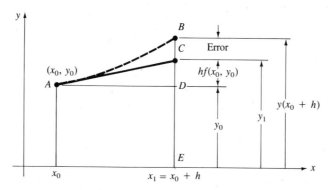

FIGURE 3–5

sequence of straight-line segments. The technique of using straight-line segments to approximate the integral curve is attributed to Leonhard Euler (1707–1783). (NOTE: Euler is pronounced "oiler.")

EULER'S METHOD

Given the initial-value problem

$$y' = f(x, y), y(x_0) = y_0$$

defined on the interval $x_0 \le x \le x_0 + h$, then at $x_1 = x_0 + h$ the approximate value of $y(x_0 + h)$, denoted by y_1, is given by

$$y_1 = y_0 + hf(x_0, y_0) \tag{3–1}$$

To understand Euler's method, consider the integral curve shown in Figure 3–5. Assume the dashed line AB represents the integral curve through the point (x_0, y_0). Line segment AC, which is tangent to the integral curve at (x_0, y_0), has a slope equal to $f(x_0, y_0)$. The exact value of y at $x_0 + h$ is represented by BE, and the value of y_1 by CE. From the figure

$$y_1 = CE = DE + CD = y_0 + hf(x_0, y_0)$$

The error in using y_1 to approximate $y(x_0 + h)$ is represented by BC.

Example 1

Given the initial-value problem

$$y' = 2x + y, y(0) = 1$$

use Euler's method to approximate the value of $y(0.4)$.

SOLUTION Here we use Equation 3–1 with $x_0 = 0$, $y_0 = 1$, and $h = 0.4$. Thus

$$y_1 = 1.0 + 0.4[2(0) + 1.0] = 1.4$$

is the approximate value of $y(0.4)$.

Notice that $y' = 2x + y$ is a linear equation, so we can solve the initial-value problem by the methods of Section 2–5. Applying the integrating factor e^{-x} and using the initial values, we find that

$$y = 3e^x - 2x - 2$$

The value of this function at $x = 0.4$ is $y = 1.675$. The percentage error in using y_1 to approximate y can be calculated by

$$\% \text{ Error } = \frac{\text{Actual value } - \text{ Approximate value}}{\text{Actual value}} \times 100$$

The percentage error in using 1.4 to approximate $y = 1.675$ is about 16.4%.   ■

Euler's method is simple to apply, but it may produce large errors if we are not careful. One way to improve the accuracy of this method is to partition a given interval into k subintervals and then use Equation 3–1 over each subinterval to obtain a succession of points (x_1, y_1), (x_2, y_2), . . . , (x_k, y_k). See Figure 3–6.

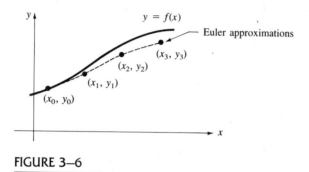

FIGURE 3–6

EULER'S METHOD: AN ITERATIVE PROCESS

For a fixed, constant value h Euler's method becomes

$$y_{n+1} = y_n + hf(x_n, y_n) \tag{3-2}$$

where $x_n = x_0 + nh$.

Example 2

Approximate the solution of the initial-value problem

$$y' = 2x + y, \, y(0) = 1$$

on the interval $0 \leq x \leq 0.4$ by using four equal subintervals. Calculate the percentage error in the approximation for $y(0.4)$.

SOLUTION   Dividing the interval [0, 0.4] into four equal parts, we get

$$h = \frac{0.4 - 0}{4} = 0.1$$

Using $f(x, y) = 2x + y$ and $x_0 = 0$, $y_0 = 1$, the required computation is conveniently arranged in Table 3–1.

TABLE 3–1 Euler's method for $y' = 2x + y$, $y(0) = 1$

x_n	y_n	$y_n + 0.1(2x_n + y_n) = y_{n+1}$
0	1.0	$1.0 + 0.1[2(0) + 1.0] = 1.1$
0.1	1.1	$1.1 + 0.1[2(0.1) + 1.1] = 1.23$
0.2	1.23	$1.23 + 0.1[2(0.2) + 1.23] = 1.39$
0.3	1.39	$1.39 + 0.1[2(0.3) + 1.39] = 1.59$
0.4	1.59	∎

From Example 1 we know that the exact solution is $y = 3e^x - 2x - 2$. The integral curve is shown in Figure 3–7 along with its approximation by straight-line segments.

The approximate value of $y(0.4)$, obtained from Table 3–1, is 1.59. Hence there is a percentage error of about 5%, a considerable improvement over the estimate in Example 1 in which only one step was used.

The computations involved in the Euler method are of a type that can be programmed and performed by a computer. For this reason we have included a flow diagram for Euler's method (Figure 3–8).

The flow diagram was used as a model for the BASIC program to solve the initial-value problem in Example 2. Table 3–2 is the computer output that parallels the computations in Table 3–1. The values of y in Table 3–2 were computed using $y = 3e^x - 2x - 2$.

TABLE 3–2 Euler's method:
$y' = 2x + y$, $y(0) = 1$;
$h = 0.1$ for $0 \leq x \leq 0.4$

x	y_1	y	% ERROR
0	1.0	1.0	0
0.1	1.1	1.1155	1.39
0.2	1.23	1.2642	2.71
0.3	1.393	1.4496	3.90
0.4	1.5923	1.6755	4.96

Example 3

The BASIC program for Euler's method was used to estimate $y(0.4)$ for $y' = 2x + y$, $y(0) = 1$, using progressively smaller values of h to show the effect of the size

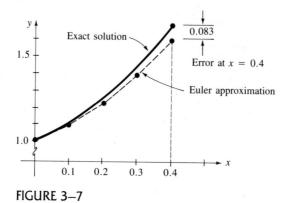

FIGURE 3–7

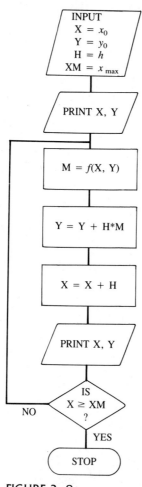

FIGURE 3–8

Flow diagram for programming Euler's method

of h on the percentage error. Table 3–3 shows the decrease in the percentage error for values of h from 0.4 to 0.005. Recall that $y(0.4) = 1.6755$.

TABLE 3–3 Estimates of $y(0.4)$, using Euler's method, for progressively smaller values of h

h	y_1	% ERROR
0.4	1.4	16.4
0.1	1.5923	4.96
0.05	1.6324	2.57
0.01	1.6666	0.53
0.005	1.6710	0.27

■

An Improved Method

The basic Euler method is rarely used in practice because other methods are available that will give a specified accuracy with proportionately fewer steps. Many of the improved methods are modifications of the basic Euler method. One improvement that can be made on the basic Euler method is to replace the slope at (x_0, y_0) with the average of the slope values at (x_0, y_0) and $(x_0 + h, y_t)$, where y_t is the estimate of $y(x_0 + h)$ obtained from the basic Euler method.

IMPROVED EULER METHOD

Given the initial-value problem

$$y' = f(x, y), \ y(x_0) = y_0$$

for a fixed, constant value of h the value of $y(x_n + h)$ can be approximated by the formula

$$y_{n+1} = y_n + hM$$

where

$$M = \frac{1}{2}[f(x_n, y_n) + f(x_{n+1}, y_t)]$$

and

$$y_t = y_n + hf(x_n, y_n)$$

Notice that $f(x_n, y_n)$ and $f(x_{n+1}, y_t)$ are approximations of the slope of the curve at (x_n, y_n) and (x_{n+1}, y_t), respectively. Figure 3–9 shows the components used in the improved Euler method relative to $x = x_0 + h$. We note that

- y is the exact value of the solution at $x_0 + h$.

- y_t is the estimate of y obtained by the basic Euler method of proceeding along the tangent line through (x_0, y_0).

- y_1 is the estimate of y obtained by proceeding along the line through (x_0, y_0) with slope $M = \frac{1}{2}[f(x_0, y_0) + f(x_1, y_t)]$.

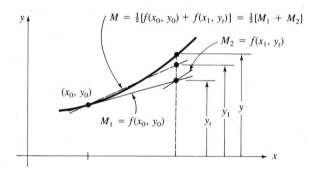

FIGURE 3–9

Example 4

Use the improved Euler method with $h = 0.4$ to estimate $y(0.4)$ if

$$y' = 2x + y, \quad y(0) = 1$$

SOLUTION   In Example 1 the basic Euler method with $h = 0.4$ was used to obtain an estimate for $y(0.4)$ of 1.4. This corresponds to y_t in the improved Euler method. Therefore

$$M = \frac{1}{2}[f(0, 1) + f(0.4, 1.4)]$$

$$= \frac{1}{2}(1 + 2.2) = 1.6$$

The value of M is now used in $y_1 = y_0 + hM$. Thus

$$y_1 = 1 + 0.4(1.6) = 1.64$$

is the estimate of $y(0.4)$. The percentage error is about 2.1%. The percentage error for the basic Euler method was found in Example 1 to be 16.4%.   ∎

The next example shows that the improved Euler method can be applied to a number of subintervals to reduce the error.

Example 5

Use the improved Euler method with step size $h = 0.1$ to estimate $y(0.4)$, if $y' = 2x + y$, $y(0) = 1$. Compare the result with $y(0.4) = 1.6755$.

SOLUTION   The computations are shown in Table 3–4.

TABLE 3–4   The improved Euler method

x_n	y_n	$y_t = y_n + 0.1(2x_n + y_n)$	$M = \frac{1}{2}[(2x_n + y_n) + (2x_{n+1} + y_t)]$	$y_{n+1} = y_n + 0.1M$
0	1	1.1	1.15	1.115
0.1	1.115	1.247	1.481	1.263
0.2	1.263	1.429	1.846	1.448
0.3	1.448	1.653	2.250	1.673
0.4	1.673			

Compared to the exact value of 1.675, the percentage error is about 0.1%. Recall from Example 2 that the percentage error using the basic Euler method with $h = 0.1$ was 5.4%. ∎

The flow diagram for the improved Euler method (Figure 3–10) is included for easy reference. Table 3–5 shows the solution of $y' = 2x + y$, $y(0) = 1$ at $x = 0.4$, for various step sizes. The values from Table 3–2 (Euler's method for the same problem) are included so the two methods can be compared.

TABLE 3–5   Estimates of $y(0.4)$, using the improved Euler method for progressively smaller values of h. [$y(0.4) = 1.6755$]

h	$y(0.4)$ IMP. EULER	% ERROR	$y(0.4)$ EULER	% ERROR
0.4	1.64	2.13	1.4	16.4
0.1	1.6727	0.17	1.5923	4.96
0.05	1.6748	0.04	1.6324	2.57
0.01	1.6754	0.006	1.6666	0.53

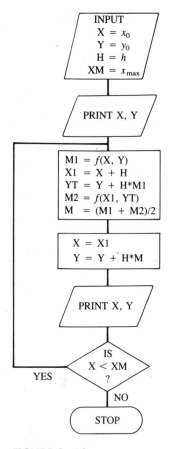

FIGURE 3–10

Flow diagram for the improved Euler method

EXERCISES FOR SECTION 3–2

Use the Euler method to approximate the indicated value of the solution function in Exercises 1–9.

1. $y' = y + 1$, $y(0) = 2$. Find $y(0.8)$, using $h = 0.2$.

2. $y' = 2x + 3$, $y(0) = 1$. Find $y(1)$, using $h = 0.25$.

3. $y' = x + y$, $y(0) = 0$. Find $y(2)$, using $h = 0.25$.

4. $y' = x^2 + y$, $y(1) = 2$. Find $y(1.5)$, using $h = 0.1$.

5. $xy' - y = 0$, $y(1) = 1$. Find $y(0.5)$, using $h = -0.1$.

6. $xy' + y = 2$, $y(1) = 0$. Find $y(0.6)$, using $h = -0.1$.

7. $s' = 2ts + 1$, $s(0) = 0$. Find $s(1)$, using $h = 0.2$.

8. $s' = \sqrt{s + t}$, $s(1) = 2$. Find $s(1.7)$, using $h = 0.1$.

9. $y' = (1 - y)/x$, $y(1) = 2$. Find $y(2)$, using $h = 0.1$.

10. Compare the values of $y(2)$ obtained in Exercise 9 with the value obtained by solving the equation exactly.

Exercises 11–20: Use the improved Euler method for Exercises 1–10.

21. The velocity of a rocket starting from rest is given by $v = 3s + 2t$ ft/sec, where s is the displacement in feet and t is elapsed time in seconds. Use the improved Euler method to estimate the displacement of the rocket at $t = 0.4$ sec. Assume $s(0) = 0$.

22. An object that is hotter than the air around it will lose heat to its surroundings. The temperature of such an object is described by

$$\frac{dT}{dt} = -2(T + 10), \quad T(0) = 100$$

Use the improved Euler method with $h = 0.2$ to estimate the temperature of the object at $t = 2.0$ sec.

23. The electric current in a series *RL* circuit is described by

$$2\frac{di}{dt} + 10i = 2t, \quad i(0) = 0$$

Use the improved Euler method to estimate the current at $t = 0.8$ sec. Let $h = 0.2$.

3–3   THE RUNGE–KUTTA METHOD

The improved Euler method discussed in the previous section can itself be improved by replacing the average slope at two points with a slope that is the weighted

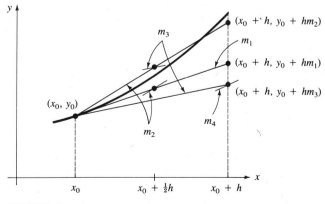

FIGURE 3–11

average of $f(x, y)$ at four points within the interval. The technique, which is attributed to C.D.T. Runge (1856–1927) and M. W. Kutta (1867–1944), is based on the following direction-field computations. See Figure 3–11.

The four slopes indicated in the figure are

$m_1 = f(x_0, y_0)$	The slope at (x_0, y_0)
$m_2 = f(x_0 + \frac{1}{2}h, y_0 + \frac{1}{2}hm_1)$	The slope at the midpoint of the interval along the line connecting (x_0, y_0) and $(x_0 + h, y_0 + hm_1)$
$m_3 = f(x_0 + \frac{1}{2}h, y_0 + \frac{1}{2}hm_2)$	The slope at the midpoint of the interval along the line connecting (x_0, y_0) and $(x_0 + h, y_0 + hm_2)$
$m_4 = f(x_0 + h, y_0 + hm_3)$	The slope at $(x_0 + h, y_0 + hm_3)$

The Runge–Kutta method uses a weighted average of m_1, m_2, m_3, and m_4 in the basic Euler formula to estimate $y(x_0 + h)$. The derivation of these formulas and the required weighted average can be found in more advanced texts.

RUNGE–KUTTA METHOD

Given the initial-value problem

$$y' = f(x, y), \ y(x_0) = y_0$$

for a fixed, constant value of h, $y(x_n + h)$ can be approximated by

$$y_{n+1} = y_n + \frac{1}{6}h(m_1 + 2m_2 + 2m_3 + m_4)$$

where

$$m_1 = f(x_n, y_n)$$

$$m_2 = f(x_n + \frac{1}{2}h, y_n + \frac{1}{2}hm_1)$$

$$m_3 = f(x_n + \frac{1}{2}h, y_n + \frac{1}{2}hm_2)$$

$$m_4 = f(x_n + h, y_n + hm_3)$$

The Runge–Kutta method is surprisingly accurate for values of $h < 1$. The next example shows the technique for the initial-value problem solved earlier by the Euler method and the improved Euler method.

Example 1

Use the Runge–Kutta method to estimate $y(0.4)$ if

$$y' = 2x + y, y(0) = 1$$

SOLUTION   In using the Runge–Kutta formulas, we note that $f(x, y) = 2x + y$, $x_0 = 0$, and $y_0 = 1$. Choosing $h = 0.4$, we have

$$m_1 = [2(0) + 1] = 1.0$$

$$m_2 = [2(0 + 0.4/2) + (1 + 0.4(1.0)/2)] = 1.6$$

$$m_3 = [2(0 + 0.4/2) + (1 + 0.4(1.6)/2)] = 1.72$$

$$m_4 = [2(0 + 0.4) + (1 + 0.4(1.72)] = 2.488$$

Hence

$$y(0.4) = 1 + \frac{1}{6}(0.4)[1.0 + 2(1.6) + 2(1.72) + 2.488]$$

$$= 1.675 \qquad \blacksquare$$

The estimate of $y(0.4)$, using the Euler method with $h = 0.4$, was 1.4, and with the improved Euler method it was 1.64. (See Examples 1 and 4, Section 3–2.) The improved accuracy of the Runge–Kutta method in the solution of this problem is obvious. (Recall that the exact value of $y(0.4)$ is 1.675 to three decimal places.)

Example 2

A certain chemical reaction takes place such that the time-rate of change of the amount of the unconverted substance q is equal to $-2q$. If the initial mass is 50 grams, use the Runge–Kutta method to estimate the amount of unconverted substance at $t = 0.8$ sec.

SOLUTION   The initial-value problem is

$$\frac{dq}{dt} = -2q, q(0) = 50$$

Using $h = 0.8$ in the Runge–Kutta formulas,

$$m_1 = -2(50) = -100$$

$$m_2 = -2[50 + 0.8(-100)/2] = -20$$

$$m_3 = -2[50 + 0.8(-20)/2] = -84$$

$$m_4 = -2[50 + 0.8(-84)] = 34.4$$

Therefore our estimate of the mass of unconverted substance at $t = 0.8$ is

$$q(0.8) = 50 + \frac{1}{6}(0.8) [-100 + 2(-20) + 2(-84) + 34.4] = 13.5 \text{ g} \qquad \blacksquare$$

In practice the Runge–Kutta formulas, like the Euler formula, are applied multiple times over a specified interval. The flow diagram in Figure 3–12 will be useful when writing a computer program for the Runge–Kutta method.

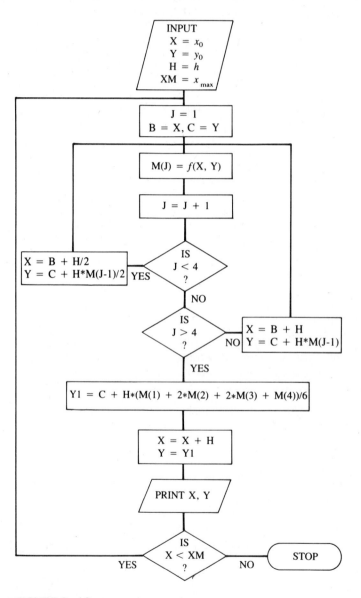

FIGURE 3–12

Flow diagram for the Runge–Kutta method

Example 3

Use the Runge–Kutta method with $h = 0.3$ to estimate the solution of $y' = 2x + y$, $y(0) = 1$, on the interval $0 \leq x \leq 1.2$.

SOLUTION Using the BASIC program for the Runge–Kutta method and the function $y = 3e^x - 2x - 2$ for the true values, Table 3–6 shows the results. ∎

TABLE 3–6 The Runge–Kutta method with $h = 0.3$

x_n	y_n (RUNGE–KUTTA APPROXIMATION)	$y(x_n)$ (TRUE VALUE)	% ERROR $\times 10^{-3}$
0	1	1	0
0.3	1.44951250	1.44957642	4.41
0.6	2.26618383	2.26635640	7.61
0.9	3.57845992	3.57880933	9.76
1.2	5.55972189	5.56035077	11.31

From a practical standpoint the determination of the actual error can be very difficult, if not impossible. The advanced methods available for making estimates of the error are beyond our scope of interest. One way of deciding when the error is sufficiently small is to decrease the size of h until the solution value shows little change. Then we "hope" this steady-state value is close to the true value.

Example 4

Use the Runge–Kutta method with progressively smaller values of h to approximate $y(2.0)$ if

$$y' = x^2 - y^2, y(1.5) = 1.8$$

SOLUTION See Table 3–7.

TABLE 3–7

h	$y(2.0)$
0.5	1.86331690
0.1	1.82119247
0.05	1.82116547
0.01	1.82116420
0.005	1.82116420

The fact that the last two values are equal suggests that the solution is close to this number. ∎

EXERCISES FOR SECTION 3–3

Use the Runge–Kutta method once to solve the initial-value problems in Exercises 1–8.

1. $y' = y + 1$, $y(0) = 2$; find $y(0.8)$.   2. $y' = 2x + 3$, $y(0) = 1$; find $y(1)$.

3. $y' = x + y$, $y(0) = 0$; find $y(2)$.   4. $y' = x^2 + y$, $y(1) = 2$; find $y(1.5)$.

5. $xy' - y = 0$, $y(1) = 1$; find $y(0.5)$.   6. $xy' + y = 2$, $y(1) = 0$; find $y(0.6)$.

7. $s' = 2ts + 1$, $s(0) = 0$; find $s(1)$.   8. $s' = \sqrt{s + t}$, $s(1) = 2$; find $s(1.7)$.

9. $\dfrac{dy}{dx} = x^2 - y$, $y(0) = 0$; find $y(1)$ by applying the Runge–Kutta method twice; that is, use $h = 0.5$.

10. Solve the differential equation in Exercise 3 by using $h = 0.5$. This requires four applications of the Runge–Kutta method. Write a computer program to solve this problem.

11. Use the Runge–Kutta method to estimate the displacement s of an object when $t = 0.5$ sec if its velocity is given by $v = 2(2t - s)$ ft/sec. Assume $s(0) = 1$ ft.

12. Show that if f is a function of x only, the Runge–Kutta formulas reduce to Simpson's rule from elementary calculus.

3–4   TWO- AND THREE-TERM TAYLOR APPROXIMATIONS

Another way to approximate the solution of the initial-value problem

$$y' = f(x, y), \ y(x_0) = y_0$$

on the interval $x_0 \leq x \leq x_0 + h$ is to consider the Taylor series of the solution about $x = x_0$. Recall that the Taylor series for a function $y(x)$ is given by

$$y(x) = y(x_0) + y'(x_0)(x - x_0) + \frac{y''(x_0)}{2}(x - x_0)^2 + \cdots$$

Then if the series converges to the function on $(x_0, x_0 + h)$, we can consider any truncation of the Taylor series to be an approximation of $y(x)$.

TAYLOR SERIES METHOD

Given the initial-value problem

$$y' = f(x, y), \ y(x_0) = y_0$$

the solution $y(x)$ is approximated by the three Taylor terms

$$y(x) = y(x_0) + y'(x_0)(x - x_0) + \frac{y''(x_0)}{2}(x - x_0)^2$$

Notice that the two-term Taylor approximation

$$y = y(x_0) + y'(x_0)(x - x_0)$$

is equivalent to the Euler approximation

$$y = y_0 + hf(x_0, y_0)$$

where $x - x_0 = h$ and $y'(x_0) = f(x_0, y_0)$. The three-term Taylor approximation is a parabolic approximation that can be applied to a number of subintervals just as with the linear approximations of Euler and Runge–Kutta.

Using the same notation as in the linear approximation, we can write the three-term Taylor approximation in the form

$$y = y_0 + hf(x_0, y_0) + \frac{1}{2}h^2 f_x(x_0, y_0)$$

or, more generally, in iterative form:

$$y_{n+1} = y_n + hf(x_n, y_n) + \frac{1}{2}h^2 f_x(x_n, y_n) \quad \text{wrong}$$

Example 1

Given the initial-value problem

$$y' = x - 3y, \quad y(2) = 1$$

use the three-term Taylor approximation to find $y(2.1)$.

SOLUTION   To use the three-term Taylor approximation, the values of $y(2)$, $y'(2)$, and $y''(2)$ must be given or computed. The value $y(2) = 1$ is given. Since $y' = x - 3y$, we find $y'(2) = 2 - 3(1) = -1$. To find y'', we differentiate the expression for y'. Thus $y'' = 1 - 3y'$, and $y''(2) = 1 - 3(-1) = 4$. The three-term Taylor approximation is then

$$y = 1 - (x - 2) + 2(x - 2)^2$$

Substituting $x = 2.1$, we get

$$y = 1 - 0.1 + 2(0.1)^2 = 0.92$$

as an estimate of $y(2.1)$. The exact solution of the given initial-value problem is

$$y = \frac{4}{9}e^{-3(x-2)} + \frac{1}{3}x - \frac{1}{9}$$

From this we get $y(2.1) = 0.9181$. ∎

COMMENT: *Since $y' = f(x, y)$, it follows that $y'' = f_x(x, y) +$*
$f_y(x, y) \cdot f(x, y)$. Using this fact, the three-term Taylor expression for y
has the form

$$y = y(x_0) + f(x_0, y_0)(x - x_0)$$

$$+ \frac{1}{2}[f_x(x_0, y_0) + f_y(x_0, y_0) \cdot f(x_0, y_0)] (x - x_0)^2 \quad (3\text{--}3)$$

$\boxed{Right}$

Example 2

Determine the three-term Taylor approximation for the solution of

$$y' = 2x + y, y(0) = 1$$

at $x = 0.4$. Use the three-term Taylor approximation with $h = 0.1$ to estimate the
values of the solution on the interval $[0, 0.4]$.

SOLUTION   Here $f(x, y) = 2x + y$, so $f_x(x, y) = 2$ and $f_y(x, y) = 1$. Since $x_0 = 0$
and $y_0 = 1$, we see that $f(0, 1) = 1, f_x(0, 1) = 2$, and $f_y(0, 1) = 1$. Using Equation
3–3, the three-term Taylor approximation about $x_0 = 0$ is

$$y = 1 + x + \frac{1}{2}[2 + 1(1)]x^2$$

$$y = 1 + x + \frac{3}{2}x^2$$

To estimate the values of the solution on $[0, 0.4]$ with a step size of 0.1, we used a
BASIC program. (See the flow diagram in Figure 3–13.) Table 3–8 shows the results
and compares the approximate values with the true values.

TABLE 3–8   Three-term Taylor approximations for
$y' = 2x + y, y(0) = 1$, on $[0, 0.4]; h = 0.1$

x_n	y_n	TRUE y	% ERROR
0	1	1	0
0.1	1.1150	1.1155	0.05
0.2	1.2631	1.2642	0.09
0.3	1.4477	1.4496	0.13
0.4	1.6727	1.6755	0.20

Example 3

The value of the step size may be varied to show the effect of step size on the
accuracy of the approximation. The BASIC program for the initial-value problem

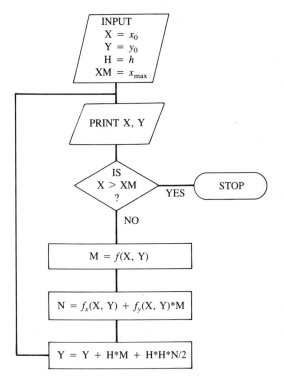

FIGURE 3–13

Flow diagram for the three-term Taylor method

$y' = 2x + y$, $y(0) = 1$, was used to evaluate $y(0.4)$ for progressively smaller values of h. The results are shown in Table 3–9.

TABLE 3–9 Three-term Taylor approximation for progressively smaller values of the step size h.

h	$y(0.4)$	% ERROR
0.4	1.64	2.12
0.1	1.6727	0.20
0.05	1.6747	0.04
0.01	1.6754	0.002

Higher-order Taylor approximations are possible but are computationally difficult because of the necessity of calculating f and its partial derivatives at each point. Although the Taylor method is not used in practice, it is the basis of the terminology used in error analysis. For example, when we say that the Runge–Kutta method is a fourth-order technique, we mean that the step error is the same order as a four-term Taylor approximation.

EXERCISES FOR SECTION 3–4   1–7 odd

In Exercises 1–6 solve the given initial-value problem by using a three-term Taylor approximation.

1. $y' = 2xy$, $y(0) = 1$; estimate $y(0.5)$, using $h = 0.5$.

2. $y' = x$, $y(0) = 0$; estimate $y(1)$, using $h = 1$.

3. $y' = x^2 + y$, $y(0) = -1$; estimate $y(1)$, using $h = 0.5$.

4. $y' = 4 - y$, $y(1) = 2$; estimate $y(1.5)$, using $h = 0.5$.

5. $y' = x^2 + 4$, $y(2) = 0$; estimate $y(2.4)$, using $h = 0.2$.

6. $y' = x^2 - y^2$, $y(1) = 2$; estimate $y(1.6)$, using $h = 0.2$.

7. Solve the initial-value problem given in Exercise 1. Compare the true value of $y(0.5)$ with the estimated value.

8. Solve the initial-value problem given in Exercise 2. Compare the true value of $y(1)$ with the estimated value.

9. Solve the initial-value problem given in Exercise 3. Compare the true value of $y(1)$ with the estimated value.

10. Solve the initial-value problem given in Exercise 4. Compare the true value of $y(1.5)$ with the estimated value.

3–5   PICARD APPROXIMATIONS (OPTIONAL)

In more theoretical discussions of differential equations a fundamental proof is often given for constructing a solution to the initial-value problem

$$y' = f(x, y), \, y(x_0) = y_0$$

which is a limit of a sequence of functions. The construction is distinctively different from the numerical techniques used earlier in this chapter in that it does not lend itself to stepwise approximations.

We can write, at least formally, the solution to the indicated initial-value problem in implicit form as

$$y = y_0 + \int_{x_0}^{x} f(t, y(t)) \, dt \tag{3–4}$$

Example 1

(a) The solution of $y' = xy^2$, $y(1) = 2$ is

$$y = 2 + \int_1^x t[y]^2 \, dt$$

(b) The solution of $y' = 2x + y$, $y(0) = 1$ is

$$y = 1 + \int_0^x [2t + y] \, dt$$ ∎

Solutions written in the form of Equation 3–4 yield little pertinent information about the integral curve. Try evaluating each of the solutions written in Example 1 at $x = 2$. The problem is that the solution is written in the form of an integral that involves the solution itself. Although there appears to be little value in writing such solutions, implicit forms such as these give rise to a method known as **iteration.** This technique permits the computation of a sequence of functions that under rather general conditions on the function $f(x, y)$, converges to a solution function. The approach shown here is based on one attributed to E. Picard (1856–1941); thus the elements of the sequence are called **Picard approximations.**

We begin an iteration by assuming that the function $y(x)$ on the right-hand side of Equation 3–4 is some known function y_0, and from this we compute a y_1. Then we use y_1 in Equation 3–4 to compute y_2 and continue in this manner to compute an entire sequence of functions. It is customary to choose y_0 to be the initial value. In general the iteration is

$$y_{n+1} = y_0 + \int_{x_0}^x f(t, y_n(t)) \, dt \tag{3–5}$$

Example 2

(a) Find the Picard approximations y_1, y_2, and y_3 to the solution of the initial-value problem

$$y' = y, \; y(0) = 2$$

(b) Use y_3 to estimate the value of $y(0.8)$.

SOLUTION   (a) Letting $y_0 = 2$, the value of y_1 is

$$y_1 = 2 + \int_0^x (2) \, dt = 2 + 2x$$

$$y_2 = 2 + \int_0^x (2 + 2t) \, dt = 2 + 2x + x^2$$

$$y_3 = 2 + \int_0^x (2 + 2t + t^2) \, dt = 2 + 2x + x^2 + \frac{1}{3}x^3$$

(b) At $x = 0.8$

$$y_3 = 2 + 2(0.8) + (0.8)^2 + \frac{1}{3}(0.8)^3 = 4.41$$

The solution of the initial-value problem, found by separation of variables, is $y = 2e^x$. At $x = 0.8$

$$y = 2e^{0.8} = 4.45$$ ■

Example 3

Find the Picard approximations y_1 and y_2 to the solution of the nonlinear initial-value problem

$$y' = 1 + y^2, \, y(2) = 1$$

SOLUTION   The Picard approximations are given by

$$y_{n+1} = y_0 + \int_2^x [1 + y_n^2(t)] \, dt$$

Using $y_0 = 1$ yields

$$y_1 = 1 + \int_2^x [1 + (1)^2] \, dt = 2x - 3$$

Substituting $y_1 = 2x - 3$,

$$y_2 = 1 + \int_2^x [1 + (2t - 3)^2] \, dt$$

$$= 1 + \int_2^x (10 - 12t + 4t^2) \, dt$$

$$= \left[1 + 10t - 6t^2 + \frac{4}{3}t^3 \right]\Big|_2^x = -\frac{20}{3} + 10x - 6x^2 + \frac{4}{3}x^3$$ ■

The method of Picard approximations may appear easier than it really is because we have selected initial-value problems for which the indefinite integrals are easy to evaluate. In fact, one reason Picard approximations are not used very much is that the required iterations cannot be evaluated.

EXERCISES FOR SECTION 3–5

In Exercises 1–5 find the first three Picard approximations to the solution of the given initial-value problems.

1. $y' = x + y$, $y(0) = -1$

2. $y' = x + y$, $y(0) = 0$

3. $y' = x^2$, $y(0) = 1$

4. $y' = x^2 - 4$, $y(1) = 2$

5. $y' = y$, $y(0) = -2$

6. Find the first two Picard approximations for $y' = x - y^2$, $y(0) = 3$.

7. Use the Picard method to find the first three approximations to the solution of $y' = e^y$, $y(0) = 1$. The difficulty encountered with y_3 is typical of what happens with Picard iterations.

REVIEW EXERCISES FOR CHAPTER 3

In Exercises 1–4 sketch the direction field associated with each initial-value problem and draw the integral curve for the given point.

1. $y' = 2 - x$, $(0, 1)$

2. $y' = 1/y$, $(1, 0)$

3. $y' = x - y$, $(0, 2)$

4. $y' = 2x + y$, $(1, 1)$

In Exercises 5–8 use the Euler method to approximate the indicated value of the solution function.

5. $y' = 2x + 3y$, $y(1) = -1$; estimate $y(1.4)$, using four increments.

6. $y' = x - y$, $y(0) = 2$; estimate $y(0.2)$, using four increments.

7. $yy' = x - y^2$, $y(1) = 1$; estimate $y(0.4)$, using three increments.

8. $t (ds/dt) = 2s - t$, $s(1) = 0$; estimate $s(0.5)$, using five increments.

Exercises 9–12: Use the improved Euler method to approximate the solutions in Exercises 5–8.

In Exercises 13–16 use the Runge–Kutta method to approximate the indicated value of the solution function.

13. $y' = 2x + 3y$, $y(1) = -1$; estimate $y(1.4)$.

14. $y' = x - y$, $y(0) = 2$; estimate $y(0.2)$.

15. $yy' = x - y^2$, $y(1) = 1$; estimate $y(0.4)$.

16. $t (ds/dt) = 2s - t$, $s(1) = 0$; estimate $s(0.5)$.

17. Given $xy' = 2y - x^2$, $y(1) = 1$. Approximate $y(1.8)$ by using
    (a) The Euler method with $h = 0.2$.
    (b) The improved Euler method with $h = 0.2$.
    (c) The Runge–Kutta method with $h = 0.8$.
    Solve the equation as a first-order linear differential equation and determine the value of $y(1.8)$. Compare this value with those obtained in (a), (b), and (c).

18. Repeat Exercise 17 with $y' = y - e^x$, $y(0) = 2$ and $y(0.8)$ to be approximated.

In Exercises 19–22 use a three-term Taylor series to approximate the indicated value of the solution function.

19. $y' = 2x + 3y$, $y(1) = -1$; estimate $y(1.4)$ with $h = 0.4$.

20. $y' = x - y$, $y(0) = 2$; estimate $y(0.2)$ with $h = 0.2$.

21. $yy' = x - y^2$, $y(1) = 1$; estimate $y(0.4)$ with $h = -0.6$.

22. $t(ds/dt) = 2s - t$, $s(1) = 0$; estimate $s(0.5)$ with $h = -0.5$.

In Exercises 23–26 find the first three Picard approximations to the solution of the given initial-value problem.

23. $y' = x + 1$, $y(0) = 1$ **24.** $y' = y$, $y(0) = 1$

25. $y' = y + 2$, $y(0) = -1$ **26.** $y' = x^2$, $y(0) = 0$

4

Homogeneous Linear Differential Equations

Sugar diabetes is a disease of the metabolism that is characterized by too much sugar (glucose) in the blood. Diabetes is usually diagnosed by means of a test called the glucose tolerance test (GTT). The identification of diabetes from the results of the GTT depends upon the solution of a differential equation of the form

$$\frac{d^2G}{dt^2} + 2\alpha \frac{dG}{dt} + \omega^2 G = 0$$

where G is the glucose in the blood at any time t, and α and ω are parameters that characterize the metabolism of glucose. The basic methods for solving this differential equation are developed in this chapter.

Ordinary differential equations are divided into two broad classes or types: **linear** equations and **nonlinear** equations. Both types occur frequently in science and engineering, but linear equations are by far the most important in elementary applications. In this chapter we begin the study of methods for solving nth-order linear differential equations.

4–1   SOME TERMINOLOGY AND AN EXISTENCE THEOREM

Recall from Section 1–1 that the general **linear differential equation of order n** may be written

$$a_n(x)y^{(n)} + a_{n-1}(x)y^{(n-1)} + \cdots + a_1(x)y' + a_0(x)y = f(x) \tag{4–1}$$

This equation is linear in y and its derivatives. The functions $f, a_0, a_1, \ldots, a_n$ may be any functions of x and, as the notation is intended to suggest, cannot depend on y. We assume that $a_n(x) \neq 0$ for x in $[a, b]$.

Example 1

The equations

$$y''' + 3y'' + 4y' = 0$$

and

$$\frac{d^2y}{dx^2} + x^2\frac{dy}{dx} + y = \cos x$$

are linear equations.   ∎

Example 2

The equation

$$(2 + y)\frac{d^2y}{dx^2} + x^2\frac{dy}{dx} + y = \cos x$$

is a nonlinear equation because the coefficient of the second derivative is a function of the dependent variable y. The second-order equation

$$y'' + y^3 = 0$$

is nonlinear because the dependent variable is cubed.   ∎

The function $f(x)$ on the right-hand side of Equation 4–1 is frequently called the **input,** or **driving,** function. If $f(x) \equiv 0$, Equation 4–1 becomes

$$a_n(x)y^{(n)} + a_{n-1}(x)y^{(n-1)} + \cdots + a_1(x)y' + a_0(x)y = 0 \tag{4–2}$$

and is called a **homogeneous** linear differential equation. If $f(x) \neq 0$ for x in $[a, b]$, the equation is called **nonhomogeneous.** (NOTE: The word *homogeneous* is being used here in a different context than in Chapter 2.)

Example 3

The equations

$$y'' + xy' - y = 0$$

and

$$y'' = 3y$$

are homogeneous linear second-order differential equations because neither contains a nonzero driving function of the independent variable. ∎

Example 4

The equations

$$y'' + y = e^x$$

and

$$y'' + y' + 3x = 0$$

are nonhomogeneous linear second-order differential equations—the first because of the nonzero driving function e^x, and the second because of the nonzero driving function $-3x$. ∎

Initial-Value Problems

In Section 1–4 we stated an existence-and-uniqueness theorem for first-order initial-value problems. The following theorem generalizes this to the nth-order linear initial-value problem.

THEOREM 4–1   *An Existence-and-Uniqueness Theorem for Linear Initial-Value Problems*

Consider the linear differential equation

$$a_n(x)y^{(n)} + a_{n-1}(x)y^{(n-1)} + \cdots + a_1(x)y' + a_0(x)y = f(x) \qquad (4\text{–}3)$$

where a_n, a_{n-1}, . . . , a_1, a_0, and f are continuous functions on the interval $[a, b]$, and $a_n(x) \neq 0$ on this interval. Further, let x_0 be any point in $[a, b]$ and let c_0, c_1, . . . , c_{n-1} be arbitrary real constants. Then there exists a unique solution function $g(x)$ of Equation 4–3 on $[a, b]$ satisfying the initial conditions $g(x_0) = c_0$, $g'(x_0) = c_1$, . . . , $g^{(n-1)}(x_0) = c_{n-1}$.

COMMENT: The differential equation in Theorem 4–1, along with the constraints $g(x_0) = c_0$, $g'(x_1) = c_1$, . . . , $g^{(n)}(x_n) = c_n$, *is called an* **nth-order initial-value problem.**

CAUTION: For the theorem to apply to Equation 4–3 the leading coefficient, $a_n(x)$, *must be nonzero for every x in* [a, b]. *Also, Theorem 4–1 does not apply to boundary-value problems or nonlinear initial-value problems.*

Example 5

(a) Theorem 4–1 does *not* guarantee a unique solution for the initial-value problem $x^2y'' + y' = 3$, $y(0) = 1$, $y'(0) = 2$ on $-\infty < x < \infty$ because the leading coefficient is 0 when $x = 0$.

(b) Theorem 4–1 does *not* apply to the *boundary-value problem* $y' + y = 0$, $y(0) = 1$, $y(\pi) = 2$.

(c) Theorem 4–1 does *not* apply to the *nonlinear* initial-value problem $y'' + y^2 = 1$, $y(0) = 1$, $y'(0) = 2$.   ■

Example 6

Use Theorem 4–1 to show that $y'' - 2y' + xy = \sin x$, $y(\pi) = 0$, $y'(\pi) = 3$ has a unique solution on $-\infty < x < \infty$.

SOLUTION   The coefficients 1, -2, x, and the driving function $\sin x$ are all continuous for all values of x in $-\infty < x < \infty$. The leading coefficient, 1, is nonzero on this interval. The point $x_0 = \pi$ is in the given interval, and 0 and 3 are real numbers. Therefore Theorem 4–1 assures us that there is a unique solution of the given initial-value problem.   ■

Example 7

(a) Show that the initial-value problem

$$(x + 1)y'' + 4y' = x^2 + 1, \; y(1) = 2, \; y'(1) = -5$$

on the interval $-\infty < x < \infty$ does not satisfy the hypothesis of Theorem 4–1.

(b) Indicate an interval for which a unique solution will exist.

SOLUTION   (a) The coefficients $(x + 1)$, 4, 0, and the driving function $x^2 + 1$ are all continuous on the given interval. However, since the coefficient of y'' is zero when $x = -1$, the conditions of the theorem are not satisfied. Since the hypothesis is not satisfied, we cannot use Theorem 4–1 to draw any conclusions about the existence of a solution. A unique solution may or may not exist since the theorem does not tell us anything about the solution if the hypothesis is not satisfied.

(b) Since the leading coefficient, $x + 1$, is zero for $x = -1$, we may eliminate this violation of the hypothesis of Theorem 4–1 by solving the equation on an interval that does not include $x = -1$, such as $-1 < x < \infty$. ■

EXERCISES FOR SECTION 4–1

In Exercises 1–14 identify each of the following equations as being linear or nonlinear and homogeneous or nonhomogeneous.

1. $y'' + y' = \cos x$

2. $\dfrac{d^3y}{dx^3} + x\dfrac{dy}{dx} + x^2y = 0$

3. $3\dfrac{d^2s}{dt^2} + 4\dfrac{ds}{dt} + 6s = 50$

4. $x''(t) + x(t) \sin 2t = \cos 2t$

5. $y'' + yy' = 0$

6. $y''' + xy'' + x^2y' = x^3$

7. $s\, ds/dt + 5s = t$

8. $y^{(4)} + (y')^2 + y = 0$

9. $y^{(4)} + 3y'' + y = 0$

10. $d^2i/dt^2 = di/dt - 3i$

11. $x^2y'' + 2y' + y = \sin x$

12. $y'' + ye^x = 0$

13. $xy''' - y'' = e^{2y}$

14. $y^{(5)} - y^{(3)} = \tan 2y$

In Exercises 15–26 use Theorem 4–1 to determine if the given differential equation with constraints has a unique solution on the interval $-\infty < x < \infty$. If it does not, try to find an interval over which such a solution exists.

15. $y'' - y' = x^2,\ y(0) = 1,\ y'(0) = 2$

16. $y'' - xy = 0,\ y(0) = 0,\ y'(0) = 1$

17. $(x - 1)y'' - y' + y = 1,\ y(0) = 2,\ y'(0) = 0$

18. $x^2y'' + y' = y,\ y(1) = 1,\ y'(1) = 0$

19. $yy'' + xy = 1,\ y(0) = 2,\ y'(0) = 0$

20. $y'' - yy' = 0,\ y(0) = 0,\ y'(0) = 0$

21. $(1 - x^2)y'' - 2xy' + 6y = 0,\ y(1) = 1,\ y'(1) = 3$

22. $x^2y'' + xy' + (x^2 - 9)y = 0,\ y(0) = 0,\ y'(0) = 1$

23. $(x + 4)y'' + xy' + y = x,\ y(0) = 0,\ y'(0) = 0$

24. $y'' + y = 0,\ y(0) = 0,\ y(\pi) = 0$

25. $y''' - xy'' = \cos x,\ y(0) = 0,\ y(\tfrac{1}{2}\pi) = 1,\ y(\pi) = 0$

26. $y''' + y'' + y' = e^{3x},\ y(0) = y'(0) = y''(0) = 0$

4–2   THE GENERAL SOLUTION OF HOMOGENEOUS LINEAR DIFFERENTIAL EQUATIONS

Suppose y_1 and y_2 are solutions of a given differential equation and c_1 and c_2 are arbitrary constants; then the sum

$$c_1y_1 + c_2y_2$$

is called a **linear combination** of the two solution functions. The next example shows that a linear combination of solutions of the homogeneous linear differential equation $y'' + y = 0$ is also a solution.

Example 1

(a) Show that $y_1 = \cos x$ and $y_2 = \sin x$ are solutions of $y'' + y = 0$.
(b) Show that $y = c_1 \cos x + c_2 \sin x$ is also a solution, where c_1 and c_2 are arbitrary constants.

SOLUTION   (a) The function $y_1 = \cos x$ is a solution since

$$y_1'' + y_1 = \frac{d^2}{dx^2}(\cos x) + \cos x = -\cos x + \cos x = 0$$

Similarly, for $y_2 = \sin x$

$$y_2'' + y_2 = \frac{d^2}{dx^2}(\sin x) + \sin x = -\sin x + \sin x = 0$$

(b) The following substitution shows that $y = c_1 \cos x + c_2 \sin x$ is also a solution of $y'' + y = 0$.

$$y'' + y = \frac{d^2}{dx^2}(c_1 \cos x + c_2 \sin x) + (c_1 \cos x + c_2 \sin x)$$
$$= -c_1 \cos x - c_2 \sin x + c_1 \cos x + c_2 \sin x = 0 \qquad\blacksquare$$

Theorem 4–2 generalizes the implications of the previous example. This theorem is also known as the **superposition principle.**

THEOREM 4–2

If $y_1, y_2, \ldots, y_R$ are solutions of the homogeneous linear equation

$$a_n(x)y^{(n)} + a_{n-1}(x)y^{(n-1)} + \cdots + a_1(x)y' + a_0(x)y = 0 \qquad (4\text{–}4)$$

and if $c_1, c_2, \ldots, c_R$ are constants, then the linear combination

$$y = c_1y_1 + c_2y_2 + \cdots + c_Ry_R$$

is also a solution of Equation 4–4.

PROOF   We prove the theorem for $R = 2$. Since y_1 and y_2 are solutions of Equation 4–4, we can write

$$a_n(x)y_1^{(n)} + a_{n-1}(x)y_1^{(n-1)} + \cdots + a_1(x)y_1' + a_0(x)y_1 = 0 \qquad (4\text{–}5)$$

and

$$a_n(x)y_2^{(n)} + a_{n-1}(x)y_2^{(n-1)} + \cdots + a_1(x)y_2' + a_0(x)y_2 = 0 \qquad (4\text{–}6)$$

Multiplying each member of Equation 4–5 by c_1 and each member of Equation 4–6 by c_2 and adding the results, we get

$$a_n(x)[c_1y_1^{(n)} + c_2y_2^{(n)}] + \cdots + a_0(x)[c_1y_1 + c_2y_2] = 0 \qquad (4\text{–}7)$$

By the linearity property of derivatives we have

$$a_1y_1^{(n)} + c_2y_2^{(n)} = (c_1y_1 + c_2y_2)^{(n)}$$

So Equation 4–7 is a statement that $y = c_1y_1 + c_2y_2$ is a solution of Equation 4–4. This completes the proof.

> *COMMENT: Theorem 4–2 applies only to homogeneous linear equations; it does* not *hold for nonhomogeneous linear equations or nonlinear equations. Note also that the theorem does* not *say that any multiple of a solution is again a solution,* only *constant multiples.*

Example 2

The functions $y_1 = e^{2x}$ and $y_2 = e^{-x}$ are both solutions of the second-order equation $y'' - y' - 2y = 0$. By Theorem 4–2 the linear combination

$$y = c_1e^{2x} + c_2e^{-x}$$

is also a solution of this equation.   ∎

Example 3

The functions $y_1 = e^x$ and $y_2 = e^x + e^{-3x}$ are both solutions of the nonhomogeneous equation $y'' + 3y' = 4e^x$. However, Theorem 4–2 does not apply to nonhomogeneous equations, so we cannot conclude that a linear combination of the functions is a solution.   ∎

Example 4

The function $y = x^{-1}$ is a solution of the second-order nonlinear equation $xy'' - 2y^2 = 0$. Since Theorem 4–2 does not apply to nonlinear equations, we cannot conclude that $y = cx^{-1}$ is also a solution. In fact, as the following substitution shows, it is not a solution. We observe that if $y = cx^{-1}$, then $y' = -cx^{-2}$ and $y'' = 2cx^{-3}$, so that

$$xy'' - 2y^2 = x\left(\frac{2c}{x^3}\right) - 2\left(\frac{c^2}{x^2}\right) \neq 0$$

except for $c = 0$ and $c = 1$. ∎

Linearly Independent Solutions

When solving a differential equation, we would like to know when we are finished—that is, when do we have all solutions? From Theorem 4–2 we know that additional solutions may be obtained by taking linear combinations of known solutions. But such solutions are not necessarily "new" solutions. In this section we develop criteria for establishing when two functions are truly different and, also, when a set of solutions constitutes the general solution of a linear differential equation.

DEFINITION

> A solution of an nth-order differential equation is said to have n **essential arbitrary constants** if it cannot be reduced algebraically to a form containing fewer than n arbitrary constants.

Example 5

The functions $y_1 = c_1 e^{2x}$ and $y_2 = c_2 e^{2x}$ are both solutions to the second-order equation

$$y'' - 4y' + 4y = 0$$

By Theorem 4–2, $y = c_1 e^{2x} + c_2 e^{2x}$ is also a solution of this equation. However, $y = c_1 e^{2x} + c_2 e^{2x}$ does not constitute a solution in which there are two arbitrary constants because it can be reduced to

$$y = (c_1 + c_2)e^{2x}$$

from which, if we let $c_1 + c_2 = c_3$, we have

$$y = c_3 e^{2x}$$

a form in which the solution has only one arbitrary constant. The question raised by this example is, how do we know when n arbitrary constants are reducible to fewer than n constants? The answer is contained in the concept of linear independence as defined next. ∎

DEFINITION

1. The n functions $y_1, y_2, \ldots, y_n$ are said to be **linearly dependent** on some interval $[a, b]$ if constants $c_1, c_2, \ldots, c_n$, not all zero, can be found such that

$$c_1 y_1 + c_2 y_2 + \cdots + c_n y_n = 0 \qquad (4\text{-}8)$$

for all x on $[a, b]$.

2. If the relation in Equation 4–8 is true only when $c_1 = c_2 = \cdots = c_n = 0$, then the n functions $y_1, y_2, \ldots, y_n$ are said to be **linearly independent** on $[a, b]$.

COMMENT: If two functions are linearly dependent, then one of them is a constant multiple of the other; conversely, if the functions are linearly independent, then it is impossible to express any of the functions as a constant multiple of the other. The next three examples illustrate the concept.

Example 6

The functions $y_1 = e^x$ and $y_2 = 4e^x$ are linearly dependent on the interval $(-\infty, \infty)$ since $-4y_1 + y_2 = -4e^x + 4e^x = 0$. That is, y_2 can be expressed as a constant multiple of y_1.   ∎

Example 7

The functions $y_1 = e^x$, $y_2 = e^{-x}$, $y_3 = \sinh x$ are linearly dependent on the interval $-\infty < x < \infty$ since

$$c_1 e^x + c_2 e^{-x} + c_3 \sinh x = 0$$

for $c_1 = 1$, $c_2 = -1$, and $c_3 = -2$. [Recall that $\sinh x = (e^x - e^{-x})/2$.]   ∎

Example 8

To show that $y_1 = x$ and $y_2 = x^3$ are linearly independent on $(-\infty, \infty)$, we note that $c_1 x + c_2 x^3$ must be identically zero for all x in the interval. That is,

$$c_1 x + c_2 x^3 \equiv 0$$

for all x. From our knowledge of polynomials we know that x and x^3 cannot be constant multiples of each other; hence we conclude that these two functions are linearly independent.   ∎

It is impractical to search for constants, not all zero, for which $c_1y_1 + \cdots + c_ny_n = 0$, or to show that such constants do not exist, in order to identify linearly independent functions. Fortunately, there is a test to determine whether or not a set of solutions of a homogeneous linear differential equation is linearly independent.

THEOREM 4–3

Suppose the coefficients a_0, a_1, . . . , a_n are continuous functions of x on the interval $a \le x \le b$, and y_1, y_2, . . . , y_n are solutions of the homogeneous linear differential equation

$$a_n(x)y^{(n)} + a_{n-1}(x)y^{(n-1)} + \cdots + a_1(x)y' + a_0(x)y = 0$$

Then the functions y_1, y_2, . . . , y_n are linearly independent on $[a, b]$ if, and only if, the determinant

$$W(x) = \begin{vmatrix} y_1(x) & y_2(x) & \cdots & y_n(x) \\ y_1'(x) & y_2'(x) & \cdots & y_n'(x) \\ y_1''(x) & y_2''(x) & \cdots & y_n''(x) \\ \vdots & \vdots & & \vdots \\ y_1^{(n-1)}(x) & y_2^{(n-1)}(x) & \cdots & y_n^{(n-1)}(x) \end{vmatrix} \neq 0 \qquad (4\text{–}9)$$

for some x in $[a, b]$. The determinant in Equation 4–9 is called the **Wronskian function** *of the n functions on $[a, b]$.*

This theorem says that a set of solutions of a homogeneous differential equation is linearly independent if, and only if, the determinant in Equation 4–9 is nonzero for some x in the interval $a \le x \le b$. We reiterate that this theorem is true only if the functions y_1, y_2, . . . , y_n are solutions of a homogeneous linear differential equation.

Example 9

The functions $y_1 = x$, $y_2 = x^2$ are solutions of the differential equation $y''' = 0$. Show that this is a linearly independent set of functions on $-\infty < x < \infty$.

SOLUTION   The Wronskian of this set of functions is

$$W(x) = \begin{vmatrix} x & x^2 \\ 1 & 2x \end{vmatrix} = 2x^2 - x^2 = x^2$$

which does not equal zero except for $x = 0$. Since $W = x^2$ is nonzero for some x in $-\infty < x < \infty$, the functions are linearly independent on this interval.   ∎

Example 10

The functions $y_1 = \sin 2x$ and $y_2 = \cos 2x$ are solutions of the second-order equation $y'' + 4y = 0$. Show that they form a linearly independent set of functions.

SOLUTION   The Wronskian for these functions is

$$W(x) = \begin{vmatrix} \sin 2x & \cos 2x \\ 2\cos 2x & -2\sin 2x \end{vmatrix} = -2\sin^2 2x - 2\cos^2 2x = -2 \neq 0$$

Since $W(x) \neq 0$ for all x, the functions $\sin 2x$ and $\cos 2x$ are linearly independent. ■

Example 11

Use the Wronskian to show that the functions e^x, e^{-x}, and $\sinh x$ are linearly dependent. Note that each of these functions is a solution of $y'' - y = 0$.

SOLUTION   The Wronskian of e^x, e^{-x}, $\sinh x$ is

$$
\begin{aligned}
W(x) &= \begin{vmatrix} e^x & e^{-x} & \sinh x \\ e^x & -e^{-x} & \cosh x \\ e^x & e^{-x} & \sinh x \end{vmatrix} \\
&= e^x[-e^{-x}\sinh x - e^{-x}\cosh x] - e^x[e^{-x}\sinh x - e^{-x}\sinh x] \\
&\quad + e^x[e^{-x}\cosh x + e^{-x}\sinh x] \\
&= -\sinh x - \cosh x - \sinh x + \sinh x + \cosh x + \sinh x = 0
\end{aligned}
$$

Since $W(x) \equiv 0$, we conclude that e^x, e^{-x}, and $\sinh x$ are linearly dependent functions. (NOTE: This agrees with the conclusion reached in Example 7.) ■

We conclude this section with a theorem that specifies the form of the general solution of an nth-order homogeneous linear differential equation.

THEOREM 4–4

If $a_0, a_1, \ldots, a_n$ are continuous functions of x and if $a_n(x) \neq 0$ on the interval $[a, b]$, then the nth-order homogeneous linear differential equation

$$a_n(x)y^{(n)} + a_{n-1}(x)y^{(n-1)} + \cdots + a_1(x)y' + a_0(x)y = 0 \qquad (4\text{–}10)$$

has n linearly independent solutions $y_1, y_2, \ldots, y_n$ on $[a, b]$ and, by the proper choice of constants $c_1, c_2, \ldots, c_n$ every solution of Equation 4–10 can be expressed as

$$c_1 y_1 + c_2 y_2 + \cdots + c_n y_n$$

Theorem 4–4 assures us that a linear combination of the n linearly independent solutions $y_1, y_2, \ldots, y_n$ must include all solutions of Equation 4–10; therefore
we call

$$y = c_1y_1 + c_2y_2 + \cdots + c_ny_n$$

the **general solution** of Equation 4–10.

Example 12

The functions $y_1 = \sin 2x$ and $y_2 = \cos 2x$ are linearly independent solutions of
$y'' + 4y = 0$. (See Example 9.) Therefore

$$y = c_1 \sin 2x + c_2 \cos 2x$$

is the general solution of $y'' + 4y = 0$. ∎

EXERCISES FOR SECTION 4–2

1. (a) Show that $y = e^{2x}$ and $y = e^{-3x}$ are solutions of $y'' + y' - 6y = 0$.
    (b) Show that $y = 3e^{2x} + 5e^{-3x}$ is also a solution of this equation.
    (c) Show that $y = xe^{2x}$ is not a solution.

2. (a) Show that $y = e^{2x}$ and $y = xe^{2x}$ are solutions of $y'' - 4y' + 4y = 0$.
    (b) Show that $y = e^{2x} + xe^{2x}$ is also a solution of this equation.
    (c) Show that $y = x^2e^{2x}$ is not a solution.

3. (a) Show that $y = \cos 2x$ and $y = \sin 2x$ are solutions of $y'' + 4y = 0$.
    (b) Show that $y = c_1 \cos 2x + c_2 \sin 2x$ is also a solution of this equation.
    (c) Show that $e^x \cos 2x$ is not a solution.

4. (a) Show that $y = x + \cos x$ and $y = x + \sin x$ are solutions of $y'' + y = x$.
    (b) Show that $y = (x + \cos x) + (x + \sin x)$ is not a solution of this equation.
    (c) Why does Theorem 4–2 fail here?

*In Exercises 5–16 determine whether the following functions are linearly dependent
or independent by using the Wronskian test. Assume (correctly) that each set of
functions is a solution set of some homogeneous linear differential equation.*

5. e^x, e^{2x}

6. $1, x$

7. $\sin 3x, \cos 3x$

8. $e^{2x}, \cosh 2x$

9. $e^x, \sin x$

10. x^2, x^3

11. e^{2x}, xe^{2x}

12. $1, 5, x$

13. $\sin 2x, \sin x \cos x$

14. $\sin^2 x, 1 - \cos 2x$

15. $x, e^x, \sin x$

16. x, e^{-x}, xe^{-x}

17. By determining constants c_1, c_2, c_3, c_4, not all zero, show that the functions $y_1 = e^x$, $y_2 = xe^x$, $y_3 = (2 + 8x)e^x$, $y_4 = x^2$ are linearly dependent. That is, show that $c_1 y_1 + c_2 y_2 + c_3 y_3 + c_4 y_4 = 0$ for cs not all zero.

18. By determining constants c_1, c_2, c_3, c_4, not all zero, show that the functions 5, e^x, $\sin^2 x$, $\cos^2 x$ are linearly dependent.

19. Show graphically that the functions x and $|x|$ are linearly dependent on the interval $0 \le x \le 1$ but not on $-1 \le x \le 1$.

20. Show that if two functions y_1 [with $y_1(x) \ne 0$] and y_2 are linearly dependent on $a \le x \le b$, then they are proportional on that interval.

21. Consider the following three functions:

$$y_1(x) = \begin{cases} 1 + x^3, & x \le 0 \\ 1, & x \ge 0 \end{cases} \qquad y_2(x) = \begin{cases} 1, & x \le 0 \\ 1 + x^3, & x \ge 0 \end{cases}$$

$$y_3(x) = 3 + x^3 \text{ for all } x$$

Show that the three functions are linearly independent even though their Wronskian vanishes for all x. What conclusion can you draw about y_1, y_2, and y_3?

22. The functions $\sin \frac{1}{2}x$ and $\cos \frac{1}{2}x$ are solutions of $4y'' + y = 0$.

(a) Show that these functions are linearly independent on $-\infty < x < \infty$.

(b) Write the general solution of the given differential equation.

23. The functions e^{-x} and xe^{-x} are solutions of $y'' + 2y' + y = 0$.

(a) Show that these functions are linearly independent on $-\infty < x < \infty$.

(b) Write the general solution of the given differential equation.

24. The functions 1, x, and e^{2x} are solutions of $y''' - 2y'' = 0$.

(a) Show that these functions are linearly independent on $-\infty < x < \infty$.

(b) Write the general solution of the given differential equation.

4–3 REDUCTION OF ORDER

In the previous section we showed that the general solution of an nth-order homogeneous linear differential equation

$$a_n(x)y^{(n)} + a_{n-1}(x)y^{(n-1)} + \cdots + a_1(x)y' + a_0(x)y = 0 \qquad (4\text{–}11)$$

can be written as the linear combination

$$c_1 y_1 + c_2 y_2 + \cdots + c_n y_n$$

in which y_1, y_2, . . . , y_n are linearly independent solutions of Equation 4–11 and c_1, c_2, . . . , c_n are arbitrary constants. However, that discussion gave no hint of how to obtain the n linearly independent solutions. In this section we introduce an important technique that permits us to construct a second linearly independent solution of a second-order linear differential equation from a known solution. The method, which is called **reduction of order,** uses the known solution to reduce the order of the differential equation. Theorem 4–5 describes the technique.

THEOREM 4–5   *Reduction of Order*

If y_1 is a nontrivial solution of the nth-order homogeneous linear differential equation

$$a_n(x)y^{(n)} + a_{n-1}(x)y^{(n-1)} + \cdots + a_1(x)y' + a_0(x)y = 0 \qquad (4\text{--}12)$$

then the substitution $y_2 = y_1 v$, followed by the substitution $w = v'$, reduces Equation 4–12 to an $(n-1)$st-order equation.

PROOF   We demonstrate the proof for second-order homogeneous equations; the extension to higher-order equations parallels this proof. Consider the second-order homogeneous equation

$$a_2(x)y'' + a_1(x)y' + a_0(x)y = 0 \qquad (4\text{--}13)$$

where a_0, a_1, and a_2 are continuous and $a_2(x) \neq 0$ for every x in some interval $a \leq x \leq b$. Let y_1 be a nontrivial solution of Equation 4–13 and let $y_2 = y_1 v$, where v is a function of x to be determined. Differentiating y_2 twice gives

$$y_2' = y_1 v' + y_1' v$$
$$y_2'' = y_1 v'' + 2y_1' v' + y_1'' v$$

Substituting y_2, y_2', and y_2'' into Equation 4–13 yields

$$a_2 y_1 v'' + 2a_2 y_1' v' + a_2 y_1'' v + a_1 y_1 v' + a_1 y_1' v + a_0 y_1 v = 0$$

Writing this in the form of a linear differential equation in v, we have

$$a_2 y_1 v'' + (2a_2 y_1' + a_1 y_1)v' + (a_2 y_1'' + a_1 y_1' + a_0 y_1)v = 0 \qquad (4\text{--}14)$$

Since y_1 is a known solution of Equation 4–13, the coefficient of v is zero. Hence Equation 4–14 reduces to

$$a_2 y_1 v'' + (2a_2 y_1' + a_1 y_1)v' = 0$$

Finally, letting $w = v'$, we obtain the first-order equation

$$a_2 y_1 w' + (2a_2 y_1' + a_1 y_1)w = 0 \qquad (4\text{--}15)$$

which demonstrates the reduction of order. Since $y_2 = vy_1$, where v is a function of x, it follows that y_2 is not a constant multiple of y_1 and consequently y_1 and y_2 are linearly independent solutions.

COMMENT: *(a) The method of reduction of order is generally attributed to the French mathematician Jean D'Alembert (1717–1783).*

(b) In the case of a second-order equation, the reduction of order yields a first-order equation that can be solved by computing an integrating factor.

Example 1

(a) Show that $y_1 = e^{-x}$ is a solution of $y'' + 3y' + 2y = 0$.
(b) Use the method of reduction of order to find a second linearly independent solution of this differential equation and write the general solution.

SOLUTION   (a) Observing that $y_1 = e^{-x}$, $y_1' = -e^{-x}$, and $y_1'' = e^{-x}$, we substitute these values into the given differential equation to obtain

$$e^{-x} + 3(-e^{-x}) + 2(e^{-x}) = 0$$

which shows that e^{-x} is a solution of the given differential equation.
(b) Using the method of reduction of order, we let $y_2 = ve^{-x}$, which, when differentiated twice, yields

$$y_2' = v'e^{-x} - ve^{-x}$$
$$y_2'' = v''e^{-x} - 2v'e^{-x} + ve^{-x}$$

Substituting into the given differential equation, we get

$$(v''e^{-x} - 2v'e^{-x} + ve^{-x}) + 3(v'e^{-x} - ve^{-x}) + 2ve^{-x} = 0$$

Expanding and collecting terms yields

$$v''e^{-x} + v'e^{-x} = 0 \qquad \text{or} \qquad v'' + v' = 0$$

Letting $w = v'$, this becomes

$$w' + w = 0$$

Separating variables on this equation yields

$$\frac{dw}{w} = -dx$$
$$\ln|Cw| = -x$$
$$w = Ce^{-x}$$

Since $w = v'$, it follows by taking the antiderivative of e^{-x} that

$$v = ce^{-x}$$

Ignoring the coefficient, a second solution is

$$y_2 = ve^{-x} = e^{-x}e^{-x} = e^{-2x}$$

The general solution of the given second-order equation is then

$$y = c_1e^{-x} + c_2e^{-2x} \qquad \blacksquare$$

The method of reduction of order yields an explicit formula for the second solution when Equation 4–15 is solved for y_2. The solution of Equation 4–15 is given in the following steps. Divide (4–15) by a_2y_1

$$w' + \left(2\frac{y_1'}{y_1} + \frac{a_1}{a_2}\right)w = 0$$

Separate variables

$$\frac{dw}{w} = -\left(2\frac{y_1'}{y_1} + \frac{a_1}{a_2}\right) dx$$

Integrate

$$\ln w = -2 \ln y_1 - \int \frac{a_1}{a_2} dx$$

Solve for w

$$w = e^{\ln y_1^{-2} - \int(a_1/a_2)dx} = y_1^{-2} e^{-\int(a_1/a_2)dx}$$

Substitute $w = v'$ and integrate

$$v = \int \frac{1}{y_1^2} e^{-\int(a_1/a_2)dx}dx$$

Substituting $v = y_2/y_1$ and solving for y_2, we obtain a second solution

$$y_2 = y_1 \int \frac{1}{y_1^2} e^{-\int(a_1/a_2)dx}dx \qquad (4\text{--}16)$$

Example 2

(a) Show that $y_1 = x$ is a solution of $2x^2y'' + xy' - y = 0$.
(b) Use the method of reduction of order to find a second linearly independent solution of this differential equation and write the general solution.

SOLUTION   (a) Substituting $y_1 = x$, $y_1' = 1$, and $y_1'' = 0$ into the given differential equation yields

$$2x^2(0) + x(1) - x = 0$$

which shows that $y_1 = x$ is a solution.

(b) Using Equation 4–16 with $y_1 = x$, $a_2 = 2x^2$ and $a_1 = x$, we have

$$y_2 = x \int \frac{1}{x^2} e^{-\int(x/2x^2)dx}dx \;\; = x \int \frac{1}{x^2} e^{-(\ln x)/2}dx$$

$$= x \int x^{-5/2} dx = -\frac{2}{3}x^{-1/2}$$

Ignoring the coefficient, the general solution is

$$y = c_1 x + c_2 x^{-\frac{1}{2}}$$

∎

EXERCISES FOR SECTION 4-3

In Exercises 1–18, show that the given function is a solution of the differential equation, use the method of reduction of order to find a second linearly independent solution, and write the general solution.

1. $y'' - 4y = 0$; $y_1 = e^{2x}$

2. $y'' - 4y = 0$; $y_1 = \cosh 2x$

3. $y'' - 9y = 0$; $y_1 = e^{3x}$

4. $y'' + y' - 6y = 0$; $y_1 = e^{-3x}$

5. $y'' - 7y' + 6y = 0$; $y_1 = e^{6x}$

6. $y'' + 4y = 0$; $y_1 = \sin 2x$

7. $y'' + 4y' + 4y = 0$; $y_1 = e^{-2x}$

8. $y'' + y = 0$; $y_1 = \cos x$

9. $\frac{1}{2}x^2 y'' - y = 0$; $y_1 = x^2$

10. $xy'' + 2y' = 0$; $y_1 = 1$

11. $xy'' + y' = 0$; $y_1 = 1$

12. $x^2 y'' - 6y = 0$; $y_1 = 1/x^2$

13. $y'' - (2 \cot x)y' = 0$; $y_1 = 1$

14. $x^2 y'' - 3xy' + 4y = 0$; $y_1 = x^2$

15. $(1 - x)y'' + xy' - y = 0$; $y_1 = e^x$

16. $(1 - x)y'' + xy' = 0$; $y_1 = 1$

17. $x^2 y'' - xy' + y = 0$; $y_1 = x$

18. $x^2 y'' - xy' + 2y = 0$; $y_1 = x \sin(\ln x)$

19. The second-order equation $(1 - x^2)y'' - 2xy' + n(n + 1)y = 0$ is called **Legendre's differential equation.** Verify that $y = x$, $|x| < 1$, is a solution of the equation for $n = 1$ and use this fact to find a second linearly independent solution.

20. The second-order equation $x^2 y'' + xy' + (x^2 - p^2)y = 0$ is called **Bessel's differential equation of order p.** Verify that $y = x^{-1/2} \sin x$ is a solution of Bessel's equation of order $\frac{1}{2}$, and use this fact to find a second linearly independent solution.

4-4   HOMOGENEOUS LINEAR EQUATIONS WITH CONSTANT COEFFICIENTS

Many applications of linear differential equations involve homogeneous equations with constant coefficients—that is, equations of the form

$$b_n y^{(n)} + b_{n-1} y^{(n-1)} + \cdots + b_1 y' + b_0 y = 0 \tag{4-17}$$

where $b_0, b_1, \ldots, b_n$ are real constants. To find the general solution of this equation, we note that any solution of Equation 4–17 must combine linearly with its derivatives to give zero. Although there may be many functions that have this property, a function whose derivatives are constant multiples of itself would be a likely candidate. One such function is e^{mx}. The following example shows that e^{mx} occurs in the general solution of $y'' + 5y' - 6y = 0$. To verify this, we substitute $y = e^{mx}$ into the equation to obtain

$$m^2 e^{mx} + 5m e^{mx} - 6e^{mx} = 0$$

Since e^{mx} is never zero, we only need to solve

$$m^2 + 5m - 6 = 0$$

This quadratic equation factors to yield roots of 1 and -6. Therefore, e^x and e^{-6x} are linearly independent solutions of $y'' + 5y' - 6y = 0$ and its general solution may be written as

$$y = c_1 e^x + c_2 e^{-6x}$$

The technique used to find a solution of $y'' + 5y' - 6y = 0$ may be extended to nth-order homogeneous differential equations. Substituting $y = e^{mx}$ into Equation 4–17 yields

$$b_n m^n e^{mx} + b_{n-1} m^{n-1} e^{mx} + \cdots + b_1 m e^{mx} + b_0 e^{mx} = 0$$

or

$$(b_n m^n + b_{n-1} m^{n-1} + \cdots + b_1 m + b_0)\, e^{mx} = 0$$

Since e^{mx} is never zero, this equation is satisfied for those values of m that are zeros of the polynomial

$$P(m) = b_n m^n + b_{n-1} m^{n-1} + \cdots + b_1 m + b_0$$

This, in turn, means that Equation 4–17 can be solved by finding the roots of $P(m) = 0$.

If m_1 is a root of the equation $P(m) = 0$, then $y = e^{m_1 x}$ is a solution of Equation 4–17. The polynomial equation $P(m) = 0$ is called the **auxiliary equation** of the differential equation.

Example 1

(a) The auxiliary equation for $y' - 3y = 0$ is $m - 3 = 0$.
(b) The auxiliary equation for $y'' + 5y' - 7y = 0$ is $m^2 + 5m - 7 = 0$.
(c) Equations such as $y'' + yy' = 0$, $y'' + y + x^2 = 0$, or $x^2 y'' + y' + xy = 0$ do not have auxiliary equations because the auxiliary-equation concept applies only to linear homogeneous equations with constant coefficients.   ∎

COMMENT: *The process of solving a homogeneous linear differential equation with constant coefficients reduces to finding the roots of the polynomial equation $P(m) = 0$. The roots of this polynomial equation must take one of the following forms:* **1.** *Distinct real roots.* **2.** *Repeated real roots.* **3.** *Distinct complex conjugate roots.* **4.** *Repeated complex conjugate roots.*

Auxiliary Equation with Distinct Real Roots

If the auxiliary equation of Equation 4–17 has n distinct real roots, denoted by m_1, $m_2, \ldots, m_n$, the n solutions

$$e^{m_1 x}, e^{m_2 x}, \ldots, e^{m_n x}$$

are linearly independent. We have the following theorem.

THEOREM 4–6

If the auxiliary equation for

$$b_n y^{(n)} + b_{n-1} y^{(n-1)} + \cdots + b_1 y' + b_0 y = 0$$

has n distinct real roots $m_1, m_2, \ldots, m_n$, then the general solution is given by

$$y = c_1 e^{m_1 x} + c_2 e^{m_2 x} + \cdots + c_n e^{m_n x} \qquad (4\text{–}18)$$

where $c_1, c_2, \ldots, c_n$ are arbitrary constants.

Example 2

Solve the differential equation

$$2 \frac{d^3 y}{dx^3} - 9 \frac{d^2 y}{dx^2} - 5 \frac{dy}{dx} = 0$$

SOLUTION   The auxiliary equation for the given differential equation is

$$2m^3 - 9m^2 - 5m = 0 \qquad \text{or} \qquad m(2m + 1)(m - 5) = 0$$

The roots of this equation are $m = 0$, $-\frac{1}{2}$, and 5; therefore the general solution is

$$y = c_1 + c_2 e^{-x/2} + c_3 e^{5x} \qquad \blacksquare$$

Example 3

Solve the initial-value problem

$$y'' + 3y' + 2y = 0, \; y(0) = 1, \; y'(0) = 2$$

SOLUTION   In this case the auxiliary equation is

$$m^2 + 3m + 2 = 0$$

whose roots are $m = -1$ and $m = -2$. Therefore the general solution is a linear combination of the functions e^{-x} and e^{-2x}, that is,

$$y = c_1 e^{-x} + c_2 e^{-2x} \qquad (4\text{–}19)$$

To find values for c_1 and c_2, we use the conditions $x = 0$, $y = 1$ in Equation 4–19 to obtain

$$1 = c_1 + c_2 \tag{4–20}$$

Then differentiation of Equation 4–19 yields

$$y' = -c_1 e^{-x} - 2c_2 e^{-2x}$$

Using $x = 0$, $y' = 2$ in this equation,

$$2 = -c_1 - 2c_2 \tag{4–21}$$

Solving Equations 4–20 and 4–21 simultaneously for c_1 and c_2, we get $c_1 = 4$ and $c_2 = -3$. Hence the desired solution is

$$y = 4e^{-x} - 3e^{-2x} \qquad \blacksquare$$

If the auxiliary equation is quadratic, the roots can always be found by using the quadratic formula. However, if the degree of $P(m)$ is greater than two, finding the roots can be quite difficult. The following two theorems from the theory of equations are useful if the auxiliary equation has rational coefficients (rational coefficients include integers).

1.   If $m = r$ is a root of $P(m) = 0$, then $m - r$ is a factor of $P(m)$. Thus $P(m) = (m - r)Q(m)$, where $Q(m)$ is a polynomial of degree $n - 1$.

2.   If $P(m)$ has integer coefficients and if $P(m) = 0$ has a rational root p/q, then p must be a factor of b_0, and q must be a factor of b_n.

Example 4

The potential rational roots of $3m^4 - 5m^2 + 7m + 6 = 0$ are

$$\pm\frac{1}{1}, \ \pm\frac{2}{1}, \ \pm\frac{3}{1}, \ \pm\frac{6}{1}, \ \pm\frac{1}{3}, \ \pm\frac{2}{3}, \ \pm\frac{3}{3}, \ \pm\frac{6}{3}$$

or, eliminating duplicate values,

$$\pm 1, \ \pm 2, \ \pm 3, \ \pm 6, \ \pm\frac{1}{3}, \ \pm\frac{2}{3} \qquad \blacksquare$$

Example 5

Solve the third-order equation

$$y''' - 5y' - 2y = 0$$

SOLUTION   The auxiliary equation is the third-degree polynomial equation

$$m^3 - 5m - 2 = 0$$

A third-degree polynomial has three roots, *one of which must be real*. We know that any rational root of this equation must be a factor of 2 divided by a factor of 1. Thus the potential rational roots are ± 2 and ± 1. To verify that $m = -2$ is a root, we use synthetic division to show that $m + 2$ is a factor of $m^3 - 5m - 2$. The format for the division is

$$m + 2 \longrightarrow \underline{-2 \rvert} \quad \begin{array}{ccccc} 1 & 0 & -5 & -2 \\ & -2 & 4 & 2 \\ \hline 1 & -2 & -1 & 0 \end{array} \qquad \begin{array}{l} \longleftarrow \quad m^3 - 5m - 2 \\[2ex] \longleftarrow \quad m^2 - 2m - 1 \end{array}$$

From the results of the division the given polynomial may be factored as

$$m^3 - 5m - 2 = (m + 2)(m^2 - 2m - 1)$$

The remaining two roots are obtained from the quadratic factor by using the quadratic formula. Thus

$$m = \frac{2 \pm \sqrt{8}}{2} = 1 \pm \sqrt{2}$$

We note that the three roots of the auxiliary equation are

$$m_1 = -2, \, m_2 = 1 + \sqrt{2}, \, m_3 = 1 - \sqrt{2}$$

so the general solution of the differential equation is

$$y = c_1 e^{-2x} + c_2 e^{(1+\sqrt{2})x} + c_3 e^{(1-\sqrt{2})x} \qquad \blacksquare$$

EXERCISES FOR SECTION 4–4

In Exercises 1–16 find the general solution.

1. $y'' - 3y' + 2y = 0$

2. $\dfrac{d^2y}{dx^2} + 5\dfrac{dy}{dx} + 6y = 0$

3. $\dfrac{d^2s}{dt^2} + \dfrac{ds}{dt} = 0$

4. $\dfrac{d^2y}{dx^2} + 5\dfrac{dy}{dx} + 4y = 0$

5. $2y'' - 3y' = 0$

6. $3y' - 4y = 0$

7. $\dfrac{d^2y}{dx^2} - 4y = 0$

8. $\dfrac{d^2i}{dt^2} - 9i = 0$

9. $y''' - 16y' = 0$

10. $y''' - 4y' = 0$

11. $\dfrac{d^3y}{dx^3} - \dfrac{d^2y}{dx^2} - 20\dfrac{dy}{dx} = 0$

12. $y''' + 9y'' + 8y' = 0$

13. $3y''' + 5y'' - 2y' = 0$

14. $2\dfrac{d^3s}{dt^3} + 5\dfrac{d^2s}{dt^2} + 3\dfrac{ds}{dt} = 0$

15. $y''' + 6y'' + 11y' + 6y = 0$

16. $9\dfrac{d^3y}{dx^3} - 7\dfrac{dy}{dx} + 2y = 0$

In Exercises 17–21 find the particular solution corresponding to the given conditions.

17. $\dfrac{d^2s}{dt^2} - 4s = 0$; when $t = 0$, $s = 0$ and $\dfrac{ds}{dt} = 2$

18. $y'' - 2y' - 3y = 0$, $y(0) = 0$, $y'(0) = -4$

19. $y'' - y = 0$, $y(0) = 1$, $y'(0) = 1$

20. $y'' + 3y' = 0$, $y(0) = 2$, $y'(0) = 6$

21. $y'' - y' - 2y = 0$, $y(0) = 2$, and $y \to 0$ as $x \to \infty$

4–5   REPEATED REAL ROOTS AND COMPLEX ROOTS

Consider the second-order equation

$$y'' - 6y' + 9y = 0$$

The auxiliary equation

$$m^2 - 6m + 9 = 0$$

has repeated real roots $m = 3, 3$. Using the method of the previous section to find the general solution,

$$y = c_1e^{3x} + c_2e^{3x}$$

$$= (c_1 + c_2)e^{3x}$$

But $c_1 + c_2$ is an arbitrary constant, so

$$y = c_3e^{3x}$$

This cannot be the *general* solution of the given equation of order *two* since it has but *one* arbitrary constant or, putting it another way, because c_1e^{3x} and c_2e^{3x} are linearly *dependent* functions, and a second-order differential equation has two linearly *independent* functions in the general solution.

To give a hint of the correct form of the solution of the differential equation $y'' - 6y' + 9y = 0$, we use the method of reduction of order to construct a second linearly independent solution from $y = c_1e^{3x}$. We assume that the second solution is $y = ve^{3x}$, so that

$$y' = v'e^{3x} + 3ve^{3x} \qquad \text{and} \qquad y'' = v''e^{3x} + 6v'e^{3x} + 9ve^{3x}$$

Substituting into the given equation, we have

$$(v''e^{3x} + 6v'e^{3x} + 9ve^{3x}) - 6(v'e^{3x} + 3ve^{3x}) + 9(ve^{3x}) = 0$$

which simplifies into

$$v''e^{3x} = 0$$

Multiplying by e^{-3x} and integrating yields

$$v' = c$$

Integrating again,

$$v = cx$$

The desired solution is then $y = ve^{3x} = xe^{3x}$. Recall that no arbitrary constant is required since we seek only a second linearly independent solution. Since e^{3x} and xe^{3x} are linearly independent solutions, the general solution of $y'' - 6y' + 9y = 0$ is

$$y = c_1 e^{3x} + c_2 x e^{3x}$$

The solution obtained above for a differential equation whose auxiliary equation has repeated roots is generalized in Theorem 4.7.

THEOREM 4–7

If the auxiliary equation for

$$b_n y^{(n)} + b_{n-1} y^{(n-1)} + \cdots + b_1 y' + b_0 y = 0$$

has n repeated real roots $m_1 = m_2 = \cdots = m_n = m$, then the general solution is given by

$$y = (c_1 + c_2 x + c_3 x^2 + \cdots + c_n x^{n-1}) e^{mx} \qquad (4\text{--}22)$$

where $c_1, c_2, \ldots, c_n$ are arbitrary constants.

Example 1

Solve the equation $y''' = 0$.

SOLUTION   The auxiliary equation is $m^3 = 0$, which gives roots $m = 0, 0, 0$. Therefore the general solution is

$$y = c_1 + c_2 x + c_3 x^2 \qquad \blacksquare$$

More generally, the auxiliary equation can have both distinct and repeated real roots. In this case the general solution consists of terms resulting from the distinct roots as well as those resulting from repeated roots.

Example 2

Solve the equation $y''' + 4y'' + 4y' = 0$.

SOLUTION   The auxiliary equation is

$$m^3 + 4m^2 + 4m = 0$$

or, in factored form,

$$m(m + 2)^2 = 0$$

from which $m = 0, -2, -2$. Hence the desired solution is

$$y = c_1 + c_2 e^{-2x} + c_3 x e^{-2x}$$

■

Complex Roots

Consider the case for which the auxiliary equation

$$P(m) = 0$$

has complex roots of the form $m = a \pm ib$. Recall that complex roots of a real polynomial equation occur in conjugate pairs; that is, if $m_1 = a + ib$ is a root of $P(m) = 0$, then $m_2 = a - ib$ is also a root.

To find a solution to Equation 4–17 when the auxiliary equation has complex roots, we assume that the method used for distinct real roots is applicable. Then if $m_1 = a + ib$ and $m_2 = a - ib$ are roots of $P(m) = 0$, the implied general solution is

$$\begin{aligned}
y &= c_3 e^{m_1 x} + c_4 e^{m_2 x} \\
&= c_3 e^{(a+ib)x} + c_4 e^{(a-ib)x} \\
&= e^{ax}(c_3 e^{ibx} + c_4 e^{-ibx})
\end{aligned} \tag{4–23}$$

The exponential function to a complex power may be expressed in terms of trigonometric functions using the **Euler formula:**

$$e^{i\theta} = \cos \theta + i \sin \theta \tag{4–24}$$

*COMMENT: The Euler identity can be verified by letting $x = i\theta$ in the **Maclaurin series** for e^x and identifying the resulting series for $\cos \theta$ and $\sin \theta$.*

$$\begin{aligned}
e^{i\theta} &= 1 + i\theta - \frac{\theta^2}{2!} - i\frac{\theta^3}{3!} + \frac{\theta^4}{4!} + i\frac{\theta^5}{5!} - \cdots \\
&= \left(1 - \frac{\theta^2}{2!} + \frac{\theta^4}{4!} - \cdots\right) + i\left(\theta - \frac{\theta^3}{3!} + \frac{\theta^5}{5!} - \cdots\right) \\
&= \cos \theta + i \sin \theta
\end{aligned}$$

Rewriting Equation 4–23, using Euler's formula,

$$\begin{aligned}
y &= e^{ax}[c_3(\cos bx + i \sin bx) + c_4(\cos bx - i \sin bx)] \\
&= e^{ax}[(c_3 + c_4)\cos bx + (c_3 - c_4)i \sin bx]
\end{aligned}$$

By letting $c_1 = c_3 + c_4$ and $c_2 = (c_3 - c_4)i$, we have the following result:

THEOREM 4–8

If the auxiliary equation for

$$b_n y^{(n)} + b_{n-1} y^{(n-1)} + \cdots + b_1 y' + b_0 y = 0$$

has the complex roots $m = a \pm ib$, then for each such pair of roots the general solution contains terms of the form

$$y = e^{ax}(c_1 \cos bx + c_2 \sin bx) \tag{4–25}$$

Sometimes we write Equation 4–25 in an alternate form

$$y = Ce^{ax} \cos (bx - \delta) \tag{4–26}$$

where C and δ are arbitrary constants, dependent on auxiliary conditions and related to the constants c_1 and c_2 by

$$C = \sqrt{c_1^2 + c_2^2} \qquad \text{and} \qquad \tan \delta = \frac{c_2}{c_1}$$

COMMENT: When confronted with complex roots to the auxiliary equation, use either Equation 4–25 or Equation 4–26 rather than Equation 4–23. The solution in the form of Equation 4–23 is not incorrect, but to be of practical value, an expression in terms of real functions is preferred. Thus complex roots generate solutions of the form of a product of an exponential and a trigonometric function.

Example 3

(a) Solve the initial-value problem

$$y'' + 2y' + 4y = 0, \, y(0) = 1, \, y'(0) = -1 + 2\sqrt{3}$$

(b) Write the solution in the form of Equation 4–26.

SOLUTION   (a) The auxiliary equation is $m^2 + 2m + 4 = 0$. The roots of this equation are $m_1 = -1 + i\sqrt{3}$ and $m_2 = -1 - i\sqrt{3}$. Hence the general solution is given by Equation 4–25 to be

$$y = e^{-x}(c_1 \cos \sqrt{3}x + c_2 \sin \sqrt{3}x)$$

To find the particular solution corresponding to $y(0) = 1$ and $y'(0) = -1 + 2\sqrt{3}$, we note that

$$1 = e^0(c_1 \cos 0 + c_2 \sin 0) \qquad \text{or} \qquad c_1 = 1$$

Since the derivative of $e^{-x}(c_1 \cos \sqrt{3}x + c_2 \sin \sqrt{3}x)$ is

$$y' = -e^{-x}(c_1 \cos \sqrt{3}x + c_2 \sin \sqrt{3}x)$$
$$+ \sqrt{3}e^{-x}(c_2 \cos \sqrt{3}x - c_1 \sin \sqrt{3}x)$$

it follows, using $y'(0) = -1 + 2\sqrt{3}$, that

$$-1 + 2\sqrt{3} = -e^0(c_1 \cos 0 + c_2 \sin 0) + \sqrt{3}e^0(c_2 \cos 0 - c_1 \sin 0)$$
$$-1 + 2\sqrt{3} = -c_1 + \sqrt{3}c_2$$

Since $c_1 = 1$, we get $c_2 = 2$. Therefore the particular solution is

$$y = e^{-x}(\cos\sqrt{3}x + 2 \sin\sqrt{3}x)$$

(b) Using Equation 4–26, we have

$$C = \sqrt{c_1^2 + c_2^2} = \sqrt{1^2 + 2^2} = \sqrt{5} \qquad \text{and} \qquad \tan \delta = \frac{2}{1}$$

NOTE: Since $c_1 > 0$, $c_2 > 0$, δ is a first quadrant angle.

$$\delta = \text{Arctan } 2 \approx 1.11$$

Thus, $y = \sqrt{5}e^{-x} \cos(\sqrt{3}x - 1.11)$.   ■

CAUTION: When evaluating δ be sure that it represents the given coefficients. For instance, if $y = -\cos 2t + \sin 2t$, we get $\tan \delta = 1/-1 = -1$. Note that $\text{Arctan }(-1) = -\pi/4$ is incorrect since $\cos 2t < 0$ and $\sin 2t > 0$ requires δ to be in the second quadrant. Hence the correct value is $\delta = \pi - \pi/4 = 3\pi/4$.

Example 4

Solve the equation $\dfrac{d^3s}{dt^3} + 4\dfrac{ds}{dt} = 0$

SOLUTION   The auxiliary equation is $m^3 + 4m = 0$, which has the roots $m_1 = 0$, $m_2 = 2i$, and $m_3 = -2i$. Therefore the general solution is

$$y = c_1 + c_2 \cos 2x + c_3 \sin 2x$$ ■

Example 5

Solve the equation $y^{(4)} + 8y'' + 16y = 0$

SOLUTION   The auxiliary equation is

$$m^4 + 8m^2 + 16 = 0$$

or, in factored form

$$(m^2 + 4)^2 = 0$$

The roots are $m_1 = m_2 = 2i$, $m_3 = m_4 = -2i$. The *formal* solution, if the roots were real, would be

$$y = C_1 e^{i2x} + C_2 e^{-i2x} + C_3 x e^{i2x} + C_4 x e^{-i2x}$$

Euler's formula can be used to express the first two terms of this expression in the form $c_1 \cos 2x + c_2 \sin 2x$, and the last two terms of the expression in the form $x(c_3 \cos 2x + c_4 \sin 2x)$. Hence the solution is

$$y = (c_1 + c_3 x) \cos 2x + (c_2 + c_4 x) \sin 2x$$ ∎

Solutions of a Homogeneous Linear Differential Equation with Constant Coefficients

The form of the solution of a homogeneous linear differential equation with constant coefficients depends on the roots of the auxiliary equation.

1. If there are k distinct real roots $m_1, m_2, \ldots, m_k$, then there are k linearly independent functions of the form $e^{m_1 x}, e^{m_2 x}, \ldots, e^{m_k x}$ in the general solution.

2. If there are k repeated real roots m, then there are k linearly independent functions of the form $e^{mx}, xe^{mx}, x^2 e^{mx}, \ldots, x^{k-1} e^{mx}$ in the general solution.

3. If there are complex roots of the form $a \pm bi$, then there are two linearly independent solutions of the form $e^{ax} \sin bx$ and $e^{ax} \cos bx$ in the general solution.

4. If there are k repeated complex conjugate roots $a \pm bi$, then there are $2k$ linearly independent functions of the form $e^{ax} \sin bx$, $xe^{ax} \sin bx$, $\ldots, x^{k-1} e^{ax} \sin bx$ and $e^{ax} \cos bx$, $xe^{ax} \cos bx, \ldots, x^{k-1} e^{ax} \cos bx$ in the general solution.

COMMENT: Note that the functions that are solutions of homogeneous linear differential equations with constant coefficients have a simple form. They are all sums of functions of the form $x^k e^{ax} \cos bx$ and $x^k e^{ax} \sin bx$, where k is a nonnegative integer and a and b are real numbers.

EXERCISES FOR SECTION 4–5

In Exercises 1–30 find the general solution of the given differential equation.

1. $y'' + 8y' + 16y = 0$

2. $4y'' - 4y' + y = 0$

3. $y'' + \dfrac{2}{3}y' + \dfrac{1}{9}y = 0$

4. $y'' + 5y' + 4y = 0$

5. $y^{(5)} - y^{(3)} = 0$

6. $d^4 s/dt^4 - d^2 s/dt^2 = 0$

7. $y^{(4)} + 18y''' + 81y'' = 0$

8. $9y^{(4)} + 6y''' + y'' = 0$

9. $4y''' - 3y' + y = 0$

10. $y''' - 3y' - 2y = 0$

11. $y^{(5)} - 3y''' - 2y'' = 0$

12. $y^{(4)} - 7y''' + 18y'' - 20y' + 8y = 0$

13. $y'' + 9y = 0$ 14. $y''' + 16y' = 0$

15. $2y''' + 50y' = 0$ 16. $y'' - 2y' + 5y = 0$

17. $\ddot{s} - 6\dot{s} + 25s = 0$ 18. $y''' + 2y'' + 4y' = 0$

19. $y^{(4)} + y'' = 0$ 20. $y^{(4)} + 4y''' + 8y'' + 16y' + 16y = 0$

21. $y''' + 18y'' + 81y' = 0$ 22. $y^{(4)} - 16y = 0$

23. $2\, d^2y/dx^2 + dy/dx + y = 0$ 24. $y^{(4)} + y''' + 2y'' + y' + y = 0$

25. $y''' + y'' + 4y' + 4y = 0$ 26. $y''' - y'' - y' + y = 0$

27. $y''' - 3y'' + 9y' + 13y = 0$ 28. $y^{(4)} - 5y'' + 12y' + 28y = 0$

29. $y^{(5)} + y^{(4)} + 18y''' + 18y'' + 81y' + 81y = 0$

30. $y^{(8)} - 2y^{(4)} + y = 0$

Solve the initial-value problems in Exercises 31–38.

31. $y'' - 8y' + 16y = 0$, $y(0) = 0$, $y'(0) = 1$

32. $y'' - 2y' + y = 0$, $y(0) = 1$, $y'(0) = 2$

33. $y'' - 6y' + 9y = 0$, $y(0) = 1$, $y'(0) = 1$

34. $y''' + 3y'' = 0$, $y(0) = 3$, $y'(0) = 0$, $y''(0) = 9$

35. $y'' + 4y = 0$, $y(0) = 0$, $y'(0) = 1$

36. $y'' + y = 0$, $y(0) = 3$, $y'(0) = 0$

37. $y'' + 4y' + 5y = 0$, $y(0) = 1$, $y'(0) = 0$

38. $y'' - 6y' + 10y = 0$, $y(0) = 2$, $y'(0) = 1$

In Exercises 39–43 solve the given boundary-value problem or show that no non-trivial solution exists.

39. $y'' + 4y = 0$, $y(0) = 0$, $y(\pi) = 0$ 40. $y'' + 4y = 0$, $y(0) = 0$, $y'(\pi) = 1$

41. $y'' - 4y = 0$, $y(0) = 0$, $y'(\pi) = 0$ 42. $y'' = 0$, $y(0) = 0$, $y'(\pi) = 0$

43. $y'' = 0$, $y(0) = 0$, $y'(\pi) = 1$

44. Determine the values of λ for which the boundary-value problem
     $y'' + \lambda^2 y = 0$, $y(0) = 0$, $y'(\pi) = 0$ has real solutions. Find the solutions.

4–6   UNDAMPED VIBRATIONS

Vibratory motion exists to some degree in all mechanical systems due to noise, wind, moving parts, and so forth. For the most part mechanical vibration is considered undesirable. For instance, the suspension system of an automobile is designed to absorb vibrations due to irregularities in roadways; booster rockets, used in launching space probes, are designed to withstand the tremendous amount of vibrational energy produced by the noise of their motors.

The simplest system available for studying vibratory motion is a spring supporting a weight W, as shown in Figure 4–1. We assume the weight of the spring is negligible. This is not unrealistic since the weight of the supported object is usually much greater than that of the spring. When the weight is hanging at rest, we say it is in the **equilibrium** position. A weight that is pulled down a certain distance and released will oscillate about its equilibrium position. Our objective is to describe this oscillatory motion analytically.

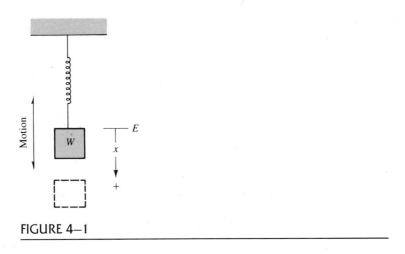

FIGURE 4–1

Figure 4–1 shows the forces acting on a weight displaced from its equilibrium position with an initial assumption that there is no air resistance or friction. The resulting motion is called **undamped motion.** The supporting spring exerts an external restoring force on the weight that, within certain elastic limits, is described by **Hooke's law.**

HOOKE'S LAW: When a spring is stretched or compressed, its restoring force is directly proportional to its change in length.

If x represents the displacement of the weight from its equilibrium position and F represents the corresponding restoring force of the spring, then by Hooke's law

$$F = -kx$$

where the minus sign indicates that the restoring force F is always opposite in direction to the displacement. The constant k, called the **spring constant,** may be found by taking the ratio of F to x for a given spring.

We establish a coordinate system with the origin as the equilibrium position and measure the vertical displacement x from this point, with the convention that x is positive when measured below the point of equilibrium and negative when above. Force is measured in pounds, mass in slugs, displacement in feet, and time in seconds.

Since the system is undamped, the unbalanced force acting on the weight is equal to $-kx$. Applying Newton's second law with $m = w/g$ and $a = d^2x/dt^2$, we have

$$m\frac{d^2x}{dt^2} = -kx$$

or

$$m\frac{d^2x}{dt^2} + kx = 0 \qquad (4\text{-}27)$$

Thus the undamped spring-mass system in Figure 4–1 is described by a second-order linear homogeneous differential equation with constant coefficients.

The general solution to this equation for undamped vibrations can be shown to be

$$x(t) = c_1 \cos \omega_0 t + c_2 \sin \omega_0 t$$

where $\omega_0 = \sqrt{k/m}$ and is called the **natural angular frequency** of the system. Note its dependence on the stiffness of the spring and on the mass.

Using Equation 4–26, the equation of motion for undamped vibration may be written as

$$x(t) = C \cos (\omega_0 t - \delta)$$

where $C = \sqrt{c_1^2 + c_2^2}$ and $\tan \delta = c_2/c_1$.

COMMENT: The graph of $C \cos (\omega_0 t - \delta)$ is shown in Figure 4–2. Note that the amplitude of the cosine wave is C, the period is $2\pi/\omega_0$, and the phase shift is δ/ω_0.

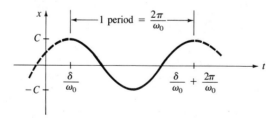

FIGURE 4–2

The motion of an undamped vibration is called **harmonic oscillation.** Figure 4–3 shows graphs of a few typical oscillations with different values of initial velocity. The frequency (the reciprocal of the period) of the harmonic oscillation is $\omega_0/2\pi$ and thus

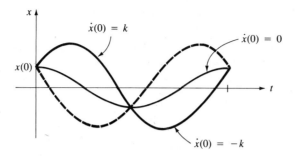

$\dot{x}(0) = k$

$\dot{x}(0) = 0$

$x(0)$

t

$\dot{x}(0) = -k$

FIGURE 4–3

1. The larger the k (stiffer spring), the higher the frequency.

2. The smaller the mass, the higher the frequency.

Example 1

It is found that a 2-lb weight stretches a certain spring 6 inches. If the weight is pulled down 3 inches and released, find the amplitude and period of the resulting motion.

SOLUTION   To find the value of k, we use Hooke's law with $x = 6$ inches $= \frac{1}{2}$ ft and $F = 2$ lb:

$$k = \frac{2}{1/2} = 4 \text{ lb/ft}$$

The differential equation of this system is then

$$\frac{2}{32}\ddot{x} + 4x = 0$$

or

$$\ddot{x} + 64x = 0$$

The auxiliary equation is $r^2 + 64 = 0$, which has roots of $r = \pm 8i$. Therefore the general solution is

$$x = c_1 \cos 8t + c_2 \sin 8t \qquad (4\text{–}28)$$

To find the required particular solution, note that the weight is initially pulled down 3 in. below the equilibrium position, so

$$x(0) = 3 \text{ in.} = \frac{1}{4} \text{ ft} \qquad (4\text{–}29)$$

Further, when the weight is released, its initial velocity is zero; that is,

$$\dot{x}(0) = 0 \qquad\qquad (4\text{-}30)$$

Using Condition 4–29 in Equation 4–28, we get

$$c_1 = \frac{1}{4}$$

Differentiation of Equation 4–28 yields

$$\dot{x}(t) = -8c_1 \sin 8t + 8c_2 \cos 8t$$

Substituting Condition 4–30 into this equation,

$$c_2 = 0$$

The required solution is then

$$x = \frac{1}{4} \cos 8t$$

This equation shows that the amplitude of the oscillation is $\frac{1}{4}$ ft, and the period is $t = 2\pi/8 = \pi/4$ sec, as shown in Figure 4–4.

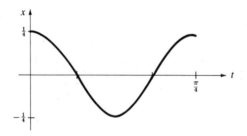

FIGURE 4–4   ∎

Example 2

Find the position, velocity, and acceleration of the weight in the previous example when $t = \frac{1}{4}$ sec.

SOLUTION   From Example 1, the displacement of the weight is given by

$$x = \frac{1}{4} \cos 8t$$

Differentiation of this with respect to t yields

$$v = \dot{x} = -2 \sin 8t$$

and

$$a = \ddot{x} = -16 \cos 8t$$

Letting $t = \frac{1}{4}$ sec, we get

$$x = \frac{1}{4} \cos 2 = \frac{1}{4}(-0.415) = -0.104 \text{ ft}$$

$$v = -2 \sin 2 = -2(0.910) = -1.82 \text{ fps}$$

$$a = -16 \cos 2 = -16(-0.415) = 6.65 \text{ fps}^2$$

The above results mean that the weight is 0.104 ft above the equilibrium position and moving upward with a velocity of 1.82 fps. Further, it has a downward acceleration of 6.65 fps². Since the acceleration is opposite to the velocity, we know the weight is slowing up at the time in question. ■

Example 3

Assume that the weight in Example 1 is given an upward velocity of 2 fps instead of simply being released. Determine the amplitude and period of the resulting oscillation.

SOLUTION   From Example 1 the solution of the differential equation

$$\ddot{x} + 64x = 0$$

is

$$x = c_1 \cos 8t + c_2 \sin 8t \qquad (4\text{--}31)$$

Thus

$$\dot{x} = -8c_1 \sin 8t + 8c_2 \cos 8t \qquad (4\text{--}32)$$

The initial conditions are $x(0) = \frac{1}{4}$ ft and $\dot{x}(0) = -2$ fps. Substituting the first condition into Equation 4–31 yields

$$c_1 = \frac{1}{4}$$

Then substituting the second condition into Equation 4–32 yields

$$c_2 = -\frac{1}{4}$$

Hence the desired solution is

$$x = \frac{1}{4} \cos 8t - \frac{1}{4} \sin 8t$$

To find the amplitude and period of this expression, we use the alternate form $x = C \cos(\omega_0 t - \delta)$. Thus

$$x = \sqrt{(1/4)^2 + (1/4)^2} \cos (8t - \delta) \tag{4–33}$$
$$= 0.354 \cos (8t - \delta)$$

where, to find δ, we use

$$\tan \delta = -1$$

so that

$$\delta = -\frac{\pi}{4} \qquad \text{(NOTE: } \delta \text{ is a fourth quadrant angle because } c_1 > 0, c_2 < 0.)$$

Substituting this value into Equation 4–33, the desired equation is

$$x = 0.354 \cos 8\left(t + \frac{1}{32}\pi \right)$$

Thus the amplitude is $C = 0.354$, and the period is $T = \frac{1}{4}\pi$. Further, there is a phase shift of $-\frac{1}{32}\pi$. The graph of the function is shown in Figure 4–5.

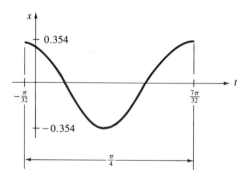

FIGURE 4–5

EXERCISES FOR SECTION 4–6

1. An object weighing 32 lb is suspended from a spring with a spring constant of 4 lb/ft. The object is started from the equilibrium position with a downward velocity of 1 ft/sec.
    (a) Determine the displacement equation of the object.
    (b) Determine the natural angular frequency of the object.
    (c) Determine the amplitude and period of the oscillation.

2. A spring with $k = 1$ lb/ft is attached to a 32-lb weight. Find the amplitude and period of the motion of the weight if it is pulled down 3 ft below the equilibrium point and released.

3. An object weighing 2 lb is attached to a spring suspended from a fixed point. The spring was stretched 1.5 inches at the time that the weight was attached. The object is set in motion by lifting it 0.25 ft above equilibrium and releasing it.
   (a) Determine the displacement, velocity, and acceleration of the object as a function of time.
   (b) Find the position, velocity, and acceleration of the object when $t = \frac{1}{8}$ sec.

4. A spring is stretched $\frac{1}{2}$ ft by a 6-lb weight in coming to rest in its equilibrium position. The weight is pulled down $\frac{1}{3}$ ft below the equilibrium position and then given a downward velocity of 2 ft/sec.
   (a) Determine the displacement of the weight as a function of time.
   (b) Determine the amplitude, period, and phase shift of the motion.
   (c) Draw the graph of the displacement of the object as a function of time.

5. A spring is stretched 24 inches by a 4-lb weight in coming to rest in its equilibrium position. Determine the displacement of the weight as a function of time if it is initially lifted 0.5 ft above the equilibrium position and then given a downward velocity of 2 ft/sec.

6. A 32-lb weight hangs from a spring having a spring constant of 9 lb/ft. It is known that $x(\pi/6) = 0$ and $x'(\pi/6) = 2$ ft/sec.
   (a) Determine the displacement of the weight as a function of time.
   (b) Determine $x(0)$ and $x'(0)$.
   (c) Determine the period of the motion.

7. A steel ball suspended from a spring is set into motion and oscillates with a period of 2.5 sec. The motion is stopped, and an additional weight of 4 lb is added to the ball. The system is again set into motion, and now it oscillates with a period of 3.5 sec. Determine the weight of the steel ball.

4–7 DAMPED VIBRATIONS

The oscillatory motion described in the preceding section would continue undiminished without some retarding force. In practice, however, several forces tend to dissipate the vibration. We assume the existence of a force proportional to the velocity of the object. Such a force is called a **damping force** or, more specifically, **viscous damping.** (There is empirical evidence to support this assumption provided the velocity is not too great.)

Mechanical systems that involve damping are usually diagrammed by connecting the mass to a "dashpot," shown schematically in Figure 4–6. The mechanical system is called a **spring-mass-damper system.**

Since the damping force is proportional to the velocity and opposes the movement of the mass through the resisting medium, we denote it by

$$-c\,\frac{dx}{dt}$$

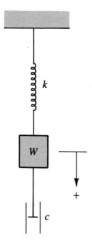

FIGURE 4–6

where c is called the **damping constant.** Both the spring force and the force due to the viscous friction oppose the movement. From Newton's second law

$$m\frac{d^2x}{dt^2} = -c\frac{dx}{dt} - kx$$

or

$$m\frac{d^2x}{dt^2} + c\frac{dx}{dt} + kx = 0 \tag{4–34}$$

The auxiliary equation written in terms of the variable r is

$$mr^2 + cr + k = 0$$

whose roots are

$$-\frac{c}{2m} \pm \frac{1}{2m}\sqrt{c^2 - 4mk}$$

Letting $\alpha = \dfrac{c}{2m}$ and $\beta = \dfrac{1}{2m}\sqrt{c^2 - 4mk}$, the roots are

$$-\alpha \pm \beta$$

The form of the solution depends on the damping, and we distinguish three cases that come from the value of $\sqrt{c^2 - 4mk}$.

Case 1 $c^2 < 4mk$: In this case β is imaginary—that is,

$$\beta = i\omega^*$$

$$\text{where } \omega^* = \frac{1}{2m}\sqrt{4mk - c^2}$$

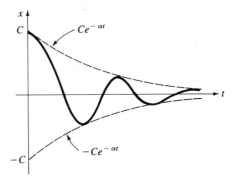

FIGURE 4–7

Thus the auxiliary equation has the complex roots $-\alpha \pm i\omega^*$ and the general solution of Equation 4–34 is

$$x = e^{-\alpha t}(c_1 \cos \omega^* t + c_2 \sin \omega^* t)$$

or

$$x(t) = Ce^{-\alpha t} \cos (\omega^* t - \delta)$$

Thus there will be oscillatory motion when the weight is released. The factor $e^{-\alpha t}$ will "damp out" the oscillation; that is, the amplitude of the motion approaches 0 as $t \to \infty$. We say that the spring-mass-damper system is **underdamped** for this case.

The graph of the solution lies between $y = Ce^{-\alpha t}$ and $y = -Ce^{-\alpha t}$, as shown in Figure 4–7. The frequency of the oscillation is $\omega^*/2\pi$ cycles per second. Note that as the damping constant c approaches zero, ω^* approaches $\sqrt{k/m}$, which is the natural frequency of the system and the frequency of harmonic oscillation. Thus underdamped motion approaches harmonic motion as the viscous damping decreases.

Case 2 $c^2 = 4mk$: If $c^2 = 4mk$, then $\beta = 0$, and therefore the auxiliary equation has the double root $m = -\alpha$. Hence the solution is

$$x = e^{-\alpha t}(c_1 + c_2 t)$$

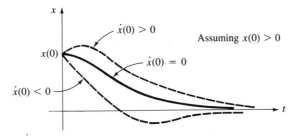

FIGURE 4–8

This solution, which does not contain sin x or cos x, indicates that the weight will return to equilibrium without oscillating. Since any decrease in damping constant c will cause underdamping, we say the spring-mass system is **critically** damped. Under these conditions the weight will return to its equilibrium position in the **shortest possible time** without oscillation. Typical graphs of a critically damped system are shown in Figure 4–8.

Case 3 $c^2 > 4mk$: If $c^2 > 4mk$, then $c^2 - 4mk > 0$, and therefore the auxiliary equation has two distinct **real** roots $m = -\alpha \pm \beta$. The solution is

$$x = c_1 e^{-(\alpha + \beta)t} + c_2 e^{-(\alpha - \beta)t}$$

This is the same basic nonoscillatory form as for the critically damped case. It can be shown that the time required for the weight to return to its equilibrium position in this case is always greater than for the critically damped case. For this reason we say the spring-mass system is **overdamped.** Figure 4–9 shows some typical overdamped motions.

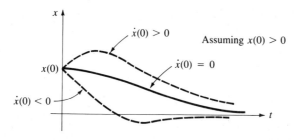

FIGURE 4–9

Example 1

A spring is stretched 0.4 ft by an 8-lb weight. The 8-lb weight is pushed up 0.1 ft and released. The motion takes place in a medium that furnishes a damping force equal to twice the instantaneous velocity. Find the displacement equation and draw its graph.

SOLUTION   The damping force in this system is $-2\, dx/dt$, the restoring force of the spring is $\frac{8}{0.4}x$, and the mass is $m = \frac{8}{32}$. Therefore the differential equation is

$$\frac{8}{32}\ddot{x} + 2\dot{x} + \frac{8}{0.4}x = 0$$

or

$$\ddot{x} + 8\dot{x} + 80x = 0$$

The auxiliary equation $r^2 + 8r + 80 = 0$ has roots

$$r = \frac{-8 \pm \sqrt{8^2 - 4(80)}}{2} = -4 \pm 8i \qquad \text{(The system is underdamped.)}$$

Therefore the general solution is

$$x = e^{-4t}(c_1 \cos 8t + c_2 \sin 8t) \qquad (4\text{--}35)$$

To find c_1 and c_2, use the initial conditions $x(0) = -0.1$ and $\dot{x}(0) = 0$. The derivative of Equation 4–35 is

$$\dot{x} = e^{-4t}[(8c_2 - 4c_1) \cos 8t - (8c_1 + 4c_2) \sin 8t]$$

Thus $c_1 = -0.1$ and $c_2 = -0.05$. Using these values in Equation 4–35, we get

$$x = -e^{-4t}(0.1 \cos 8t + 0.05 \sin 8t)$$

or

$$x = -0.112e^{-4t} \cos (8t - 0.46)$$

The graph of this function is bounded above by the graph of $x = 0.112e^{-4t}$ and below by $x = -0.112e^{-4t}$, as shown in Figure 4–10. The cosine function has a period of $\pi/4$ sec and a phase shift of 0.058 sec.

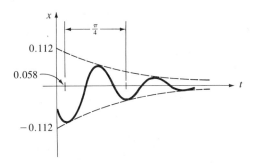

FIGURE 4–10 ■

Example 2

A spring has a spring constant of 8 lb/ft. A 4-lb weight is pulled down 0.5 ft below equilibrium and then given an initial upward velocity of 5 fps. The damping force is equal to twice the instantaneous velocity. Find the equation of motion.

SOLUTION   Here we have

$$\frac{4}{32}\ddot{x} + 2\dot{x} + 8x = 0$$

or

$$\ddot{x} + 16\,\dot{x} + 64x = 0$$

with the initial conditions $x(0) = \frac{1}{2}$ and $\dot{x}(0) = -5$.

The auxiliary equation is $r^2 + 16r + 64 = 0$, which has the double root $r = -8$, so the general solution for the critically damped system is

$$x = e^{-8t}(c_1 + c_2 t)$$

The derivative of the displacement function is

$$\dot{x} = (c_2 - 8c_1 - 8c_2 t)e^{-8t}$$

Therefore we have $c_1 = \frac{1}{2}$ and $c_2 = -1$ and

$$x = \left(\frac{1}{2} - t\right)e^{-8t}$$

The graph of this function is shown in Figure 4–11.

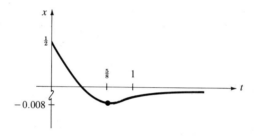

FIGURE 4–11

EXERCISES FOR SECTION 4–7

1. A 2-lb weight attached to a spring with a spring constant of 4 lb/ft experiences a damping force equal to its velocity. The weight is set into motion by pulling it 3 inches below the equilibrium position and releasing it.
   (a) Determine the displacement of the weight as a function of time.
   (b) Draw the graph of the displacement function.

2. A spring is such that a 32-lb weight stretches it 2 ft in coming to the equilibrium position. When in motion, the weight experiences a damping force equal to eight times its velocity. The system is set into motion from the equilibrium position by a downward velocity of 1 ft/sec.
   (a) Determine the displacement of the weight as a function of time.
   (b) Find the maximum displacement of the weight from the equilibrium position.
   (c) Draw the graph of the displacement function.

3. A spring is stretched 0.4 ft by a 4-lb weight in coming to its equilibrium position. A damping force equal to the velocity of the object acts on the system. The weight

is set into motion from the equilibrium position by an initial upward velocity of 2 ft/sec.

(a) Determine the displacement of the weight as a function of time.

(b) Determine the displacement of the weight at the first three stops in its motion.

(c) Draw the graph of the displacement function.

4. An object having a mass of 1 slug is suspended from a spring having a spring constant of 5 lb/ft. Assume the damping force to be equal to twice the velocity of the object. Determine the displacement of the object as a function of time if it is pulled down 24 inches below the equilibrium position and released.

5. Indicate the type of damping associated with each of the following equations:

(a) $\ddot{x} + 4\dot{x} + 4x = 0$

(b) $\ddot{x} + 10\dot{x} + 9x = 0$

(c) $3\ddot{x} + 6\dot{x} + 3x = 0$

(d) $\ddot{x} + 2\dot{x} + 3x = 0$

(e) $\ddot{x} + \dot{x} + x = 0$

6. How does the frequency of a damped oscillation depend on the damping force?

7. Determine where the maxima and minima of an underdamped oscillation occur and compare to the points where the graph is tangent to Ce^{-at} and $-Ce^{-at}$.

8. In the case of an underdamped motion show that the time between two successive positive maxima is $2\pi/\omega^*$.

9. Prove that, in the overdamped case, the mass cannot pass through equilibrium more than once.

REVIEW EXERCISES FOR CHAPTER 4

In Exercises 1–6 identify the given equation as linear or nonlinear. If the equation is linear, use Theorem 4–1 to determine if it has a unique solution on $-\infty < x < \infty$.

1. $xy'' + 3y = y^2$, $y(0) = 1$, $y'(0) = 2$

2. $3yy' = 2 - \sin x$, $y(0) = 0$

3. $(x + 2)y'' - xy' = 0$, $y(1) = -1$, $y'(1) = 0$

4. $3y'' + 4y' - 6y = e^{2x}$, $y(1) = 3$, $y'(-1) = 2$

5. $x^2y'' - 3y = 10$, $y(0) = -3$, $y'(0) = 0$

6. $y'' - xy' + \tan x = 0$, $y(0) = y'(0) = 1$

In Exercises 7–10 show that the indicated functions are linearly independent on $-\infty < x < \infty$.

7. $1, x, x^2$

8. e^{-2x}, e^{3x}

9. $\sin x, x \sin x$

10. $\cos 2x, \cos 3x$

11. By determining constants c_1, c_2, c_3, not all zero, show that the functions $x, 2e^{-x}$, and $5e^{-x}$ are linearly dependent.

12. By determining constants c_1, c_2, c_3, c_4, not all zero, show that the functions e^x, e^{-x}, $\cos^2 x$, and $1 + \cos 2x$ are linearly dependent.

In Exercises 13–16 use the method of reduction of order to find a second linearly independent solution of the given differential equation.

13. $y'' - 5y' + 4y = 0$, $y_1 = e^x$

14. $y'' + 3y' + 2y = 0$, $y_1 = e^{-x}$

15. $2x^2y'' - xy' + y = 0$, $y_1 = x$

16. $(x - 1)y'' - xy' + y = 0$, $y_1 = e^x$

In Exercises 17–26 find the general solution of the given homogeneous differential equation.

17. $2y''' - 5y'' - 3y' = 0$

18. $y''' - 12y'' = 0$

19. $y'' + 5y = 0$

20. $y'' + 2y' + 2y = 0$

21. $y'' + 16y' + 64y = 0$

22. $y'' - 3y = 0$

23. $2y'' + 2y' - y = 0$

24. $y'' + y' + y = 0$

25. $y''' + 2y'' + 4y' + 8y = 0$

26. $2y''' - 3y'' - 8y' - 3y = 0$

Solve the initial-value problems in Exercises 27–30.

27. $y'' + 2y = 0$, $y(0) = 1$, $y'(0) = -2$

28. $y'' + 2y' = 0$, $y(0) = -5$, $y'(0) = -1$

29. $y'' + 4y' + 4y = 0$, $y(0) = 1$, $y'(0) = 1$

30. $2y'' - 3y' - 2y = 0$, $y(0) = 0$, $y'(0) = 2$

31. A 16-lb weight is attached to a spring with a spring constant of 8 lb/ft. The weight is set in motion by pushing it up $\frac{1}{3}$ ft above the equilibrium position and imparting to it an upward velocity of 4 ft/sec.

   (a) Determine the displacement and the velocity as functions of time.

   (b) Determine the displacement and velocity of the weight when $t = 1$ sec.

   (c) Determine the amplitude and period of the motion.

32. A 3.2-lb weight is attached to a spring with a spring constant of 2 lb/ft. The weight experiences a damping force equal to 0.4 of its velocity. To set the system in motion, the weight is raised 1 ft above the equilibrium position and released.

   (a) Determine the displacement of the weight as a function of time.

   (b) Describe the motion as underdamped, critically damped, or overdamped.

33. When a 2-lb weight is attached to a spring, it stretches the spring 0.5 ft in coming to rest. A damping force equal to the velocity of the object acts on the system. The weight is set in motion from the equilibrium position by a downward velocity of 10 ft/sec.

   (a) Determine the displacement of the weight as a function of time.

   (b) Determine the maximum displacement of the weight from the equilibrium position.

5

Nonhomogeneous Linear Differential Equations

In a simple electrical circuit, such as shown in Figure 5–1, the electric charge $q(t)$ will satisfy the differential equation

$$L\frac{d^2q}{dt^2} + R\frac{dq}{dt} + \frac{1}{C}q = v(t)$$

where L, R, and C are constants and $v(t)$ is the voltage supplied to the circuit. Some techniques for solving this differential equation are developed in this chapter.

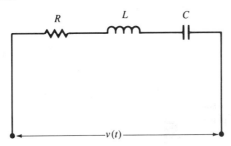

FIGURE 5–1

5–1   THE GENERAL APPROACH

The nth-order nonhomogeneous linear differential equation can be expressed in the form

$$a_n(x)y^{(n)} + a_{n-1}(x)y^{(n-1)} + \cdots + a_1(x)y' + a_0(x)y = f(x) \qquad (5\text{–}1)$$

where $a_n(x)$ and $f(x)$ are not identically zero on some interval $a \leq x \leq b$. The function $f(x)$ is often called the **driving function** of the equation. In this chapter we define the general solution of Equation 5–1 and develop methods for finding such solutions.

Any function y_p that is free of arbitrary constants and satisfies Equation 5–1 is called a **particular solution** of the equation.

Example 1

(a) A particular solution of

$$y'' + y' = 5$$

is $y_p = 5x$ since substituting $y_p' = 5$ and $y_p'' = 0$ into $y'' + y' = 5$ yields the identity $0 + 5 = 5$.

(b) $y_p = 2e^{3x}$ is a particular solution of

$$y'' - 2y' + y = 8e^{3x}$$

since substituting $y_p = 2e^{3x}$, $y_p' = 6e^{3x}$, and $y_p'' = 18e^{3x}$ into the given equation yields

$$18e^{3x} - 2(6e^{3x}) + 2e^{3x} = 8e^{3x} \qquad \blacksquare$$

Associated with Equation 5–1 is the homogeneous linear differential equation

$$a_n(x)y^{(n)} + a_{n-1}y^{(n-1)} + \cdots + a_1(x)y' + a_0y = 0 \qquad (5\text{–}2)$$

which is called the **corresponding homogeneous equation.**

Example 2

(a) The corresponding homogeneous equation for $y'' - 2y' - 3y = \sin x$ is

$$y'' - 2y' - 3y = 0$$

(b) The corresponding homogeneous equation for $y'' + y = 25$ is

$$y'' + y = 0 \qquad \blacksquare$$

The general solution of the corresponding homogeneous equation, denoted by y_c, is called the **complementary solution** of the nonhomogeneous equation. The complementary solution is found by using the methods of Chapter 4.

Example 3

(a) The general solution of $y'' - 2y' - 3y = 0$ is $y = c_1 e^{3x} + c_2 e^{-x}$. Therefore the complementary solution of $y'' - 2y' - 3y = \sin x$ is

$$y_c = c_1 e^{3x} + c_2 e^{-x}$$

(b) The complementary solution of $y'' + y = 25$ is

$$y_c = c_1 \cos x + c_2 \sin x$$

since this is the general solution of $y'' + y = 0$. ∎

The general solution of a nonhomogeneous linear differential equation is composed of a particular solution and the complementary solution, as proved in Theorem 5–1. The theorem is stated and proved for the case in which the coefficient functions are constants.

THEOREM 5–1

Let y_p be any particular solution of the nth-order constant-coefficient linear differential equation

$$b_n y^{(n)} + b_{n-1} y^{(n-1)} + \cdots + b_1 y' + b_0 y = f(x) \tag{5-3}$$

and let y_c be the general solution of the corresponding homogeneous equation

$$b_n y^{(n)} + b_{n-1} y^{(n-1)} + \cdots + b_1 y' + b_0 y = 0$$

Then the general solution of Equation 5–3 is

$$y = y_c + y_p$$

PROOF   For the sake of simplicity we prove the theorem for the second-order equation. The extension to the nth-order case is merely a matter of detail. Let y_p be a particular solution of $b_2 y'' + b_1 y' + b_0 y = f(x)$, and let y_c be the complementary solution. Then $y = y_c + y_p$ is a function that has two arbitrary constants. Also, we have

$$b_2(y_c + y_p)'' + b_1(y_c + y_p)' + b_0(y_c + y_p)$$
$$= (b_2 y_c'' + b_1 y_c' + b_0 y_c) + (b_2 y_p'' + b_1 y_p' + b_0 y_p)$$
$$= 0 + f(x)$$

The expression $b_2 y_c'' + b_1 y_c' + b_0 y_c$ is zero because y_c is the complementary solution; the expression $b_2 y_p'' + b_1 y_p' + b_0 y_p$ is equal to $f(x)$ because y_p is a particular solution of the given nonhomogeneous differential equation. Hence the function $y_c + y_p$ is a solution of Equation 5–3.

To complete the proof we need to show that if g is any solution to Equation 5–3 then it is derivable from $y_c + y_p$ where $y_c = c_1 y_1 + c_2 y_2$ and y_1 and y_2 are linearly independent solutions of the corresponding second-order homogeneous equation. That is, for some choice of c_1 and c_2, $g = y_c + y_p$. Since g and y_p are both solutions of Equation 5–3, their difference, $g - y_p$, is a solution of the corresponding homogeneous equation and, hence, by Theorem 4–4 for some particular choice of constants, say c_1^* and c_2^*,

$$g - y_p = c_1^* y_1 + c_2^* y_2$$

Thus g may be written in terms of y_c and y_p, which is what we wished to show. Hence, $y_c + y_p$ is the general solution of Equation 5–3.

> *COMMENT: (a) Although Theorem 5–1 is stated for equations with constant coefficients, nothing in the proof requires such a restriction. In fact, the theorem is valid for linear differential equations with continuous coefficients.*
>
> *(b) Theorem 5–1 says that if any particular solution of the differential equation is known, then finding the general solution is a matter of solving the corresponding homogeneous equation.*

Example 4

Show that $x - 1$ is a particular solution of $y'' - y = 1 - x$ and then find the general solution.

SOLUTION   To verify that $y = x - 1$ is a solution of the differential equation, substitute $y = x - 1$ and $y'' = 0$ into the equation. Next we note that the corresponding homogeneous equation is $y'' - y = 0$, and the roots of the auxiliary equation $m^2 - 1 = 0$ are 1 and -1. Thus the general solution of the corresponding homogeneous equation is $y_c = c_1 e^x + c_2 e^{-x}$. The general solution of the given equation is

$$y = c_1 e^x + c_2 e^{-x} + x - 1 \qquad\blacksquare$$

The driving function is often written as the sum of two or more elementary functions, say $f(x) = f_1(x) + f_2(x)$. The next theorem shows that the solution of a nonhomogeneous differential equation with the driving function $f_1(x) + f_2(x)$ is found by considering $f_1(x)$ and $f_2(x)$ separately.

THEOREM 5–2

Let y_{p1} be a particular solution of
$$b_n y^{(n)} + \cdots + b_1 y' + b_0 y = f_1(x)$$

and y_{p2} be a particular solution of
$$b_n y^{(n)} + \cdots + b_1 y' + b_0 y = f_2(x)$$

Then $y_{p1} + y_{p2}$ is a particular solution of
$$b_n y^{(n)} + \cdots + b_1 y' + b_0 y = f_1(x) + f_2(x)$$

Example 5

The general solution of the nonhomogeneous linear equation

$$3y'' - 5y' + 2y = \sin x + x^2$$

is a matter of finding

1. The complementary solution, y_c, of $3y'' - 5y' + 2y = 0$.

2. A particular solution, y_p of $3y'' - 5y' + 2y = \sin x$.

3. A particular solution, y_q of $3y'' - 5y' + 2y = x^2$.

Then $y_c + y_p + y_q$ is the general solution of the given differential equation. The complementary solution, y_c, is found by the methods of Chapter 4; a systematic method for finding the particular solutions y_p and y_q is explained next. ∎

Using the Method of Reduction of Order to Find a Particular Solution (Optional)

Finding a particular solution y_p is at the heart of the method of solving a non-homogeneous equation. But how do we find such a y_p? Several methods are available for this purpose, with their applicability dependent upon the differential equation under consideration.

One systematic approach to finding a y_p is based on the method of reduction of order that was introduced in Section 4–3 in connection with homogeneous differential equations. The method can also be used to find a particular solution of a nonhomogeneous equation; all we need is a particular solution of the corresponding homogeneous equation. The examples show the technique.

Example 6

Solve the differential equation

$$y'' + 2y' + y = e^{3x}$$

using the method of reduction of order to find a y_p.

SOLUTION The corresponding homogeneous equation is $y'' + 2y' + y = 0$. The auxiliary equation $m^2 + 2m + 1 = 0$, which has repeated real roots $-1, -1$, yields the complementary solution

$$y_c = (c_1 + c_2 x)e^{-x}$$

To find y_p by reduction of order, we let $y_p = vy_1$, where y_1 is any particular solution of the corresponding homogeneous equation. For simplicity we choose $y_1 = e^{-x}$. Thus

$$y_p = v e^{-x}$$
$$y_p' = v' e^{-x} - v e^{-x}$$
$$y_p'' = v'' e^{-x} - 2v' e^{-x} + v e^{-x}$$

Substituting into the given differential equation, we get

$$(v'' e^{-x} - 2v' e^{-x} + v e^{-x}) + 2(v' e^{-x} - v e^{-x}) + v e^{-x} = e^{3x}$$

This reduces to

$$v'' e^{-x} = e^{3x}$$

from which

$$v'' = e^{4x}$$

Integrating twice, we get

$$v = \tfrac{1}{16} e^{4x}$$

(NOTE: The arbitrary constants are omitted in the function v because we need only one solution.)

Substituting the value of v in $y_p = v e^{-x}$, a particular solution of the given differential equation is

$$y_p = \tfrac{1}{16} e^{4x} e^{-x} = \tfrac{1}{16} e^{3x}$$

Finally, the general solution is

$$y = \underbrace{(c_1 + c_2 x)e^{-x}}_{y_c} + \underbrace{\tfrac{1}{16} e^{3x}}_{y_p}$$ ■

Example 7

Solve the differential equation

$$y'' + 2y' = 3x$$

using the method of reduction of order to find a y_p.

SOLUTION   The corresponding homogeneous equation is $y'' + 2y' = 0$. The auxiliary equation $m^2 + 2m = 0$ has the roots 0 and -2, so

$$y_c = c_1 + c_2 e^{-2x}$$

To find y_p, we let $y_p = v y_1$. Here the simplest choice for y_1 is $y_1 = 1$, so

$$y_p = v$$
$$y_p' = v'$$
$$y_p'' = v''$$

Substituting into the given differential equation, we get

$$v'' + 2v' = 3x$$

Or, letting $w = v'$, this becomes the first-order linear equation

$$w' + 2w = 3x$$

An integrating factor for this equation is

$$e^{\int 2\,dx} = e^{2x}$$

Multiplying both sides by this factor yields

$$e^{2x}w' + 2we^{2x} = 3xe^{2x}$$

which may be written

$$\frac{d}{dx}(we^{2x}) = 3xe^{2x}$$

Integrating, we have

$$\begin{aligned}
we^{2x} &= \int 3xe^{2x}\,dx \\
&= \tfrac{3}{2}xe^{2x} - \tfrac{3}{4}e^{2x} \quad \text{(integration by parts)}
\end{aligned}$$

Solving for w yields

$$w = \tfrac{3}{2}x - \tfrac{3}{4}$$

Recalling that $w = v'$,

$$v' = \tfrac{3}{2}x - \tfrac{3}{4} \quad \text{and} \quad v = \tfrac{3}{4}x^2 - \tfrac{3}{4}x$$

Since $y_p = 1 \cdot v$, this is also the expression for y_p. Consequently, the general solution is

$$y = \underbrace{c_1 + c_2 e^{-2x}}_{y_c} + \underbrace{\tfrac{3}{4}x^2 - \tfrac{3}{4}x}_{y_p} \qquad \blacksquare$$

EXERCISES FOR SECTION 5–1

In Exercises 1–6 determine the complementary solution of the nonhomogeneous equation.

1. $y'' + 3y' + 2y = 12e^x$

2. $y'' + 2y' - 8y = 4$

3. $y'' + 6y' + 9y = 9x + 2$

4. $y'' - 4y' + 4y = 5x^2 + e^{-x}$

5. $y'' + 4y = e^{-x}$

6. $y'' + 16y = 2\sin 3x$

7. Verify that $y_p = 2e^x$ is a particular solution of $y'' + 3y' + 2y = 12e^x$ and then find the general solution.

8. Verify that $y_p = x - \frac{4}{9}$ is a particular solution of $y'' + 6y' + 9y = 9x + 2$ and then find the general solution.

9. Verify that $y_p = \frac{1}{5}e^{-x}$ is a particular solution of $y'' + 4y = e^{-x}$ and then find the general solution.

10. Verify that $y_p = \frac{2}{7}\sin 3x$ is a particular solution of $y'' + 16y = 2\sin 3x$ and find the general solution.

11. Prove Theorem 5–2 for second-order linear differential equations.

12. Show that if y_1 is a solution of $a_2(x)y'' + a_1(x)y' + a_0y = f(x)$ and c is a constant, then cy_1 is a solution of $a_2(x)y'' + a_1(x)y' + a_0(x)y = cf(x)$.

Optional

Find the general solution of the differential equations in Exercises 13–26. Find the y_p by the method of reduction of order.

13. $y'' - 4y = 3$ **14.** $y'' - y = e^{3x}$

15. $y'' - y = x$ **16.** $y'' - y' - 6y = 5$

17. $y'' + 4y' + 4y = e^{-2x}$ **18.** $y'' + y' = x$

19. $y'' + 3y' = e^x$ **20.** $y'' - 2y' = e^{2x}$

21. $y'' + 3y' + 2y = 25$ **22.** $y'' + 4y' + 4y = 1$

23. $y'' + 4y' = e^{2x}$ **24.** $y'' - y' = e^{-x}$

25. $y'' - y' = \sin x$ **26.** $y'' + 3y' = \cos 2x$

5–2   THE METHOD OF UNDETERMINED COEFFICIENTS: DIFFERENTIAL FAMILIES

Two popular methods are used in the approach to undetermined coefficients: the differential family method explained in this section and the annihilator method explained in Section 5–2*. As noted in the Preface, only one of the two methods needs to be covered.

In this section we consider a method for determining y_p for nonhomogeneous differential equations that is based on the assumption that y_p has the same general form as the driving function $f(x)$. The basic idea of the method is to construct the general form of the particular solution from the driving function and then determine coefficients for y_p that allow it to satisfy the given differential equation. The process of evaluating the coefficients for y_p is called the **method of undetermined coefficients.** Before discussing the general approach, we consider some examples that illustrate the concept.

Example 1

Find a particular solution of $y'' + 3y' + 2y = e^{2x}$.

SOLUTION   Since the driving function is e^{2x}, we anticipate that perhaps there is a particular solution of the form $y_p = Ae^{2x}$. To see if there is such a y_p, we substitute $y_p = Ae^{2x}$, $y_p' = 2Ae^{2x}$, and $y_p'' = 4Ae^{2x}$ into the given equation and solve for A. Thus

$$4Ae^{2x} + 3(2Ae^{2x}) + 2(Ae^{2x}) = e^{2x}$$

yields $A = \frac{1}{12}$. We conclude that $y_p = \frac{1}{12}e^{2x}$ is a particular solution of $y'' + 3y' + 2y = e^{2x}$.   ∎

Example 2

Show that the method employed to find y_p in Example 1 fails for the equation

$$y'' + 3y' + 2y = 5x^2$$

SOLUTION   Since the driving function is $5x^2$, we assume that there is a particular solution of the form $y_p = Ax^2$. Substituting $y_p = Ax^2$, $y_p' = 2Ax$, and $y_p'' = 2A$ into the given equation yields

$$2A + 3(2Ax) + 2(Ax^2) = 5x^2$$

which is impossible to solve for A, so the method fails to give a particular solution.   ∎

Example 2 indicates that our method of choosing the general form for the y_p needs refinement. Toward this end, we introduce the concept of the differential family of a function. The technique is limited to driving functions that are sums of functions of the form $x^k e^{ax} \cos bx$ and $x^k e^{ax} \sin bx$, where k is a nonnegative integer and a and b are real numbers. Significantly, this same class of functions represents all possible solutions to homogeneous linear differential equations with constant coefficients.

DEFINITION

The **differential family** of a function is the set of functions consisting of the function itself and all of its linearly independent derivatives. Numerical coefficients are deleted in the differential family.

Example 3

Find the differential family of $f(x) = 2 \sin 3x$.

SOLUTION   Computing the derivatives of $f(x)$, we get

$$f'(x) = 6 \cos 3x$$
$$f''(x) = -18 \sin 3x$$
$$f'''(x) = -54 \cos 3x$$
$$f^{(4)}(x) = 162 \sin 3x$$
$$\vdots$$

We note that f and f' are linearly independent functions and that the other derivatives are constant multiples of one of these two functions. Thus the differential family of $2 \sin 3x$ is the set $\{\sin 3x, \cos 3x\}$.   ∎

We could continue to derive differential families for various functions, but we choose instead to list the differential families of some of the important driving functions in Table 5–1.

TABLE 5–1   Differential Families

FUNCTION	DIFFERENTIAL FAMILY
x^n (n a positive integer)	$\{1, x, x^2, \ldots, x^n\}$
e^{ax}	$\{e^{ax}\}$
$\sin bx$ or $\cos bx$	$\{\sin bx, \cos bx\}$
$x^n e^{ax}$	$\{e^{ax}, xe^{ax}, x^2 e^{ax}, \ldots, x^n e^{ax}\}$
$e^{ax} \sin bx$ or $e^{ax} \cos bx$	$\{e^{ax} \sin bx, e^{ax} \cos bx\}$

The significance of the concept of a differential family in solving nonhomogeneous linear differential equations is that **it can be shown that under appropriate conditions, the general form of y_p consists of a linear combination of the members of the differential family of the driving function.**

Example 4

Given $3y'' + 2y = f(x)$, write the general form of y_p as a linear combination of members of the differential family of the following driving functions:

(a) $f(x) = 5e^{2x} + 2x^3$
(b) $f(x) = x^2 e^{-3x}$
(c) $f(x) = 15 \sin 2x$

SOLUTION   (a) The differential family of $5e^{2x}$ is $\{e^{2x}\}$ and of $2x^3$ is $\{1, x, x^2, x^3\}$. Expressing y_p as a linear combination of these functions, we have

$$y_p = Ae^{2x} + B + Cx + Dx^2 + Ex^3$$

(b) The differential family of $x^2 e^{-3x}$ is $\{e^{-3x}, xe^{-3x}, x^2 e^{-3x}\}$. Therefore the general form of y_p is

$$y_p = Ae^{-3x} + Bxe^{-3x} + Cx^2 e^{-3x}$$

(c) The differential family of $15 \sin 2x$ is $\{\sin 2x, \cos 2x\}$, so the general form of y_p is

$$y_p = A \sin 2x + B \cos 2x$$ ∎

We now return to Example 2 and show that a particular solution of the equation $y'' + 3y' + 2y = 5x^2$ can be found by writing y_p as a linear combination of functions from the differential family of $5x^2$.

Example 5

Find a particular solution of $y'' + 3y' + 2y = 5x^2$.

SOLUTION   The differential family of $5x^2$ is $\{1, x, x^2\}$, so the general form of y_p is $A + Bx + Cx^2$. Using the method of undetermined coefficients, we substitute

$$y_p = A + Bx + Cx^2, \ y_p' = B + 2Cx, \ y_p'' = 2C$$

into the differential equation and solve for A, B, and C. Thus

$$(2C) + 3(B + 2Cx) + 2(A + Bx + Cx^2) = 5x^2$$

Expanding,

$$2C + 3B + 6Cx + 2A + 2Bx + 2Cx^2 = 5x^2$$

Collecting like terms, we have

$$(2A + 3B + 2C) + (2B + 6C)x + 2Cx^2 = 5x^2$$

Equating the coefficients of like terms yields the system of equations

$$
\begin{aligned}
2A + 3B + 2C &= 0 \\
2B + 6C &= 0 \\
2C &= 5
\end{aligned}
$$

which has the solution $A = \frac{35}{4}$, $B = -\frac{15}{2}$, $C = \frac{5}{2}$. Hence the required particular solution is

$$y_p = \frac{35}{4} - \frac{15}{2}x + \frac{5}{2}x^2$$ ∎

Example 6

Given $y'' - 4y = 8e^{2x}$, show that $y_p = Ae^{2x}$ fails to produce a particular solution of the equation.

SOLUTION   The differential family of the driving function is $\{e^{2x}\}$, so we let $y_p = Ae^{2x}$. The method of undetermined coefficients fails here since substituting $y_p = Ae^{2x}$ and $y_p'' = 4Ae^{2x}$ into the given equation yields

$$4Ae^{2x} - 4Ae^{2x} = 8e^{2x} \qquad \text{or} \qquad 0 = 8e^{2x}$$

which is impossible. The fact that the substitution of Ae^{2x} into the given equation reduces the left-hand side to 0, instead of a multiple of e^{2x}, means that e^{2x} is a solution of the corresponding homogeneous equation. To show this, we note that the auxiliary equation of $y'' - 4y = 0$ is $m^2 - 4 = 0$, which has roots ± 2. Thus the complementary solution is

$$y_c = c_1 e^{2x} + c_2 e^{-2x}$$

Clearly, Ae^{2x} is a constant multiple of the first term of y_c, which means Ae^{2x} and y_c are linearly dependent functions. ∎

The difficulty encountered in Example 6 can be averted by modifying the form of y_p. To accommodate driving functions that are constant multiples of terms of y_c, we multiply the duplicated function in y_p by the minimum positive integer power of x necessary to make y_c and y_p linearly independent functions. Thus in the previous example we should let $y_p = Axe^{2x}$. With the understanding that y_c and y_p must be linearly independent functions we formalize a rule for choosing the general form of y_p.

THE GENERAL FORM OF y_p

(a) The general form of y_p, except for the case explained in (b), consists of a linear combination of the functions in the differential family of the driving function.

(b) If any term of y_p is duplicated in y_c, then the duplicated terms in y_p must be multiplied by the minimum positive integer power of x required to make all terms in $y_c + y_p$ linearly independent functions.

Example 7

Given $y'' - 4y = f(x)$, write the general form of y_p for the following driving functions:

(a) $f(x) = 3xe^{-2x}$
(b) $f(x) = 7 \sin 3x - 10e^{-2x}$

SOLUTION   The solution of the corresponding homogeneous equation, from Example 6, is

$$y_c = c_1 e^{2x} + c_2 e^{-2x}$$

(a) The differential family of $3xe^{-2x}$ is $\{e^{-2x}, xe^{-2x}\}$, so we assume the form of y_p to be $y_p = Ae^{-2x} + Bxe^{-2x}$. However, Ae^{-2x} is a constant multiple of the second term of y_c, so we multiply this term by x to obtain

$$y_p = Axe^{-2x} + Bxe^{-2x}$$

But now the two terms of y_p are constant multiples of one another. To make these functions linearly independent, we multiply one of them by x. Thus the correct form of y_p is

$$y_p = Axe^{-2x} + Bx^2e^{-2x}$$

(b) The general form of y_p is

$$y_p = A \sin 3x + B \cos 3x + Ce^{-2x}$$

Comparing the terms of y_p to those of y_c, we note that Ce^{-2x} is a term of y_c. Therefore we multiply this term by x and write

$$y_p = A \sin 3x + B \cos 3x + Cxe^{-2x}$$

Notice that the functions $A \sin 3x$ and $B \cos 3x$ are not multiplied by x since they are not constant multiples of either term of y_c. ∎

Example 8

Use the method of undetermined coefficients to find the general solution of $y'' + 4y = 3 \cos x$.

SOLUTION   The solution of the corresponding homogeneous equation $y'' + 4y = 0$, which has roots $\pm 2i$, is

$$y_c = c_1 \sin 2x + c_2 \cos 2x$$

To find a particular solution of the given equation, we assume that the general form of y_p is

$$y_p = A \sin x + B \cos x$$

This is the correct expression since neither of these functions is in y_c. Substituting y_p and y_p'' into the given differential equation, we obtain

$$(-A \sin x - B \cos x) + 4(A \sin x + B \cos x) = 3 \cos x$$

Expanding and collecting like terms yields

$$3A \sin x + 3B \cos x = 3 \cos x$$

Equating the coefficients of like terms, we get the system of equations

$$3A = 0, 3B = 3$$

which has the solution $A = 0$ and $B = 1$. Hence $y_p = \cos x$, and the general solution is

$$y = \underbrace{c_1 \sin 2x + c_2 \cos 2x}_{y_c} + \underbrace{\cos x}_{y_p}$$ ∎

Example 9

Solve the differential equation

$$y'' + 3y' + 2y = e^{-x}$$

SOLUTION   The general solution of the corresponding homogeneous equation $y'' + 3y' + 2y = 0$ is $y_c = c_1 e^{-x} + c_2 e^{-2x}$. Since the driving function is in the y_c, the general form of y_p is $y_p = Axe^{-x}$. Substituting

$$y_p = Axe^{-x}, y'_p = -Axe^{-x} + Ae^{-x}, y''_p = Axe^{-x} - 2Ae^{-x}$$

into the given differential equation, we obtain

$$Axe^{-x} - 2Ae^{-x} - 3Axe^{-x} + 3Ae^{-x} + 2Axe^{-x} = e^{-x}$$

This yields the result $A = 1$. Therefore the general solution is

$$y = \underbrace{c_1 e^{-x} + c_2 e^{-2x}}_{y_c} + \underbrace{xe^{-x}}_{y_p}$$ ∎

COMMENT: The method of undetermined coefficients is effective for driving functions with finite differential families, but not for those with infinite differential families. Thus, equations with driving functions such as $x^{1/2}$ and sec x cannot be solved by this method.

EXERCISES FOR SECTION 5–2

1. Consider the differential equation $y''' + 2y'' + y' = f(x)$. Determine the y_p to be used if $f(x)$ equals each of the following:

   (a) x (b) $x + 2$ (c) $\sin x + x$
   (d) e^{-x} (e) xe^{-x} (f) $\sinh x$
   (g) $\sinh x + \cosh x$ (h) $x(1 + e^{-x})$

Solve the equations in Exercises 2–26.

2. $y'' - 4y' + 4y = e^x$

3. $y'' - 4y' + 4y = e^x + 1$

4. $y'' - 4y' + 4y = e^{2x}$

5. $y'' - 4y' + 4y = \sin x$

6. $y'' - 4y' + 4y = xe^{2x} + e^{2x}$

7. $y'' - 4y' + 4y = xe^{2x}$

8. $y'' + y = \sin 2x$

9. $y'' + 4y = \sin 2x$

10. $y'' + 4y' = \sin 2x$

11. $y'' - 2y' + 5y = \sin 2x$

12. $y'' - 2y' + 5y = e^x \cos 2x$

13. $y'' - y = \cosh 2x$

14. $y'' - y = \cosh x$

15. $y'' + y = x \cos x$

16. $y'' - 3y' + 2y = e^x + e^{2x} + e^{-x}$

17. $y'' - y' = x^2$

18. $y'' - y' = 3$

19. $y'' + 3y' = x + 3$

20. $y'' + 4y = 3$ **21.** $y'' + 4y = 3x$

22. $y''' - 3y' - 2y = \sin 2x$ **23.** $y^{(4)} - y = e^{-x}$

24. $y''' + 4y'' + 9y' + 10y = -e^x$ **25.** $y''' + y'' = 1$

26. $y''' + y'' - 2y = x^2 + 10 \cos 2x$

5–2* THE METHOD OF UNDETERMINED COEFFICIENTS: DIFFERENTIAL OPERATORS

As noted in Section 5–2, this section need not be covered if the method of differential families is preferred to the method of differential operators.

Differential Polynomial Operators

To simplify writing and solving systems of nth-order linear differential equations with constant coefficients, we introduce the concept of a differential polynomial operator.

DEFINITION

> By the operator D^n is meant the nth-order derivative operator d^n/dx^n. The positive integer n is called the **power** of the operator.

The differential operator D^n must operate on some n-times differentiable function. Thus when D^2 operates on $x^3 + 3x$, the expression $6x$ is obtained and we write

$$D^2(x^3 + 3x) = 6x$$

DEFINITION

> A linear combination of differentiable operators of the form
> $$b_n D^n + b_{n-1} D^{n-1} + \cdots + b_1 D + b_0$$
> is called an **nth-order polynomial operator** and is denoted $P(D)$, where $b_0, b_1, \ldots, b_n$ are constants.

To indicate that $P(D)$ is being applied to an n-times differentiable function y, we write $P(D)y$ and observe that

$$P(D)y = (b_n D^n + b_{n-1} D^{n-1} + \cdots + b_1 D + b_0)y$$
$$= b_n D^n y + b_{n-1} D^{n-1} y + \cdots + b_1 Dy + b_0 y$$

Recalling that $D^n = d^n/dx^n$, we see that $P(D)y$ means

$$P(D)y = b_n \frac{d^n y}{dx^n} + b_{n-1} \frac{d^{n-1} y}{dx^{n-1}} + \cdots + b_1 \frac{dy}{dx} + b_0 y$$

COMMENT: *Any linear differential equation with constant coefficients can be expressed in terms of a polynomial operator. For example,*

$$y'' + 3y' - y = 0 \quad \text{may be written as} \quad (D^2 + 3D - 1)y = 0$$

and

$$y''' - 4y' = \cos x \quad \text{may be written as} \quad (D^3 - 4D)y = \cos x$$

In general the nth-order linear differential equation with constant coefficients may be written

$$P(D)y = f(x)$$

There are limitations on the kinds of functions for which $P(D)y$ has meaning, such as the existence or continuity of some derivative. Functions for which $P(D)y$ has meaning are said to be **admissible** for that operator. For example, only twice differentiable functions are admissible for the operator $(D^2 + 1)$.

The differential operator $P(D)$ has the two properties

$$P(D)(y_1 + y_2) = P(D)y_1 + P(D)y_2$$
$$P(D)(cy) = cP(D)y$$

where y, y_1, and y_2 are admissible functions for $P(D)$, and c is any constant. These two properties are called the **linearity properties.** Hence $P(D)$ is called a **linear operator.** The algebraic laws of polynomial operators are included in the following definitions:

DEFINITION

Two operators $P_1(D)$ and $P_2(D)$ are said to be **equal** if, and only if, $P_1(D)y = P_2(D)y$ for all admissible functions y.

DEFINITION

The **sum** of two operators $P_1(D) + P_2(D)$ is obtained by first expressing P_1 and P_2 as linear combinations of the D operator and adding coefficients of like powers of D.

DEFINITION

By the **product** of two operators $P_1(D)P_2(D)$ is meant the equivalent operator obtained by using the operator $P_2(D)$ followed by $P_1(D)$. Thus the product $P_1(D)P_2(D)$ is interpreted to mean

$$[P_1(D)P_2(D)]y = P_1(D)[P_2(D)y]$$

To **expand** a product of operators, determine the effect of the product operator on any admissible function and then express it as a linear combination of powers of D. The technique is illustrated in the following example:

Example 1

Let $P_1 = 2D + 3$ and $P_2 = D - 5$. Find the expansion of the product operator $P_1 P_2$.

SOLUTION   Applying the product operator $P_1 P_2$ to a function y, we have

$$P_1 P_2(y) = P_1[P_2(y)]$$

Here

$$P_2(y) = (D - 5)y = \frac{dy}{dx} - 5y$$

and

$$
\begin{aligned}
P_1[P_2(y)] &= (2D + 3)\left[\frac{dy}{dx} - 5y\right] \\
&= 2D\left[\frac{dy}{dx} - 5y\right] + 3\left[\frac{dy}{dx} - 5y\right] \\
&= 2\frac{d^2y}{dx^2} - 7\frac{dy}{dx} - 15y \\
&= (2D^2 - 7D - 15)y
\end{aligned}
$$

which means that

$$(2D + 3)(D - 5) = 2D^2 - 7D - 15 \qquad \blacksquare$$

COMMENT: It can be shown that polynomial operators satisfy all of the laws of elementary algebra with regard to addition and multiplication, which means that polynomial operators may be manipulated in the same way we manipulate algebraic expressions.

Example 2

(a) If $P_1(D) = 3D^2 + 7D - 5$ and $P_2(D) = D^3 + 6D^2 - 2D - 3$, then

$$P_1(D) + P_2(D) = D^3 + 9D^2 + 5D - 8$$

(b) The product operator $(D^2 - 4)(D + 2)$ may be expanded and written as

$$(D^2 - 4)(D + 2) = D^3 + 2D^2 - 4D - 8$$

(c) The operator $D^3 - D^2 - 6D$ may be factored and written as

$$D^3 - D^2 - 6D = D(D - 3)(D + 2)$$ ■

Annihilators

DEFINITION

> Let $f(x)$ have n derivatives. Then the polynomial operator $P(D)$ is said to **annihilate** $f(x)$ if
>
> $$P(D)f(x) = 0$$

Example 3

(a) D^2 annihilates $3x + 1$ because $D^2(3x + 1) = 0$.

(b) $(D^2 + 1)$ annihilates $\sin x$ because

$$(D^2 + 1)(\sin x) = D^2(\sin x) + \sin x = -\sin x + \sin x = 0$$

(c) $(D + 1)$ annihilates e^{-x} because

$$(D + 1)e^{-x} = D(e^{-x}) + e^{-x} = -e^{-x} + e^{-x} = 0$$ ■

We note the obvious but important fact that every linear homogeneous differential equation with constant coefficients can be written in the operator form as $P(D)y = 0$. Thus, from the viewpoint of operators, **$P(D)$ annihilates every solution of $P(D)y = 0$.** Recall from Chapter 4 that all solutions of $P(D)y = 0$ are linear combinations of terms of the form $x^k e^{ax} \sin bx$ or $x^k e^{ax} \cos bx$. Thus *these are the only functions that a polynomial operator can annihilate*.

Example 4

(a) $(D - 1)$ annihilates e^x since e^x is a solution of $(D - 1)y = 0$.

(b) $(D^2 - 3D + 2)$ annihilates any linear combination of e^x and e^{2x} because $c_1 e^x + c_2 e^{2x}$ is a solution of $(D^2 - 3D + 2)y = 0$.

(c) $(D^2 + 1)$ annihilates any linear combination of $\sin x$ and $\cos x$ because $c_1 \sin x + c_2 \cos x$ is a solution of $(D^2 + 1)y = 0$.

(d) No polynomial-operator annihilator can be found for functions such as $\ln x$, $\sec x$, or $(x^2 + 1)^{1/2}$ because none of these is a solution of a linear homogeneous differential equation with constant coefficients. ∎

The determination of a polynomial-operator annihilator for a given function is like solving in reverse a linear homogeneous differential equation with constant coefficients.

Example 5

Find a polynomial annihilator for

$$3e^{2x} + e^{-x}$$

SOLUTION   The function $3e^{2x} + e^{-x}$ is a solution of a homogeneous linear differential equation whose auxiliary equation has the roots 2 and -1. Hence the auxiliary equation, in factored form, must be

$$(m - 2)(m + 1) = 0$$

Expanding, we get

$$m^2 - m - 2 = 0$$

The corresponding linear homogeneous equation is

$$(D^2 - D - 2)y = 0$$

Therefore the annihilator for $3e^{2x} + e^{-x}$ is $D^2 - D - 2$. ∎

Example 6

Find a polynomial annihilator for $5xe^{-3x}$.

SOLUTION   The function $5xe^{-3x}$ is a solution of a linear homogeneous differential equation having the repeated real roots -3, -3. (We know this because the exponential is multiplied by x.) The required auxiliary equation is then

$$(m + 3)^2 = 0$$

From the auxiliary equation we conclude that $(D + 3)^2$ is the annihilator of the function $5xe^{-3x}$. ∎

Example 7

Find a polynomial annihilator for $x + e^x \sin 2x$.

SOLUTION   The function x arises in the solution of a linear homogeneous differential equation whose auxiliary equation has a double root of 0. Also, the function $e^x \sin 2x$ implies an auxiliary equation having roots $1 \pm 2i$. Thus the assumed auxiliary equation has the form

$$m^2(m - 1 - 2i)(m - 1 + 2i) = 0$$

or, in expanded form,

$$m^4 - 2m^3 + 5m^2 = 0$$

Therefore $D^4 - 2D^3 + 5D^2$ is an annihilator of the given function.   ■

Examination of the preceding examples leads us to make the following observations concerning polynomial annihilators:

1. $(D - a)^{k+1}$ annihilates $x^k e^{ax}$.

2. $[(D - a)^2 + b^2]^{k+1}$ annihilates both $x^k e^{ax} \cos bx$ and $x^k e^{ax} \sin bx$.

3. A sum of terms is annihilated by a product of annihilators of the individual terms.

EXERCISES

Write the given differential equations, using operator notation, in Exercises 1–6.

1. $7y'' + 3y' + 4y = x^2$

2. $\dfrac{d^2y}{dx^2} - \dfrac{dy}{dx} - y = 0$

3. $\dfrac{d^2y}{dx^2} + 4y = 0$

4. $3y''' - 7y' + 4y = 0$

5. $y''' + 6y'' - 2y' = \sin x$

6. $2y^{(4)} + 9y^{(2)} - y' = e^{2x}$

Write the indicated operators in expanded form in Exercises 7–14.

7. $(D - 2)(D + 3)$

8. $(D - 2)(D + 2)$

9. $(D + 2)(D^2 - 5)$

10. $D(D + 1)(D + 3)$

11. $D^2(D^3 + 4D)$

12. $(D + 2)^3$

13. $(D + 1)(D - 1)^2$

14. $(D + 5)(D^2 - D)$

Factor the indicated operators in Exercises 15–22.

15. $D^2 - 9$

16. $D^2 - D - 6$

17. $D^3 + 6D^2 + 9D$

18. $D^3 - 10D^2$

19. $2D^2 + 3D - 2$

20. $D^3 - 3D^2 + 4$

21. $D^4 + D^3 - 2D^2 + 4D - 24$

22. $D^3 + 6D^2 + 12D + 8$

Find a polynomial annihilator for each function in Exercises 23–44.

23. e^{3x}

24. $6e^{-4x}$

25. $7e^{-x} + e^x$

26. $25e^{x/2}$

27. $9xe^{-3x} + e^{-3x}$ **28.** $9xe^{-3x} + e^x$

29. $1 + x + x^2$ **30.** $3 + xe^{5x}$

31. $4 - x^2 + e^x$ **32.** $x^3e^x + 3x$

33. $5 \sin x + 4 \cos x$ **34.** $e^{2x} + \sin 5x$

35. $x + 3 \sin 2x$ **36.** $3x \cos 2x$

37. $x \sin x$ **38.** $e^{2x} \sin x$

39. $e^{-x} \cos 2x$ **40.** $e^{3x}(3 \cos 2x - \sin 2x)$

41. $\sin^2 x$ **42.** $6 \sin x \cos x$

43. $\cos^2 x - \sin^2 x$ **44.** $5 + \cos^2 x$

The Method of Undetermined Coefficients

As proved in Theorem 5–1, the general solution of

$$P(D)y = f(x)$$

is given by $y = y_c + y_p$, where y_c is the general solution of $P(D)y = 0$, and y_p is any particular solution of the nonhomogeneous equation. In this section we shall discuss a technique for finding y_p when $f(x)$ is a linear combination of functions of the form $x^k e^{ax} \cos bx$ and $x^k e^{ax} \sin bx$ (k a nonnegative integer). The technique assumes that $f(x)$ can be annihilated by some polynomial operator $P(D)$ and is explained in the following example.

Example 8

Solve the differential equation

$$y'' + 3y' + 2y = e^{2x}$$

SOLUTION   In operator form this equation is

$$(D + 1)(D + 2)y = e^{2x}$$

and the solution of the corresponding homogeneous equation

$$(D + 1)(D + 2)y = 0$$

is

$$y_c = c_1e^{-x} + c_2e^{-2x}$$

Since the general solution of a nonhomogeneous equation is

$$y = y_c + y_p$$

we must find *one* particular solution. Observe that the driving function of the given equation is annihilated by the operator $(D - 2)$—that is,

$$(D - 2)e^{2x} = 0$$

Using this fact, we convert the given nonhomogeneous equation into a homogeneous equation that we can solve by the methods of Chapter 4. Applying $D - 2$ to both sides of the given equation, we get

$$(D - 2)(D + 1)(D + 2)y = 0$$

which has the solution

$$y = \underbrace{c_1 e^{-x} + c_2 e^{-2x}}_{y_c} + \underbrace{c_3 e^{2x}}_{y_p} \tag{5-4}$$

We conclude from Equation 5–4 that y_p must correspond to $c_3 e^{2x}$ since the first two terms constitute y_c. Using a different notation for the coefficient of e^{2x} to indicate that it is not arbitrary, we write

$$y_p = Ae^{2x}$$

To determine the numerical value of the coefficient, we use the fact that y_p must satisfy the equation

$$y'' + 3y' + 2y = e^{2x}$$

Substituting $y_p = Ae^{2x}$, $y_p' = 2Ae^{2x}$, and $y_p'' = 4Ae^{2x}$ into this equation, we get

$$4Ae^{2x} + 3(2Ae^{2x}) + 2(Ae^{2x}) = e^{2x}$$

from which $A = \frac{1}{12}$. Therefore the specific y_p is

$$y_p = \frac{1}{12} e^{2x}$$

and the general solution is

$$y = c_1 e^{-x} + c_2 e^{-2x} + \frac{1}{12} e^{2x} \qquad ■$$

The method used in Example 8 to find y_p is outlined here under the heading of the method of undetermined coefficients.

THE METHOD OF UNDETERMINED COEFFICIENTS

Given the nonhomogeneous differential equation

$$P(D)y = f(x)$$

where $f(x)$ is a function that can be annihilated. To find a y_p, proceed as follows:

1. *Determine the complementary solution y_c.*

2. *Determine an annihilator $P_1(D)$ for $f(x)$ and apply it to both sides of the equation $P(D)y = f(x)$ to obtain $P_1(D)P(D)y = 0$.*

3. *Solve the homogeneous differential equation $P_1(D)P(D)y = 0$. This solution will consist of y_c and terms due to the factor $P_1(D)$. The terms added because of this factor give the **form** of the y_p.*

4. *To find the specific y_p, substitute the general form of y_p into the given differential equation and solve for the unknown coefficients.*

Example 9

Solve $y'' + 3y' = 6x$.

SOLUTION The auxiliary equation $m^2 + 3m = 0$ has roots of -3 and 0, so

$$y_c = c_1 e^{-3x} + c_2$$

To annihilate $f(x) = 6x$, we operate on the given differential equation with D^2. Thus

$$D^2(D^2 + 3D)y = 0$$

or

$$D^3(D + 3)y = 0$$

Since the solution of this homogeneous equation is

$$y = \underbrace{c_1 e^{-3x} + c_2}_{y_c} + \underbrace{c_3 x + c_4 x^2}_{y_p}$$

we conclude that the general form of y_p is

$$y_p = Ax + Bx^2$$

Substituting $y_p = Ax + Bx^2$, $y_p' = A + 2Bx$, and $y_p'' = 2B$ into the given differential equation, we get

$$2B + 3(A + 2Bx) = 6x$$

Collecting terms and equating coefficients of like terms yields the system of equations

$$3A + 2B = 0$$
$$6B = 6$$

The solution of this system is $A = -\frac{2}{3}$, $B = 1$, which means that

$$y_p = -\frac{2}{3}x + x^2$$

The general solution of $y'' + 3y' = 6x$ is then

$$y = c_1 e^{-3x} + c_2 - \tfrac{2}{3}x + x^2 \qquad\blacksquare$$

If we were to solve a number of nonhomogeneous differential equations by using annihilators, we would eventually notice that the general form of y_p is predictable from $f(x)$ if y_c is known. The following rule may be used to find the general form of y_p.

THE GENERAL FORM OF y_p

(a) The general form of y_p, except for the case explained in (b), is the same as the most general function that will be annihilated by the annihilator of $f(x)$. Specifically, we cite the following cases:

$f(x)$	y_p
x^n, n a nonnegative integer	$A_0 + A_1 x + A_2 x^2 + \cdots + A_n x^n$
e^{ax}	$A e^{ax}$
$x^n e^{ax}$	$(A_0 + A_1 x + A_2 x^2 + \cdots + A_n x^n) e^{ax}$
$\sin bx$ or $\cos bx$	$A \sin bx + B \cos bx$
$e^{ax} \sin bx$ or $e^{ax} \cos bx$	$e^{ax}(A \sin bx + B \cos bx)$

(b) If any term of y_p is duplicated in y_c, then the duplicated terms in y_p must be multiplied by the minimum positive integer power of x required to make all terms in $y_c + y_p$ linearly independent functions.

Example 10

Given $y'' + 3y' + 2y = f(x)$, use the preceding rule to construct the general form of y_p for

(a) $f(x) = 5e^{2x}$
(b) $f(x) = 2x^3 + 7$
(c) $f(x) = 10e^{-2x} + 3 \cos 5x$

SOLUTION   The solution of the corresponding homogeneous equation is

$$y_c = c_1 e^{-x} + c_2 e^{-2x}$$

(a) $f(x) = 5e^{2x}$: Since e^{2x} is not a term in y_c, the correct form for y_p is

$$y_p = A e^{2x}$$

(b) $f(x) = 2x^3 + 7$: Here we assume that

$$y_p = A + Bx + Cx^2 + Dx^3$$

Since none of these terms is duplicated in y_c, this is the correct form for y_p.

(c) $f(x) = 10e^{-2x} + 3 \cos 5x$: In this case we assume that $y_p = Ae^{-2x} + B \cos 5x + C \sin 5x$. However, since e^{-2x} is a term in y_c, we multiply it by x and write

$$y_p = Axe^{-2x} + B \cos 5x + C \sin 5x$$

Notice that only the duplicated term is multiplied by x. ∎

Example 11

Find the general solution of

$$y''' + 4y' = x^2 + \sin x$$

SOLUTION   In operator form this equation may be written

$$D(D^2 + 4)y = x^2 + \sin x$$

The solution of the corresponding homogeneous equation $D(D^2 + 4)y = 0$ is

$$y_c = c_1 + c_2 \cos 2x + c_3 \sin 2x$$

Initially, we take the form of y_p to be $y_p = A + Bx + Cx^2 + D \sin x + E \cos x$. However, since a constant term occurs in y_c, we must use Rule (b) and write y_p as

$$y_p = Ax + Bx^2 + Cx^3 + D \sin x + E \cos x$$

The coefficients of the terms in the expression for y_p are found by substituting into the given differential equation. Since

$$
\begin{aligned}
y_p &= Ax + Bx^2 + Cx^3 + D \sin x + E \cos x \\
y_p' &= A + 2Bx + 3Cx^2 + D \cos x - E \sin x \\
y_p'' &= 2B + 6Cx - D \sin x - E \cos x \\
y_p''' &= 6C - D \cos x + E \sin x
\end{aligned}
$$

we have that

$$
\begin{aligned}
y_p''' + 4y_p' &= 6C - D \cos x + E \sin x + 4A + 8Bx + 12Cx^2 \\
&\quad + 4D \cos x - 4E \sin x \\
&= 4A + 6C + 8Bx + 12Cx^2 + 3D \cos x - 3E \sin x
\end{aligned}
$$

Therefore

$$4A + 6C + 8Bx + 12Cx^2 + 3D \cos x - 3E \sin x = x^2 + \sin x$$

We equate coefficients of like terms on both sides of this equation:

Constant terms	$4A + 6C = 0$
Coefficient of x	$8B = 0$
Coefficient of x^2	$12C = 1$
Coefficient of $\cos x$	$3D = 0$
Coefficient of $\sin x$	$-3E = 1$

From these equations the coefficients are found to be $A = -\frac{1}{8}$, $B = 0$, $C = \frac{1}{12}$, $D = 0$, and $E = -\frac{1}{3}$. Hence the general solution is

$$y = \underbrace{c_1 + c_2 \cos 2x + c_3 \sin 2x}_{y_c} \underbrace{- \frac{x}{8} + \frac{x^3}{12} - \frac{\cos x}{3}}_{y_p} \qquad \blacksquare$$

EXERCISES FOR SECTION 5–2*

1. Consider the differential equation $y''' + 2y'' + y' = f(x)$. Determine the y_p to be used if $f(x)$ equals each of the following:

   (a) x (b) $x + 2$ (c) $\sin x + x$
   (d) e^{-x} (e) xe^{-x} (f) $\sinh x$
   (g) $\sinh x + \cosh x$ (h) $x(1 + e^{-x})$

Solve the equations in Exercises 2–26.

2. $y'' - 4y' + 4y = e^x$ 3. $y'' - 4y' + 4y = e^x + 1$

4. $y'' - 4y' + 4y = e^{2x}$ 5. $y'' - 4y' + 4y = \sin x$

6. $y'' - 4y' + 4y = xe^{2x} + e^{2x}$ 7. $y'' - 4y' + 4y = xe^{2x}$

8. $y'' + y = \sin 2x$ 9. $y'' + 4y = \sin 2x$

10. $y'' + 4y' = \sin 2x$ 11. $y'' - 2y' + 5y = \sin 2x$

12. $y'' - 2y' + 5y = e^x \cos 2x$ 13. $y'' - y = \cosh 2x$

14. $y'' - y = \cosh x$ 15. $y'' + y = x \cos x$

16. $y'' - 3y' + 2y = e^x + e^{2x} + e^{-x}$ 17. $y'' - y' = x^2$

18. $y'' - y' = 3$ 19. $y'' + 3y' = x + 3$

20. $y'' + 4y = 3$ 21. $y'' + 4y = 3x$

22. $y''' - 3y' - 2y = \sin 2x$ 23. $y^{(4)} - y = e^{-x}$

24. $y''' + 4y'' + 9y' + 10y = -e^x$ 25. $y''' + y'' = 1$

26. $y''' + y'' - 2y = x^2 + 10 \cos 2x$

(NOTE: These are the same problems used in the exercise set for Section 5–2. The solutions are given in the answer section for Section 5–2.)

5–3   THE METHOD OF VARIATION OF PARAMETERS

The method of undetermined coefficients is limited to those equations whose driving functions are finite sums of functions of the form $x^k e^{ax} \cos bx$ and $x^k e^{ax} \sin bx$, where k is a nonnegative integer and a and b are real numbers. In this section we consider a method first introduced by LaGrange in 1774 that is applicable to all linear differential equations, including those with variable coefficients. However, to keep the discussion simple, we restrict the discussion to the second-order equation with constant coefficients of the form

$$b_2 y'' + b_1 y' + b_0 y = f(x) \tag{5-5}$$

Variation of Parameters

Let y_1 and y_2 be any two linearly independent solutions of the corresponding homogeneous equation for Equation 5–5. Assume that a particular solution of the equation may be written in the form

$$y_p = u y_1 + v y_2$$

where u and v are unknown functions to be determined. Since two functions are to be determined, we impose two conditions on y_p:

1. $u y_1 + v y_2$ must satisfy the given differential equation.

2. $u' y_1 + v' y_2 = 0$. This condition helps to simplify the computations.

To impose the first condition, substitute y_p' and y_p'' into the given differential equation. Thus

$$y_p' = u' y_1 + u y_1' + v' y_2 + v y_2'$$

Since we require that $u' y_1 + v' y_2 = 0$, y_p' becomes

$$y_p' = u y_1' + v y_2'$$

From this we compute y_p'' to be

$$y_p'' = u' y_1' + v' y_2' + u y_1'' + v y_2''$$

Substituting y_p, y_p', and y_p'' for y, y', and y'' in Equation 5–5, the differential equation becomes

$$b_2(u' y_1' + v' y_2' + u y_1'' + v y_2'') + b_1(u y_1' + v y_2') + b_0(u y_1 + v y_2) = f(x)$$
$$b_2(u' y_1' + v' y_2') + u(b_2 y_1'' + b_1 y_1' + b_0 y_1) + v(b_2 y_2'' + b_1 y_2' + b_0 y_2) = f(x)$$

Since y_1 is a solution of the corresponding homogeneous equation, the expression $(b_2 y_1'' + b_1 y_1' + b_0 y_1)$ is zero. Similarly, the coefficient of v is zero, so u' and v' satisfy the system of equations

$$u' y_1 + v' y_2 = 0$$
$$u' y_1' + v' y_2' = \frac{f(x)}{b_2}$$

This system may be solved for u' and v' by **Cramer's rule.** Thus

$$u' = \frac{\begin{vmatrix} 0 & y_2 \\ f(x)/b_2 & y_2' \end{vmatrix}}{\begin{vmatrix} y_1 & y_2 \\ y_1' & y_2' \end{vmatrix}} \qquad v' = \frac{\begin{vmatrix} y_1 & 0 \\ y_1' & f(x)/b_2 \end{vmatrix}}{\begin{vmatrix} y_1 & y_2 \\ y_1' & y_2' \end{vmatrix}}$$

Note that the denominator is the Wronskian of the two functions y_1 and y_2. Since y_1 and y_2 are linearly independent solutions, it follows from Theorem 4–3 that this Wronskian is nonzero. Hence

$$u' = -\frac{y_2 f(x)/b_2}{W} \qquad \text{and} \qquad v' = \frac{y_1 f(x)/b_2}{W}$$

Integrating, we get

$$u = -\int \frac{y_2 f(x)/b_2}{W} \, dx \qquad \text{and} \qquad v = \int \frac{y_1 f(x)/b_2}{W} \, dx$$

The particular solution is then

$$y_p = uy_1 + vy_2$$

> COMMENT: *It is the integration needed to obtain u and v that can make this method technically difficult to use. Thus despite the generality of the method, we use a simpler approach, such as the method of undetermined coefficients, when appropriate.*

Example 1

Solve the differential equation

$$y'' + y = \sec x$$

SOLUTION   Since two linearly independent solutions of the corresponding homogeneous equation are $\cos x$ and $\sin x$, the general solution of the given equation is

$$y = c_1 \cos x + c_2 \sin x + y_p$$

where $y_p = u \cos x + v \sin x$, and u' and v' are given by

$$u' = \frac{\begin{vmatrix} 0 & \sin x \\ \sec x & \cos x \end{vmatrix}}{\begin{vmatrix} \cos x & \sin x \\ -\sin x & \cos x \end{vmatrix}} \qquad v' = \frac{\begin{vmatrix} \cos x & 0 \\ -\sin x & \sec x \end{vmatrix}}{\begin{vmatrix} \cos x & \sin x \\ -\sin x & \cos x \end{vmatrix}}$$

Therefore $u' = -\tan x$ and $v' = 1$, from which $u = \ln |\cos x|$ and $v = x$. The general solution is

$$y = \underbrace{c_1 \cos x + c_2 \sin x}_{y_c} + \underbrace{\ln|\cos x| \cos x + x \sin x}_{y_p}$$

Notice that the method of undetermined coefficients could not be used to obtain the y_p, because sec x is not a solution of a homogeneous linear differential equation. ∎

Example 2

Solve the differential equation

$$y'' - y = xe^x$$

SOLUTION The corresponding homogeneous equation $y'' - y = 0$ has the general solution $y_c = c_1e^x + c_2e^{-x}$. (NOTE: Because of the nature of the driving function, the y_p could be found by the method of undetermined coefficients.) In the method of variation of parameters, let $y_1 = e^x$ and $y_2 = e^{-x}$ in the expressions for u' and v' to obtain

$$u' = \frac{\begin{vmatrix} 0 & e^{-x} \\ xe^x & -e^{-x} \end{vmatrix}}{\begin{vmatrix} e^x & e^{-x} \\ e^x & -e^{-x} \end{vmatrix}} \qquad v' = \frac{\begin{vmatrix} e^x & 0 \\ e^x & xe^x \end{vmatrix}}{\begin{vmatrix} e^x & e^{-x} \\ e^x & -e^{-x} \end{vmatrix}}$$

Therefore $u' = x/2$ and $v' = -xe^{2x}/2$, from which $u = x^2/4$ and $v = -(xe^{2x}/4) + (e^{2x}/8)$. The general solution is then

$$y = c_1e^x + c_2e^{-x} + \tfrac{1}{4}x^2e^x - \tfrac{1}{4}xe^x + \tfrac{1}{8}e^x$$

Finally, we note that c_1e^x and $\tfrac{1}{8}e^x$ can be combined as $(c_1 + \tfrac{1}{8})e^x = c_1'e^x$ and the general solution written as

$$y = c_1'e^x + c_2e^{-x} - \tfrac{1}{4}xe^x + \tfrac{1}{4}x^2e^x$$ ∎

EXERCISES FOR SECTION 5–3

In Exercises 1–10 solve the indicated equation by using the method of variation of parameters.

1. $y'' - y' - 2y = e^{2x}$

2. $y'' + 2y' + y = \dfrac{e^{-x}}{x}$

3. $y'' + 4y = \tan 2x$

4. $y'' + 4y = \tan^2 2x$

5. $y'' + 4y = \sin^2 2x$

6. $y'' - 3y' + 2y = \cos(e^{-x})$

7. $y'' - 4y' + 4y = \dfrac{e^{2x}}{x}$

8. $y'' + 6y' + 9y = \dfrac{e^{-3x}}{x^2 + 1}$

9. $y'' + 2y' + y = e^{-x}\ln x$

10. $y'' + 2y' + y = \dfrac{e^{-x}}{x^3}$

11. Verify that x and $1/x$ are solutions to the differential equation $x^2y'' + xy' - y = 0$ on $(0, \infty)$. Solve the differential equation

$$x^2y'' + xy' - y = x^2 \ln x$$

12. Solve the differential equation $x^2y'' + xy' - y = x^2$. (HINT: See Exercise 11.)

13. Verify that $y_1 = x$ and $y_2 = x \ln x$ are solutions of the corresponding homogeneous equation of $x^2y'' - xy' + y = \dfrac{1}{x}$ and then solve the differential equation.

14. Verify that x and e^x are solutions to $(1 - x)y'' + xy' - y = 0$ on $(1, \infty)$. Solve

$$(1 - x)y'' + xy' - y = (x - 1)^2e^{-x}$$

15. The second-order equation $a_2x^2y'' + a_1xy' + a_0y = f(x)$, where the a_i are constants, is called the **Cauchy–Euler equation.** Show that the substitution for the independent variable $x = e^v$ (or, equivalently, $v = \ln x$) reduces the Cauchy–Euler equation to

$$a_2\frac{d^2y}{dv^2} + (a_1 - a_2)\frac{dy}{dv} + a_0y = f(e^v)$$

which is a second-order equation with constant coefficients and can be solved by the methods of the previous sections.

Solve the equations in Exercises 16–20.

16. $x^2y'' + 4xy' + 2y = 0$

17. $x^2y'' - 2xy' + 2y = 4x + \sin(\ln x)$

18. $x^2y'' - xy' - 3y = x^2 \ln x$

19. $x^2y'' + xy' + 9y = 0$

20. $x^2y'' - 2xy' + 2y = (x - 1) \ln x$

21. By using the method of variation of parameters, show that the solution to $y'' + y = f(x)$ can be written in the form

$$y = c_1 \cos x + c_2 \sin x + \int_a^x f(t) \sin(x - t)\, dt$$

22. Solve the equation $y''' + y'' = 1$ by the method of variation of parameters extended to third-order equations. HINT: Let $y_p = uy_1 + vy_2 + wy_3$, where y_1, y_2, y_3 are linearly independent solutions of the corresponding homogeneous equation. Then, impose the following three conditions on u, v, and w:

    (i) $uy_1 + vy_2 + wy_3$ must satisfy the given differential equation
    (ii) $u'y_1 + v'y_2 + w'y_3 = 0$
    (iii) $u'y_1' + v'y_2' + w'y_3' = 0$

23. Solve the equation $y''' - 3y' - 2y = \sin 2x$ by the method of variation of parameters (see Exercise 22).

24. Using the hint for Exercise 22, derive formulas for u', v', w' when using the method of variation of parameters to solve the third-order nonhomogeneous equation

$$b_3y''' + b_2y'' + b_1y' + b_0y = f(x).$$

5–4 APPLICATIONS TO MECHANICAL SYSTEMS

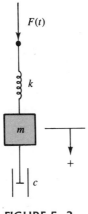

FIGURE 5–2

Nonhomogeneous linear differential equations have an immediate application to spring-mass-damper systems upon which external forces are acting. These forces may be applied to the support holding the spring or directly to the weight. In either case the applied force causes the weight to move up and down in some prescribed manner. Denoting the external applied force by $F(t)$, the differential equation of the system shown in Figure 5–2 is

$$m\frac{d^2x}{dt^2} + c\frac{dx}{dt} + kx = F(t) \qquad (5\text{–}6)$$

Periodic Forcing Functions

Consider the forcing function given by $F(t) = F_0 \cos \omega t$, where F_0 is the constant amplitude and ω is the **angular frequency.** Equation 5–6 becomes

$$m\frac{d^2x}{dt^2} + c\frac{dx}{dt} + kx = F_0 \cos \omega t \qquad (5\text{–}7)$$

Since the solution of this equation is dependent on the damping force, we consider separately the cases $c = 0$ (undamped) and $c \neq 0$ (damped).

Undamped Forced Vibration

If there is no damping force, Equation 5–7 becomes

$$m\ddot{x} + kx = F_0 \cos \omega t \qquad (5\text{–}8)$$

Assume the weight to be initially at rest and that $\omega \neq \omega_0 = \sqrt{k/m}$. It can be shown (say, by the method of undetermined coefficients) that the general solution of this equation is

$$x(t) = c_1 \cos \omega_0 t + c_2 \sin \omega_0 t + \frac{F_0}{m(\omega_0^2 - \omega^2)} \cos \omega t \qquad (5\text{–}9)$$

where $\omega_0 = \sqrt{k/m}$. Then, using $x(0) = \dot{x}(0) = 0$, we get

$$c_1 = -\frac{F_0}{m(\omega_0^2 - \omega^2)} \qquad \text{and} \qquad c_2 = 0$$

The desired solution is then

$$x(t) = \frac{F_0}{m(\omega_0^2 - \omega^2)} (\cos \omega t - \cos \omega_0 t) \qquad (5\text{–}10)$$

If we use the identity $\cos A - \cos B = -2 \sin \frac{1}{2}(A + B) \sin \frac{1}{2}(A - B)$, Equation 5–10 can be rewritten in the form

$$x(t) = \frac{2F_0}{m(\omega_0^2 - \omega^2)} \sin \tfrac{1}{2}(\omega_0 + \omega)t \sin \tfrac{1}{2}(\omega_0 - \omega)t \qquad (5\text{–}11)$$

Since the two sine functions are of different frequencies there will be times, especially when ω is close to ω_0, when their amplitudes will magnify one another and other times when they will cancel each other. (See Figure 5–3.) This magnification and cancellation occurs at regular intervals and is called **beats.** In acoustics these fluctuations can be heard when two tuning forks of different frequencies are set into vibration simultaneously. This same phenomenon occurs in electronics, where it is called **amplitude modulation.**

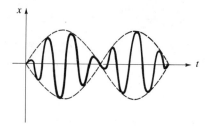

FIGURE 5–3

If the period of $F(t)$ is the same as the natural period of the system—that is, $\omega = \omega_0$—Equation 5–8 becomes

$$m\ddot{x} + kx = F_0 \cos \omega_0 t$$

The general solution is then

$$x = c_1 \cos \omega_0 t + c_2 \sin \omega_0 t + \frac{F_0}{2m\omega_0} t \sin \omega_0 t$$

Using the initial conditions $x(0) = \dot{x}(0) = 0$,

$$c_1 = 0 \qquad \text{and} \qquad c_2 = 0$$

and therefore the desired solution is

$$x = \frac{F_0}{2m\omega_0} t \sin \omega_0 t$$

The term $\sin \omega_0 t$ is periodic, but the amplitude is variable and increases without bound as $t \to \infty$. (See Figure 5–4.) A condition of this type is called **undamped resonance.**

In practice, damping forces are always present, and therefore oscillations do not build up without bound even though they may become very large. Probably the most famous case of the destructive nature of resonance occurred in 1940 when the

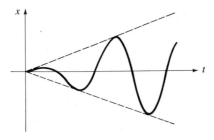

FIGURE 5–4

Tacoma Narrows suspension bridge was destroyed by resonant oscillations set up by the wind.

Example 1

A 4-lb weight stretches a spring 0.5 ft. An external force equal to $\frac{1}{2}\cos 8t$ is acting on the spring. Find the displacement equation if the weight is started from its equilibrium position with an upward velocity of 4 fps.

SOLUTION   The spring constant is $k = \frac{4}{0.5} = 8$, and the mass is $m = \frac{4}{32} = \frac{1}{8}$. Therefore the differential equation of the system is

$$\frac{1}{8}\ddot{x} + 8x = \frac{1}{2}\cos 8t$$

or

$$\ddot{x} + 64x = 4\cos 8t \qquad\qquad (5\text{--}12)$$

The complementary solution is

$$x_c = c_1 \cos 8t + c_2 \sin 8t$$

Since the driving function $4\cos 8t$ is found in x_c, we write x_p as

$$x_p = t(A\cos 8t + B\sin 8t)$$

from which

$$\dot{x}_p = A\cos 8t + B\sin 8t + t(-8A\sin 8t + 8B\cos 8t)$$

and

$$\ddot{x}_p = -16A\sin 8t + 16B\cos 8t - t(64A\cos 8t + 64B\sin 8t)$$

Substituting these values into Equation 5–12 and equating coefficients of like terms, we get $A = 0$, $B = \frac{1}{4}$, so

$$x_p = \frac{1}{4}t\sin 8t$$

Hence

$$x = c_1 \cos 8t + c_2 \sin 8t + \frac{1}{4}t \sin 8t$$

and

$$\dot{x} = -8c_1 \sin 8t + 8c_2 \cos 8t + \frac{1}{4}\sin 8t + 2t \cos 8t$$

Now using the initial conditions $x(0) = 0$ and $\dot{x}(0) = -4$, we get

$$c_1 = 0 \qquad \text{and} \qquad c_2 = -\frac{1}{2}$$

So

$$x = \frac{1}{4}(t - 2) \sin 8t$$

is the desired solution. This is shown graphically in Figure 5–5.

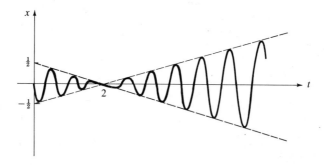

FIGURE 5–5

EXERCISES FOR SECTION 5–4

1. A 16-lb weight is suspended from a spring having a spring constant of 50. A force of $9 \cos 8t$ is applied to the spring. If the weight starts from rest in its equilibrium position, find the displacement equation.

2. An object having a mass of $\frac{3}{8}$ slug is suspended on a spring with $k = 24$. A force of $9 \sin 4t$ is applied. If the object is pulled down $\frac{1}{4}$ ft below the equilibrium point and released, what is the displacement equation?

3. An object having a mass of 1 slug is suspended from a spring for which $k = 1$. A force of $-9 \sin 2t$ is applied to the spring. If the object is pulled down 1 ft and released, what is the displacement equation?

4. A 32-lb weight is suspended from a spring having $k = 9$. The applied force varies with time in accordance with $12 \cos 3t$. If the weight is pulled down 4 ft and released, describe the ensuing motion.

5. A weight of 6 lb is attached to a spring having $k = 12$. Suppose a force equal to $3 \cos 8t$ is applied to the spring. Describe the motion if the weight starts from rest at the equilibrium point.

6. Draw a graph showing the motion of the object in Exercise 5.

7. Describe the output vibration of an undamped mechanical system with $k = 4$, $m = 1$ if the external excitation is $\sin 2t$. Assume $x(0) = \dot{x}(0) = 0$.

8. Starting with Equation 5–8, completely verify the solution given by Equation 5–10.

5–5 DAMPED FORCED VIBRATION

As noted in Section 5–4, the motion of a spring-mass system with viscous damping constant c and forcing function $F_0 \cos \omega t$ is described by the differential equation

$$m\ddot{x} + c\dot{x} + kx = F_0 \cos \omega t \qquad (5\text{--}13)$$

The corresponding homogeneous equation is

$$m\ddot{x} + c\dot{x} + kx = 0$$

which is the same as the equation describing damped vibration without a forcing function. In Section 4–7 we found that the solution of $m\ddot{x} + c\dot{x} + kx = 0$ is dependent upon the sign of $c^2 - 4mk$. Thus

if $c^2 - 4mk > 0$,

$$x_c(t) = c_1 e^{-(\alpha - \beta)t} + c_2 e^{-(\alpha + \beta)t} \qquad (5\text{--}14)$$

if $c^2 - 4mk = 0$,

$$x_c(t) = e^{-\alpha t}(c_1 t + c_2) \qquad (5\text{--}15)$$

if $c^2 - 4mk < 0$,

$$x_c(t) = e^{-\alpha t}(c_1 \cos \omega^* t + c_2 \sin \omega^* t)$$

$$= Ce^{-\alpha t} \cos(\omega^* t - \delta) \qquad (5\text{--}16)$$

where $\alpha = c/2m$, $\beta = \dfrac{1}{2m}\sqrt{c^2 - 4mk}$, $\omega^* = \dfrac{1}{2m}\sqrt{4mk - c^2}$, $C = \sqrt{c_1^2 + c_2^2}$,

and $\tan \delta = c_2/c_1$.

Since no constant multiple of the driving function $F_0 \cos \omega t$ is a term of $x_c(t)$, the particular solution is of the form

$$x_p(t) = A \cos \omega t + B \sin \omega t$$

Two differentiations yield

$$\dot{x}_p(t) = -\omega A \sin \omega t + \omega B \cos \omega t$$

$$\ddot{x}_p(t) = -\omega^2 A \cos \omega t - \omega^2 B \sin \omega t$$

Substituting into the differential equation and collecting the sine and cosine terms, we have

$$[(k - m\omega^2)A + \omega cB] \cos \omega t + [-\omega cA + (k - m\omega^2)B] \sin \omega t = F_0 \cos \omega t$$

Equating the coefficients of the sine and cosine terms on both sides of this equality yields

$$-\omega cA + (k - m\omega^2)B = 0$$

$$(k - m\omega^2)A + \omega cB = F_0$$

The solution of this system is

$$A = \frac{F_0(k - m\omega^2)}{(k - m\omega^2)^2 + \omega^2 c^2} \qquad B = \frac{F_0 \omega c}{(k - m\omega^2)^2 + \omega^2 c^2} \qquad (5\text{--}17)$$

Recalling that $\sqrt{k/m} = \omega_0$, we write A and B as

$$A = \frac{F_0 m(\omega_0^2 - \omega^2)}{m^2(\omega_0^2 - \omega^2)^2 + \omega^2 c^2} \qquad B = \frac{F_0 \omega c}{m^2(\omega_0^2 - \omega^2)^2 + \omega^2 c^2}$$

We choose to write $x_p(t)$ in the form

$$x_p(t) = C \cos(\omega t - \delta)$$

where $C = F_0 / \sqrt{m^2(\omega_0^2 - \omega^2)^2 + \omega^2 c^2}$ and $\tan \delta = \omega c / m(\omega_0^2 - \omega^2)$. Thus the general solution to Equation 5–13 is

$$x(t) = x_c(t) + x_p(t)$$

where $x_c(t)$ is one of Equations 5–14, 5–15, or 5–16.

In each case the term due to $x_c(t)$ is initially significant as time increases. For large values of t the motion is essentially described by $x_p(t)$. For this reason $x_p(t)$ is called the **steady-state solution,** and $x_c(t)$ is called the **transient solution.** After a sufficiently long time the motion of the mass will be a harmonic oscillation of the same frequency as the forcing function.

Recall that the amplitude of an **undamped** vibration becomes unbounded as the frequency ω approaches ω_0. On the other hand, a damped vibration will remain finite as ω is varied, although the maximum value of the vibration amplitude may be significant. The value of ω for which this maximum value is obtained is called **damped resonance.** (The definition of undamped resonance is given in Section 5–4.)

To determine the maximum amplitude of x_p, we should set $dC/d\omega$ equal to zero. (C is the amplitude of the oscillation.) Equivalently, we compute the derivative of the reciprocal of C^2 and equate it to zero. Thus

$$\frac{d(1/C^2)}{d\omega} = -4m^2\omega(\omega_0^2 - \omega^2) + 2\omega c^2 = 0$$

from which the condition on ω for maximum amplitude is

$$2m^2(\omega_0^2 - \omega^2) - c^2 = 0$$

Solving for ω and labeling it ω_r, we have

$$\omega_r = \sqrt{\omega_0^2 - c^2/2m^2} \qquad (5\text{–}18)$$

(NOTE: The condition for maximum amplitude is *not* $\omega = \omega_0$ as it was for un-damped resonance.)

If $c^2/2m^2 > \omega_0^2$, Formula 5–18 does not give a real value for ω_r. In this case the maximum output occurs when $\omega = 0$. Otherwise, a positive value of ω_r is obtained.

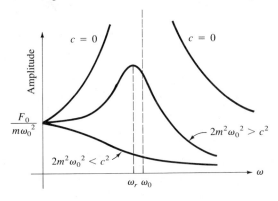

FIGURE 5–6

Figure 5–6 shows the variation of the amplitude with ω for the cases where the maximum output occurs for $\omega = 0$ and for the case of some positive value of ω_r. The values of the resonant and natural frequencies differ by a small number if the damping is small. In this case, damped resonance is roughly approximated by undamped resonance ω_0.

The Sliding Block "(skip)"

Consider a block of mass m on a horizontal plane with a constant frictional force that always opposes the movement of the block. A spring (with spring constant k) is attached to a support and connected to the block. The block is displaced from equilibrium and then released with zero initial velocity. See Figure 5–7.

The differential equation of the system is given by one of the two equations

$$m\ddot{x} + kx = \pm F$$

where F is a measure of the constant frictional force. The plus sign is chosen if the velocity dx/dt is negative, and the minus sign if dx/dt is positive. The general

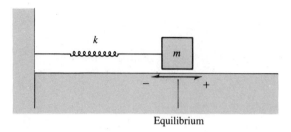

Equilibrium

FIGURE 5–7

solution of the differential equation over a specific interval will be one of the two equations

$$x(t) = c_1 \cos \omega_0 t + c_2 \sin \omega_0 t \pm \frac{F}{k}$$

where $\omega_0 = \sqrt{k/m}$ is the natural angular frequency of the system. The constants c_1 and c_2 are determined from the conditions at the time when the block changes directions. Since the velocity at the beginning of each time interval is 0 (the block stops instantaneously when changing directions), $c_2 = 0$, so that

$$x(t) = c_1 \cos \omega_0 t \pm \frac{F}{k}$$

The value of c_1 is determined at the beginning of each interval of time by using the value of the displacement of the weight from its rest position at that time.

The values of t for which the block changes direction can be found by using the condition that the instantaneous velocity is zero at these times. Since $dx/dt = \omega_0 c_1 \sin \omega_0 t = 0$, it follows that $\omega_0 t = n\pi$, $n = 0, 1, 2, \ldots$. That is, the block stops instantaneously for $t = n\pi/\omega_0$, $n = 0, 1, 2, \ldots$.

> COMMENT: The block will stop completely when the spring force at one of the extremes is insufficient to overcome the frictional force.

Example 1

A block of mass 4 slugs is attached to a spring with spring constant equal to 9 lb/ft. If the frictional force is 4.5 lb and the block is initially 3 ft from its position of equilibrium with zero initial velocity, determine the motion of the block and sketch the graph of the motion.

SOLUTION   For this system the natural frequency is $\sqrt{k/m} = \frac{3}{2}$. Hence the block will change directions at

$$\frac{3}{2}t_0 = 0, \frac{3}{2}t_1 = \pi, \frac{3}{2}t_2 = 2\pi, \ldots$$

or

$$t_0 = 0, t_1 = \frac{2\pi}{3}, t_2 = \frac{4\pi}{3}, \dots$$

For the interval $0 \le t \le 2\pi/3$ sec,

$$x(t) = c_1 \cos \frac{3}{2}t + \frac{1}{2}, \qquad x(0) = 3 \text{ ft}$$

which implies $c_1 = \frac{5}{2}$. Hence

$$x(t) = \frac{5}{2} \cos \frac{3}{2}t + \frac{1}{2}$$

Note that the value of $x(2\pi/3) = -2$ ft. Notice that the spring force is $2(9) = 18$, which is sufficient to overcome the frictional force.

For the interval $2\pi/3 \le t \le 4\pi/3$ sec,

$$x(t) = c_1 \cos \frac{3}{2}t - \frac{1}{2}, \qquad x\left(\frac{2\pi}{3}\right) = -2 \text{ ft}$$

which implies $c_1 = \frac{3}{2}$. Hence

$$x(t) = \frac{3}{2} \cos \frac{3}{2}t - \frac{1}{2}$$

Note that $x(4\pi/3) = 1$ ft. Here, the spring force is $(1)(9) = 9$ lb, which is again greater than the frictional force.

For the interval $4\pi/3 \le t \le 2\pi$ sec,

$$x(t) = c_1 \cos \frac{3}{2}t + \frac{1}{2}, \qquad x\left(\frac{4\pi}{3}\right) = 1 \text{ ft}$$

which implies $c_1 = \frac{1}{2}$. Hence

$$x(t) = \frac{1}{2} \cos \frac{3}{2}t + \frac{1}{2}$$

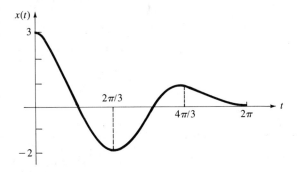

FIGURE 5–8

Note that $x(2\pi) = 0$, which is the equilibrium point. Since the spring force is zero at that point, it will be unable to overcome the frictional force, and the block will stop. (See Figure 5–8.) ■

Example 2

Repeat Example 1, but change the frictional force to 4 lb.

SOLUTION   Since the points at which the block changes direction are determined by the natural frequency, they are unchanged from Example 1.

For the interval $0 \le t \le 2\pi/3$ sec,

$$x(t) = c_1 \cos \frac{3}{2}t + \frac{4}{9}, \qquad x(0) = 3 \text{ ft},$$

which implies $c_1 = \frac{23}{9}$. Hence

$$x(t) = \frac{23}{9} \cos \frac{3}{2}t + \frac{4}{9}$$

Note that $x(2\pi/3) = -\frac{19}{9}$ ft and the spring force $= 19$ lb.

For the interval $2\pi/3 \le t \le 4\pi/3$ sec,

$$x(t) = c_1 \cos \frac{3}{2}t - \frac{4}{9}, \qquad x\left(\frac{2\pi}{3}\right) = -\frac{19}{9} \text{ ft}$$

which implies $c_1 = \frac{15}{9}$. Hence

$$x(t) = \frac{15}{9} \cos \frac{3}{2}t - \frac{4}{9}$$

Note that $x(4\pi/3) = \frac{11}{9}$ ft and the spring force $= 11$ lb.

For the interval $4\pi/3 \le t \le 2\pi$ sec,

$$x(t) = c_1 \cos \frac{3}{2}t + \frac{4}{9}, \qquad x\left(\frac{4\pi}{3}\right) = \frac{11}{9} \text{ ft}$$

which implies $c_1 = \frac{7}{9}$. Hence

$$x(t) = \frac{7}{9} \cos \frac{3}{2}t + \frac{4}{9}$$

Note that $x(2\pi) = -\frac{1}{3}$ ft. The spring force at that spot is 3 lb, which is insufficient to overcome the frictional force of 4 lb. Hence the block stops at $x = -\frac{1}{3}$ ft from the equilibrium point. ■

EXERCISES FOR SECTION 5–5

1. Find the transient and steady-state oscillations of a mechanical system with $m = 9$, $c = 5$, and $k = 16$, with external force $10 \cos t$, if the initial displacement and velocity are zero.

2. Repeat Exercise 1 with $m = 1$, $c = 2$, $k = 2$, with external force $5 \sin t$.

3. Find the resonant frequencies for the systems of Exercises 1 and 2.

4. Compute the maximum amplitude of an output vibration corresponding to the resonant frequency.

5. Compute the phase shift of the output vibration as a function of input frequency.

6. Determine the motion of a block of 1 slug on a horizontal plane attached to a spring with spring constant 1. The friction force is $\frac{3}{8}$ lb, and the block is initially displaced 2 ft from equilibrium.

7. Repeat Exercise 6 with $m = 4$, $k = 9$, $F = 4.5$, and $x(0) = 1.25$.

8. Repeat Exercise 6 with $m = 4$, $k = 9$, $F = 7$, and $x(0) = 2$.

5–6 ELECTRICAL CIRCUITS

Another application of second-order linear differential equations with constant coefficients is to electrical circuits. Every circuit has three passive elements that behave to impede the flow of current, the change in current, or the change in voltage. In elementary circuit analysis these elements are considered as being "lumped" components called resistance R, inductance L, and capacitance C. These three constants are essentially defined by the following three equations:

$$Ri_R = v_R \qquad L\frac{di_L}{dt} = v_L \qquad C\frac{dv_C}{dt} = i_C$$

where i is current and v is voltage. In this section we discuss series arrangements of R, L, and C components.

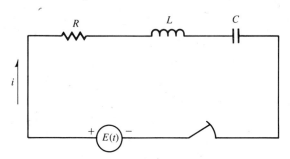

FIGURE 5–9

The governing physical law for an *RLC* series circuit is given by Kirchhoff's law, which says that the driving voltage to the circuit is equal to the sum of the voltage drops around the circuit. In equation form

$$L\frac{di}{dt} + Ri + \frac{1}{C}\int_{-\infty}^{t} i(t^*)\,dt^* = E(t) \qquad (5\text{--}19)$$

where $E(t)$ is the driving voltage. This equation relates the response (current) to the excitation (input voltage). Because $i = dq/dt$, Equation 5–19 is usually written as

$$L\frac{d^2q}{dt^2} + R\frac{dq}{dt} + \frac{1}{C}q = E(t) \qquad (5\text{--}20)$$

Example 1

The circuit shown in Figure 5–9 has $L = 1$ H, $R = 300$ Ω, and $C = 5 \times 10^{-5}$ f. At $t = 0$ the switch is closed to a 40-v battery. Find the charge $q(t)$ on the capacitor, and the current $i(t)$, for any $t > 0$. Assume $q(0) = i(0) = 0$.

SOLUTION   The differential equation governing the charge is given by

$$\frac{d^2q}{dt^2} + 300\frac{dq}{dt} + 20000\,q = 40$$

The auxiliary equation of the corresponding homogeneous equation is $m^2 + 300m + 20000 = 0$, which factors into

$$(m + 100)(m + 200) = 0$$

Thus

$$q_c = c_1e^{-100t} + c_2e^{-200t}$$

The form of q_p is $q_p = A$, so

$$20000A = 40$$
$$A = 2 \times 10^{-3} = 0.002$$

The general form of the expression for the charge is

$$q(t) = c_1e^{-100t} + c_2e^{-200t} + 0.002$$

and for the current is

$$i(t) = \frac{dq}{dt} = -100c_1e^{-100t} - 200c_2e^{-200t}$$

Imposing the initial conditions gives

$$c_1 + c_2 + 0.002 = 0$$
$$c_1 + 2c_2 = 0$$

from which $c_1 = -0.004$ and $c_2 = 0.002$. This gives the following expressions for $q(t)$ and $i(t)$:

$$q(t) = 0.002(-2e^{-100t} + e^{-200t} + 1)$$
$$i(t) = 0.4(e^{-100t} - e^{-200t})$$ ∎

We see from Equation 5–20 that the differential equation of a series *RLC* circuit is a second-order linear differential equation with constant coefficients. Thus the *RLC* circuit is the electrical analog of the spring-mass-damper system.

MECHANICAL

$$m\frac{d^2x}{dt^2} + c\frac{dx}{dt} + kx = F(t)$$

Driving force, $F(t)$

Displacement, $x(t)$

Mass, m

Damping constant, c

Spring constant, k

Damping factor, $\alpha = c/2m$

$$\beta = \frac{1}{2m}\sqrt{c^2 - 4mk}$$

Natural frequency, $\omega_0 = \sqrt{k/m}$

$$\omega^* = \frac{1}{2m}\sqrt{4mk - c^2}$$

Kinetic energy of mass, $\frac{1}{2}m\left(\frac{dx}{dt}\right)^2$

Potential energy of mass, $\frac{1}{2}kx^2$

Heat loss, $\int_0^t c\left(\frac{dx}{dt}\right)^2 dt$

ELECTRICAL

$$L\frac{d^2q}{dt^2} + R\frac{dq}{dt} + \frac{1}{C}q = E(t)$$

Excitation voltage, $E(t)$

Charge, $q(t)$

Inductance, L

Resistance, R

Reciprocal of capacitance, $1/C$

Damping factor, $a = R/2L$

$$b = \sqrt{(R/2L)^2 - 1/LC}$$

Natural frequency, $\omega_0 = \sqrt{1/LC}$

$$\omega^* = \sqrt{1/LC - (R/2L)^2}$$

Magnetic energy of inductance, $\frac{1}{2}L\left(\frac{dq}{dt}\right)^2$

Electric potential energy in C, $\frac{1}{2C}q^2$

Power loss, $\int_0^t R\left(\frac{dq}{dt}\right)^2 dt$

COMMENT: *If the excitation voltage is differentiable, then Equation 5–19 can be converted into a differential equation with the current as the unknown function.*

$$L\frac{d^2i}{dt^2} + R\frac{di}{dt} + \frac{i}{C} = \frac{dE}{dt} \qquad (5-21)$$

If, for example, the excitation voltage is $V_0 \sin \omega t$, then $dE/dt = V_0\omega \cos \omega t$, and Equation 5–21 can be compared to Equation 5–6. The electrical circuit with this type of sinusoidal input is therefore essentially solved in Sections 5–4 and 5–5.

We divide the discussion of *RLC* circuits with sinusoidal input into transient current and steady-state current.

Transient Current

Case 1   *Overdamped.* If $(R/2L)^2 - 1/LC > 0$, the transient current has the form

$$i_c = c_1 e^{-(\alpha-\beta)t} + c_2 e^{-(\alpha+\beta)t}$$

(See Equation 5–14.) Figure 5–10 shows the transient current for some typical initial conditions.

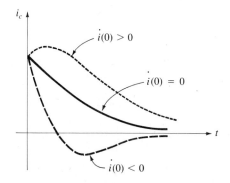

FIGURE 5–10

Case 2   *Critically Damped.* If $(R/2L)^2 - 1/LC = 0$, then the solution is

$$i_c(t) = e^{-\alpha t}(c_1 t + c_2)$$

(See Equation 5–15.) The graph of the transient current is of the same general form as for the overdamped case.

Case 3   *Underdamped.* If $(R/2L)^2 - 1/LC < 0$, then

$$i_c = e^{-\alpha t}(c_1 \cos \omega^* t + c_2 \sin \omega^* t)$$

which may be put into the form

$$i_c = Ce^{-\alpha t} \cos (\omega^* t - \delta)$$

where

$$C = \sqrt{c_1^2 + c_2^2} \qquad \text{and} \qquad \delta = \text{Tan}^{-1} c_2/c_1$$

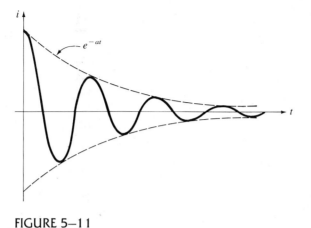

FIGURE 5–11

(See Equation 5–16.) Figure 5–11 shows a typical underdamped case for the initial conditions $i(0) = 1$, $\dot{i}(0) = 0$.

COMMENT: *The graphs of the overdamped and critically damped transients cut the t-axis at most once, while the underdamped transient is zero infinitely often.*

Steady-State Current

The particular solution to the nonhomogeneous equation corresponding to $i_c = 0$ is called the **steady-state current** and is denoted by i_{ss}.* Let

$$i_{ss} = A \cos \omega t + B \sin \omega t$$

Using the analog of Equation 5–17,

$$A = \frac{-E(\omega L - 1/\omega C)}{(\omega L - 1/\omega C)^2 + R^2} \qquad B = \frac{ER}{(\omega L - 1/\omega C)^2 + R^2}$$

The quantity $(\omega L - 1/\omega C)$ is called the **reactance** and is denoted by X. The quantity $\sqrt{X^2 + R^2}$ is called the **impedance** and is denoted by Z. Using these notations,

$$A = -\frac{EX}{Z^2} \qquad \text{and} \qquad B = \frac{ER}{Z^2}$$

and therefore

$$i_{ss} = \frac{E}{Z^2}(-X \cos \omega t + R \sin \omega t)$$

---

*In general the steady-state current is the value of the current as $t \to \infty$.

To express the steady-state current in the form $C \cos(\omega t - \delta)$, which displays the amplitude of the output oscillation C, and the phase shift δ, we have the following:

$$C = \sqrt{A^2 + B^2} = \sqrt{(E^2X^2 + E^2R^2)/Z^4} = \frac{E}{Z}; \quad \delta = \text{Tan}^{-1}\left(-\frac{R}{X}\right)$$

Total Current

In summary, the total output current is given by

Output current = Transient current + Steady-state current

$$i = c_1 e^{-(\alpha + \beta)t} + c_2 e^{-(\alpha - \beta)t} + \frac{E}{Z} \cos\left(\omega t + \text{Tan}^{-1}\frac{R}{X}\right) \quad \textbf{Overdamped}$$

$$i = e^{-\alpha t}(c_1 + c_2 t) + \frac{E}{Z} \cos\left(\omega t + \text{Tan}^{-1}\frac{R}{X}\right) \quad \textbf{Critically damped}$$

$$i = e^{-\alpha t} c_3 \cos(\omega^* t - c_4) + \frac{E}{Z} \cos\left(\omega t + \text{Tan}^{-1}\frac{R}{X}\right) \quad \textbf{Underdamped}$$

One other interesting case occurs when $R = 0$ (this corresponds to undamped forced vibration in Equation 5–8).

$$i = c_1 \cos \omega_0 t + c_2 \sin \omega_0 t + \frac{E}{X} \cos \omega t \quad \textbf{No damping}$$

The initial conditions $i(0) = i(0) = 0$ give $c_1 = -(E/X)$, $c_2 = 0$. Thus the total current for the no-damping ($R = 0$) case is

$$i = \frac{E}{X}(\cos \omega t - \cos \omega_0 t)$$

which, as in Section 5–4, may be put into the form

$$i = 2\frac{E}{X}\left[\sin\left(\frac{\omega_0 + \omega}{2}\right)t \sin\left(\frac{\omega_0 - \omega}{2}\right)t\right]$$

This is the "beats" solution shown in Figure 5–3.

Resonance

The amplitude of the steady-state current is given by $\alpha = E/Z$ and is therefore dependent upon E, L, C, R, and ω. That is,

$$\alpha = \frac{E}{Z} = \frac{E}{\sqrt{(\omega L - 1/\omega C)^2 + R^2}}$$

Conditions for which the amplitude is a maximum are called **resonance.** For a fixed value of E the output is maximized by minimizing the impedance as follows:

1. For fixed L, C, and ω, by making $R = 0$.

2. For fixed R, C, and ω, by making $L = 1/(\omega^2 C)$.

3. For fixed L, R, and ω, by making $C = 1/(\omega^2 L)$.

4. For fixed L, R, and C, by making $\omega = 1/\sqrt{LC}$.

In summary, resonance conditions occur in the RLC circuit for a resistance as small as possible and with the condition $\omega^2 LC = 1$.

Example 2

Using initial conditions $i(0) = 0$, $\dot{i}(0) = 0$, find the current in an RLC circuit if
(a) $L = 1$, $R = 0$, and $C = 1$ with an input voltage $E = \sin t$.
(b) $L = 1$, $R = 0$, $C = 1/1.1025$, and $E = (\sin 0.95t)/0.95$.
(c) $L = 1$, $R = 0.2$, $C = 1/1.01$, and $E = \sin t - 0.05 \cos t$.

SOLUTION   The three differential equations that must be solved with initial conditions $i(0) = 0$; $\dot{i}(0) = 0$ are
(a) $\ddot{i} + i = \cos t$.
(b) $\ddot{i} + 1.1025i = \cos 0.95t$.
(c) $\ddot{i} + 0.2\dot{i} + 1.01i = \cos t + 0.05 \sin t$.
We recognize Case (a) as undamped resonance since $R = 0$ and $\omega = \omega_0 = 1$. The general solution is found to be

$$i = c_1 \cos t + c_2 \sin t + \frac{1}{2}t \sin t$$

Evaluating the constants from the initial conditions, we obtain $c_1 = c_2 = 0$, so that $i = (t \sin t)/2$, which becomes unbounded with time.
The general solution to Case (b) is found to be

$$i = c_1 \cos 1.05t + c_2 \sin 1.05t + 5 \cos 0.95t$$

The initial conditions give $c_1 = -5$ and $c_2 = 0$, from which

$$i = 5(\cos 0.95t - \cos 1.05t) = 10 \sin (0.05t) \sin t.$$

This is the "beats" solution.
The general solution to Case (c) is

$$i = e^{-0.1t}(c_1 \cos t + c_2 \sin t) + 5 \sin t$$

The initial conditions give $c_1 = 0$, $c_2 = -5$, so that

$$i = (5 - 5e^{-0.1t}) \sin t = 5(1 - e^{-0.1t}) \sin t$$

Note that the amplitude of the "beats" solution in Case (b) is exactly twice that of the steady-state solution of the damped current in Case (c). ∎

EXERCISES FOR SECTION 5–6

In Exercises 1–4 find the charge $q(t)$ and the current $i(t)$ for the indicated series RLC circuits on the given time interval. Assume $q(0) = i(0) = 0$.

1. $R = 200\ \Omega$, $L = 1\ H$, $C = 10^{-4}\ f$, $E(t) = 3\ v$; $t \geq 0$

2. $R = 8000\ \Omega$, $L = 1\ H$, $C = 1/15 \times 10^{-6}\ f$, $E(t) = 30\ v$; $t \geq 0$

3. $R = 17{,}000\ \Omega$, $L = 6\ H$, $C = 1/12 \times 10^{-6}\ f$, $E(t) = 96(3 - t)\ v$; $0 \leq t \leq 36$

4. $R = 1500\ \Omega$, $L = 1\ H$, $C = 2 \times 10^{-6}\ f$, $E(t) = 20(5 - t)\ v$; $0 \leq t \leq 5$

5. If the resistance in a circuit is close to zero, the term for resistance is ignored and the circuit is called an *LC* circuit. Find the expressions for $q(t)$ and $i(t)$ in an *LC* circuit if $L = 1\ H$, $C = \frac{1}{9}\ f$, and $E(t) = 10 \sin t$. Assume $q(0) = i(0) = 0$. What is the transient current?

6. Find $q(t)$ and $i(t)$ for an *LC* circuit in which $L = 1\ H$, $C = \frac{1}{9}\ f$, and $E(t) = e^{-t}\ v$. Assume $q(0) = i(0) = 0$. What is the transient current?

Find the transient and steady-state currents in the RLC series circuits of Exercises 7–13.

7. $R = 20$, $L = 10$, $C = \frac{1}{10}$, $E = 40 \sin t$.

8. $R = 8$, $L = 2$, $C = \frac{1}{40}$, $E = 40 \sin 2t$.

9. $R = 20$, $L = 10$, $C = \frac{1}{50}$, $E = 100 \sin t$.

10. $R = 200$, $L = 100$, $C = 5 \times 10^{-3}$, $E = 500 \sin t$.

11. $R = 20$, $L = 5$, $C = 0.01$, $E = 850 \sin 4t$.

12. $R = 60$, $L = 10$, $C = 0.02$, $E = 2600 \sin t$.

13. $R = 6$, $L = 1$, $C = 0.04$, $E = 24 \cos 5t$.

14. Show that the current in a series *RLC* circuit satisfies the relation

$$i(0) = \frac{1}{L}\left[E(0) - Ri(0) - \frac{1}{C}q(0)\right]$$

where $q(0)$ is the charge in the capacitor at $t = 0$.

REVIEW EXERCISES FOR CHAPTER 5

In Exercises 1–4 use the method of reduction of order to find y_p. Write the general solution of the given nonhomogeneous equation.

1. $y'' + 3y' + 2y = e^x$ 2. $y'' + 3y' = 5$

3. $y'' - y' = \cos x$ 4. $y'' + 4y = \sin x$

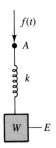

FIGURE 5–12

Use the method of undetermined coefficients to find y_p in Exercises 5–10. Write the general solution of the equation, or the particular solution if initial conditions are given.

5. $y'' + 3y' = 5e^{2x}$ **6.** $y'' + y = x^2$

7. $x'' + 2x' - 3x = 2e^t$ **8.** $y'' + y' - 6y = 10e^{2x} - 18e^{3x}$

9. $y'' + 8y' + 16y = 8e^{-2x}$, $y(0) = 2$, $y'(0) = 0$

10. $x'' - 5x' = t - 2$, $x(0) = 0$, $x'(0) = 0$

In Exercises 11–14 use the method of variation of parameters to solve the equations.

11. $y'' - y = e^x$ **12.** $y'' + y = \tan x$

13. $y'' + y = \sec^3 x$ **14.** $y'' + 2y' + 2y = e^{-x} \ln x$

15. A 32-lb weight is attached to a spring having a spring constant of 4. A driving force equal to 16 sin 2t is applied at point A (see Figure 5–12). If there is no damping force and the weight is initially at rest in the equilibrium position, determine the equation for the displacement of the weight.

16. A 16-lb object is attached to a spring whose spring constant is 10 lb/ft. The object experiences a damping force equal to twice its velocity. The system is driven by a

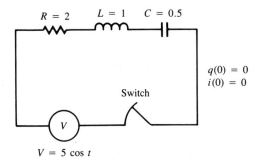

FIGURE 5–13

force of 5 cos 2*t*, applied at point *A* (see Figure 5–12). Determine the equation for the displacement of the object if it is initially at rest in the equilibrium position.

17. Find the transient and steady-state currents in the circuit of Figure 5–13.

18. Show that the current in a series *LC* circuit satisfies the relation

$$\dot{i}(0) = \frac{E(0)}{L} - \frac{q(0)}{LC}$$

where $q(0)$ is the charge in the capacitor at $t = 0$, and $E(0)$ is the initial voltage.

6

The Laplace Transform

The importance of linear differential equations should be apparent from the discussion in the preceding chapters. The Laplace transform, introduced in this chapter, is another device for solving linear differential equations which, as you will find in Chapter 7, is particularly useful if the driving function is discontinuous. In Chapter 9 the Laplace transform is used to solve systems of differential equations.

6–1   DEFINITION AND GENERAL PROPERTIES

The Laplace transform is only one of a class of operations called "transforms." The general idea of a transform is that of a pairing of functions where, to be useful, there must exist some sort of uniqueness in the pairing. For example, a function may be considered as paired with its derivative function. The derivative transform is unique in the sense that two functions with the same derivative differ at most by a constant. Another familiar transformation is the indefinite integral of a function. Recall that the indefinite integral of $f(t)$ is given by

$$I\{f(t)\} = \int_0^x f(t)\, dt$$

The Laplace transform, which can be considered as an extension of the idea of the indefinite integral transform, is defined as follows.

DEFINITION

> The **Laplace transform** of $f(t)$, if it exists, is denoted by $\mathscr{L}\{f(t)\}$ and defined by
>
> $$\mathscr{L}\{f(t)\} = \int_0^\infty e^{-st} f(t)\, dt \qquad (6\text{--}1)$$
>
> where s is a real number called a **parameter** of the transform.

We note that the Laplace transform takes a function $f(t)$ and transforms it into a function $F(s)$ of the parameter s. More generally, we represent functions of t by lowercase letters such as f, g, and h, and represent their respective Laplace transforms by the corresponding capital letters F, G, and H. Thus we also write $\mathscr{L}\{f(t)\} = F(s)$ or

$$F(s) = \int_0^\infty e^{-st} f(t)\, dt$$

The defining equation for the Laplace transform is an improper integral, the "improperness" arising because the upper limit is unbounded. Improper integrals of this kind are defined by

$$\int_0^\infty e^{-st} f(t)\, dt = \lim_{T \to \infty} \int_0^T e^{-st} f(t)\, dt \qquad (6\text{--}2)$$

Therefore, the existence of the Laplace transform of f depends upon the existence of this limit. We can find the Laplace transform of several elementary functions directly from the definition.

Example 1

Show that the Laplace transform of $f(t) = 1$ is given by

$$\mathcal{L}\{1\} = \frac{1}{s}, \qquad s > 0$$

SOLUTION   By definition we have

$$\mathcal{L}\{1\} = \int_0^\infty e^{-st}[1] \, dt = \lim_{T \to \infty} \int_0^T e^{-st} \, dt$$

To determine if this limit exists, we must consider three separate cases involving the parameter s.

1.   When $s < 0$, the exponent of e is positive for $t > 0$. Therefore

$$\lim_{T \to \infty} \int_0^T e^{-st} \, dt = \lim_{T \to \infty} \frac{e^{-st}}{-s} \Big|_0^T = \lim_{T \to \infty} \left[-\frac{1}{s} e^{-sT} + \frac{1}{s} \right] = \infty$$

which means the integral diverges.

2.   When $s = 0$, the integral becomes

$$\lim_{T \to \infty} \int_0^T dt = \lim_{T \to \infty} t \Big|_0^T = \lim_{T \to \infty} T = \infty$$

3.   When $s > 0$, the exponent is negative for $t > 0$, and therefore

$$\mathcal{L}\{1\} = \lim_{T \to \infty} \left[-\frac{1}{s} e^{-sT} + \frac{1}{s} \right] = \frac{1}{s} \qquad \blacksquare$$

The case considered in Example 1 is typical; that is, the Laplace transform is usually valid only over some interval of s. However, this limitation in no way reduces the effectiveness of the Laplace transform in applications and, in most cases, the limitation on the domain of $\mathcal{L}\{f(t)\}$ will go almost unnoticed. The Laplace transforms of some of the important elementary functions are derived in the following examples.

Example 2

Show that

$$\mathcal{L}\{e^{kt}\} = \frac{1}{s - k}, \qquad s > k \tag{6–3}$$

SOLUTION

$$\mathcal{L}\{e^{kt}\} = \int_0^\infty e^{-st}e^{kt}\, dt = \int_0^\infty e^{-(s-k)t}\, dt$$

$$= \lim_{T\to\infty} \left[\frac{-e^{-(s-k)t}}{s-k} \right]_0^T$$

This converges for $s > k$, hence

$$\mathcal{L}\{e^{kt}\} = \lim_{T\to\infty} \left[-\frac{e^{-(s-k)T}}{s-k} + \frac{1}{s-k} \right] = \frac{1}{s-k}, \qquad s > k \quad \blacksquare$$

Example 3

Show that

$$\mathcal{L}\{\sin kt\} = \frac{k}{s^2 + k^2}, \qquad s > 0 \tag{6-4}$$

SOLUTION   The Laplace transform of $\sin kt$ is given by

$$\mathcal{L}\{\sin kt\} = \int_0^\infty e^{-st} \sin kt\, dt$$

An integral formula for $\int e^{ax} \sin bx\, dx$ allows us to evaluate the integral as follows:

$$\mathcal{L}\{\sin kt\} = \int_0^\infty e^{-st} \sin kt\, dt$$

$$= \lim_{T\to\infty} \left[\frac{e^{-st}(-s \sin kt - k \cos kt)}{s^2 + k^2} \right]_0^T$$

$$= \lim_{T\to\infty} \left[\frac{e^{-sT}(-s \sin kT - k \cos kT)}{s^2 + k^2} + \frac{k}{s^2 + k^2} \right]$$

$$= \frac{k}{s^2 + k^2}, \qquad s > 0 \qquad\qquad \blacksquare$$

Example 4

Show, for n a positive integer, that

$$\mathcal{L}\{t^n\} = \frac{n!}{s^{n+1}}, \qquad s > 0 \tag{6-5}$$

SOLUTION   By definition we have

$$\mathcal{L}\{t^n\} = \int_0^\infty e^{-st}t^n\, dt$$

We proceed by applying the formula for integration by parts with

$u = t^n$	$dv = e^{-st}\,dt$
$du = nt^{n-1}\,dt$	$v = -\dfrac{1}{s}e^{-st}$

Thus

$$\int_0^\infty e^{-st}t^n\,dt = \left[\frac{-t^n e^{-st}}{s}\right]_0^\infty + \frac{n}{s}\int_0^\infty e^{-st}t^{n-1}\,dt$$

The first term on the right-hand side is equal to zero for $n > 0$ and $s > 0$, so

$$\mathcal{L}\{t^n\} = \int_0^\infty e^{-st}t^n\,dt = \frac{n}{s}\int_0^\infty e^{-st}t^{n-1}\,dt = \frac{n}{s}\,\mathcal{L}\{t^{n-1}\} \qquad (6\text{–}6)$$

Applying Equation 6–6 and replacing n with $n - 1$, we conclude that

$$\mathcal{L}\{t^{n-1}\} = \frac{n-1}{s}\,\mathcal{L}\{t^{n-2}\} \qquad (6\text{–}7)$$

Combining the results in Equations 6–6 and 6–7, we can write

$$\mathcal{L}\{t^n\} = \frac{n(n-1)}{s^2}\,\mathcal{L}\{t^{n-2}\}$$

Continued iteration of this process yields

$$\mathcal{L}\{t^n\} = \frac{n(n-1)(n-2)\cdots 3\cdot 2\cdot 1}{s^n}\,\mathcal{L}\{t^0\}$$

The numerator is $n!$, and $\mathcal{L}\{t^0\} = \mathcal{L}\{1\} = \dfrac{1}{s}$, so

$$\mathcal{L}\{t^n\} = \frac{n!}{s^n}\left(\frac{1}{s}\right) = \frac{n!}{s^{n+1}}, \qquad s > 0 \qquad\blacksquare$$

Like the derivative and the integral, the Laplace transform is a **linear operator;** that is, if $f_1(t)$ and $f_2(t)$ have Laplace transforms and if c_1 and c_2 are constants, then

$$\mathcal{L}\{c_1 f_1(t) + c_2 f_2(t)\} = c_1 \mathcal{L}\{f_1(t)\} + c_2 \mathcal{L}\{f_2(t)\}$$

To use the Laplace transform in the solution of differential equations, it is important to have a readily available list or table of elementary functions and their transforms. Table 6–1 is an example of such a table and will be large enough for the purposes of this book. This table, the linearity property, and a general formula for the Laplace transform of the derivative function are sufficient to solve a great many differential equations.

TABLE 6–1   Laplace Transforms

ORIGINAL FUNCTION	LAPLACE TRANSFORM	
$f(t)$	$F(s)$	
1	$\dfrac{1}{s}$	$s > 0$
t^n, n a positive integer	$\dfrac{n!}{s^{n+1}}$	$s > 0$
e^{kt}	$\dfrac{1}{s - k}$	$s > k$
$t^n e^{kt}$	$\dfrac{n!}{(s - k)^{n+1}}$	$s > k$
$\sin kt$	$\dfrac{k}{s^2 + k^2}$	$s > 0$
$\cos kt$	$\dfrac{s}{s^2 + k^2}$	$s > 0$
$\sinh kt$	$\dfrac{k}{s^2 - k^2}$	$s > |k|$
$\cosh kt$	$\dfrac{s}{s^2 - k^2}$	$s > |k|$

Example 5

Evaluate $\mathcal{L}\{3t + 5e^{-2t}\}$.

SOLUTION $\mathcal{L}\{3t + 5e^{-2t}\} = 3\,\mathcal{L}\{t\} + 5\,\mathcal{L}\{e^{-2t}\}$

$$= 3\left[\frac{1}{s^2}\right] + 5\left[\frac{1}{s + 2}\right] \qquad (\textit{from Table 6–1})$$

$$= \frac{5s^2 + 3s + 6}{s^2(s + 2)}$$ ∎

Example 6

Evaluate $\mathcal{L}\{\cos^2 3t\}$.

SOLUTION   Using the identity $\cos^2 x = \frac{1}{2} + \frac{1}{2}\cos 2x$, we have

$$\mathcal{L}\{\cos^2 3t\} = \mathcal{L}\{\tfrac{1}{2} + \tfrac{1}{2}\cos 6t\}$$

$$= \frac{1}{2}\left(\frac{1}{s}\right) + \frac{1}{2}\left(\frac{s}{s^2 + 6^2}\right) = \frac{s^2 + 18}{s(s^2 + 36)}$$ ∎

EXERCISES FOR SECTION 6–1

In Exercises 1–18 evaluate the Laplace transform of the function, using the linearity property and Table 6–1.

1. $\mathcal{L}\{t^3\}$

2. $\mathcal{L}\{e^{-2t}\}$

3. $\mathcal{L}\{2t^4 + t^2 + 6\}$

4. $\mathcal{L}\{e^{7t} + e^{-2t}\}$

5. $\mathcal{L}\{e^{-2t} + 4e^t\}$

6. $\mathcal{L}\{3t^2 + \cos 2t\}$

7. $\mathcal{L}\{t^5 e^{-3t}\}$

8. $\mathcal{L}\{4t^3 e^{2t}\}$

9. $\mathcal{L}\{e^{3t} + \sin 3t\}$

10. $\mathcal{L}\{3 \sin 4t - 2 \cos 4t\}$

11. $\mathcal{L}\{t^3 - \sinh 2t\}$

12. $\mathcal{L}\{1 - 2t + \cosh 3t\}$

13. $\mathcal{L}\{2te^{-t} + e^{-t}\}$

14. $\mathcal{L}\{5 + te^{-2t} - e^{-2t}\}$

15. $\mathcal{L}\{t(t - 2)e^{3t}\}$

16. $\mathcal{L}\{(t - 2)^2 e^{4t}\}$

17. $\mathcal{L}\{(1 - e^{t/2})^2\}$

18. $\mathcal{L}\{(1 + \cos 2t)^2\}$

19. The hyperbolic sine of u is denoted $\sinh u$ and defined by

$$\sinh u = \frac{1}{2}(e^u - e^{-u})$$

Using this definition and the Laplace transform for e^{kt}, show that

$$\mathcal{L}\{\sinh kt\} = \frac{k}{s^2 - k^2}, \qquad s > |k|$$

20. Using $\cosh u = \frac{1}{2}(e^u + e^{-u})$, show that

$$\mathcal{L}\{\cosh kt\} = \frac{s}{s^2 - k^2}, \qquad s > |k|$$

21. Show that

$$\mathcal{L}\{\sin kt \cos kt\} = \frac{k}{s^2 + 4k^2}, \qquad s > 0$$

22. Show that

$$\mathcal{L}\{\sin^2 kt\} = \frac{2k^2}{s(s^2 + 4k^2)}, \qquad s > 0$$

6–2   WHAT KINDS OF FUNCTIONS HAVE LAPLACE TRANSFORMS?

Piecewise Continuity

In the previous section we showed that certain elementary functions have Laplace transforms. Now we address the broader question of the existence of Laplace transforms. We will specify two conditions on a function that are sufficient to ensure that the function has a Laplace transform.

DEFINITION

A function f is said to be **piecewise continuous** over the closed interval $a \leq t \leq b$ if the interval can be divided into a finite number of open subintervals $c < t < d$, such that

1. The function f is continuous on each subinterval $c < t < d$.

2. The function f has a finite limit as t approaches each endpoint from within the interval; that is, $\lim_{t \to c^+} f(t)$ and $\lim_{t \to d^-} f(t)$ exist.

The second condition means that a piecewise continuous function f may contain **finite** or **jump** discontinuities. Figure 6–1 shows some functions that are piecewise continuous. (Incidentally, every continuous function is also piecewise continuous.) Functions such as $f(t) = 1/t$ are not piecewise continuous on any closed interval containing the origin since there is an infinite discontinuity at $t = 0$.

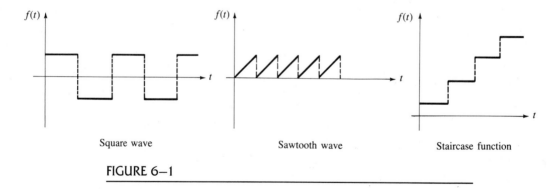

Square wave   Sawtooth wave   Staircase function

FIGURE 6–1

Example 1

The function $h(t)$ pictured in Figure 6–2 and defined by

$$h(t) = \begin{cases} t, & 0 \leq t < 2 \\ 3, & t > 2 \end{cases}$$

has a discontinuity at $t = 2$ since $h(2)$ is not defined. However, the function is

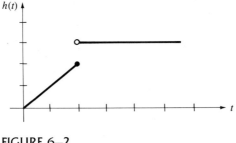

FIGURE 6–2

piecewise continuous over the interval $t \geq 0$ since it is continuous over the open subintervals $0 < t < 2$ and $t > 2$ and $\lim\limits_{t \to 2^-} h(t) = 2$ and $\lim\limits_{t \to 2^+} h(t) = 3$.   ∎

Example 2

The function

$$g(t) = \frac{\sin t}{t}$$

is discontinuous at $t = 0$. See Figure 6–3. From calculus we know that

$$\lim_{t \to 0^+} \frac{\sin t}{t} = 1$$

Hence the function is piecewise continuous for $t \geq 0$.

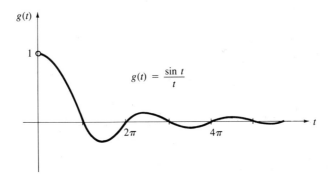

$$g(t) = \frac{\sin t}{t}$$

FIGURE 6–3   ∎

From calculus we know that a finite number of finite discontinuities of an integrand function do not affect the existence of the integral. Thus, the integral of the piecewise continuous function

$$f(t) = \begin{cases} 1, & 0 \leq t < 2 \\ 3, & 2 \leq t \leq 4 \end{cases}$$

on the interval [0, 4] is given by

$$\int_0^4 f(t)\, dt = \int_0^2 1\, dt + \int_2^4 3\, dt = \Big[t \Big]_0^2 + \Big[3t \Big]_2^4 = 8$$

Similarly, the discontinuities of a piecewise continuous function do not prevent us from finding the Laplace transform of the function. The Laplace transform of a piecewise continuous function $f(t)$ exists provided

$$\int_0^\infty e^{-st} f(t)\, dt$$

exists.

Example 3

Find the Laplace transform of the piecewise continuous function

$$g(t) = \begin{cases} 2t, & 0 \le t < 3 \\ -1, & t > 3 \end{cases}$$

SOLUTION   We use the definition of $\mathcal{L}\{g(t)\}$ and split the integral into two parts, one for the interval $0 \le t < 3$ and the other for $t > 3$. Thus

$$\mathcal{L}\{g(t)\} = \int_0^3 e^{-st}[2t]\, dt + \int_3^\infty e^{-st}[-1]\, dt$$

$$= \left[-\frac{2te^{-st}}{s} - \frac{2e^{-st}}{s^2} \right]_0^3 + \left[\frac{1}{s}e^{-st} \right]_3^\infty$$

where integration by parts was used to evaluate the integral from 0 to 3. The value of $\dfrac{1}{s}e^{-st}$ converges to zero as $t \to \infty$, if $s > 0$. Therefore

$$\mathcal{L}\{g(t)\} = \left[-\frac{6}{s}e^{-3s} - \frac{2}{s^2}e^{-3s} \right] - \left[0 - \frac{2}{s^2} \right] + \left[0 - \frac{1}{s}e^{-3s} \right]$$

$$= \frac{2}{s^2} - \frac{2e^{-3s}}{s^2} - \frac{7e^{-3s}}{s}, \qquad s > 0 \qquad \blacksquare$$

Exponential Order

In Section 6–1 we showed that e^{kt} has a Laplace transform for all $s > k$. The graph of $y = e^{kt}$ is shown in Figure 6–4. Notice from the graph that e^{kt} increases rapidly as t increases. Now consider another function g, shown in Figure 6–4 as a dashed

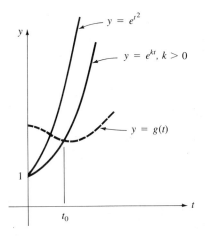

FIGURE 6–4

curve, bounded by e^{kt} for all $t \geq t_0$. If $|g|$ is bounded by e^{kt} for all $t \geq t_0$, we might speculate that since $\mathcal{L}\{e^{kt}\}$ exists, $\mathcal{L}\{g(t)\}$ should also exist. This speculation motivates us to make the following definition:

DEFINITION

A function f is of **exponential order** if there are real numbers a, M, and t_0 such that

$$|f(t)| < Me^{at} \qquad \text{for} \qquad t > t_0$$

COMMENT: (a) *We interpret the definition to mean that the graph of the function is below that of the exponential function Me^{at} for $t > t_0$.*

(b) *It follows from the definition that f is of exponential order if and only if, for some exponential function e^{at},*

$$\lim_{t \to \infty} |f(t)|e^{-at} = 0$$

We will use this latter interpretation to show that a given function is of exponential order.

(c) *Figure 6–4 also shows the graph of $y = e^{t^2}$. The graph is intended to show that there is no exponential function of the form e^{kt} that bounds e^{t^2} for $t \geq 0$, and therefore e^{t^2} is not of exponential order.*

Example 4

Show that $f(t) = t^2$ is of exponential order.

SOLUTION   We note that if a is any constant > 0, then

$$\lim_{T \to \infty} |t^2| e^{-at} = \lim_{T \to \infty} \frac{t^2}{e^{at}} = \lim_{T \to \infty} \frac{2t}{ae^{at}} = \lim_{T \to \infty} \frac{2}{a^2 e^{at}} = 0$$

where the last two limits were obtained by using **l'Hôpital's rule.** Thus t^2 is of exponential order.   ■

By the same reasoning used in Example 4 any function of the form t^n, where n is a positive integer, is of exponential order. We are now in a position to state a basic existence theorem for the Laplace transform of a function f.

THEOREM 6–1

If a function f is piecewise continuous on $[0, \infty)$ and of exponential order, then the Laplace transform exists for s greater than some a.

PROOF   Since the function f is of exponential order, we know that there are constants t_0 and a such that for any M

$$|e^{-at} f(t)| < M, \qquad t \geq t_0$$

Multiplying by $e^{-st} e^{at}$, we have

$$|e^{-st} f(t)| < M e^{-st} e^{at}$$

Hence

$$\int_0^\infty |e^{-st} f(t)| dt < \int_0^\infty M e^{-(s-a)t} \, dt = \left[-\frac{M e^{-(s-a)t}}{s - a} \right]_0^\infty$$

For $s > a$ the last member approaches zero as $t \to \infty$. Therefore

$$\int_0^\infty |e^{-st} f(t)| dt < \frac{M}{s - a}, \qquad s > a$$

which implies the existence of the improper integral defining the Laplace transform of f and completes the proof.

Theorem 6–1 does not give a necessary condition for the existence of the Laplace transform of a function f; that is, it does not say that a function must be piecewise continuous and of exponential order to have a Laplace transform. For example, the function $t^{-1/2}$, which is not piecewise continuous on $t \geq 0$, has a Laplace transform of $(\pi/s)^{1/2}$ if $s > 0$. Although functions of this type are important in the theory of Laplace transforms, we shall limit our discussion to functions that are both piecewise continuous and of exponential order. The proof of Theorem 6–1 provides us with an interesting corollary.

COROLLARY

If a function f is piecewise continuous and of exponential order, and $\mathcal{L}\{f(t)\} = F(s)$, then

$$\lim_{s \to \infty} F(s) = 0$$

PROOF

$$|\mathcal{L}\{f(t)\}| = \left| \int_0^\infty e^{-st} f(t)\, dt \right| \leq \int_0^\infty |e^{-st} f(t)|\, dt$$

From Theorem 6–1 this last quantity is $\leq M/(s - a)$. Hence

$$0 \leq |F(s)| \leq \frac{M}{s - a}$$

which shows that $\lim_{s \to \infty} F(s) = 0$.

COMMENT: This corollary makes it clear that not all functions of s are Laplace transforms. For example, $s/(s + 1)$ does not approach zero as s becomes infinite, and therefore it is not a possible Laplace transform of a function satisfying the conditions of Theorem 6–1.

EXERCISES FOR SECTION 6–2

In Exercises 1–8 show that the function is piecewise continuous on the interval $t \geq 0$.

1. $f(t) = \begin{cases} 1, & 0 \leq t < 2 \\ 2, & t > 2 \end{cases}$

2. $f(t) = \begin{cases} t, & 0 \leq t < 4 \\ 4, & t > 4 \end{cases}$

3. $g(t) = \begin{cases} t + 2, & 0 \leq t < 1 \\ 5, & t > 1 \end{cases}$

4. $h(t) = \begin{cases} t^2, & 0 \leq t < 2 \\ 6, & t > 2 \end{cases}$

5. $f(t) = \begin{cases} e^t, & 0 \le t < 2 \\ 0, & t > 2 \end{cases}$ **6.** $m(t) = \begin{cases} \sin t, & 0 \le t < \pi \\ 0, & t > \pi \end{cases}$

7. $f(t) = \dfrac{\cos t - 1}{t}$ **8.** $x(t) = \dfrac{1 - e^t}{t}$

9. Show that $f(t) = t^{-1/2}$ is not piecewise continuous on $t \ge 0$.

10–15. Find the Laplace transform of each of the functions in Exercises 1–6.

16. Find $\mathcal{L}\{x(t)\}$, given

$$x(t) = \begin{cases} 1, & 0 \le t < 2 \\ 2, & 2 < t < 4 \\ 0, & t > 4 \end{cases}$$

In Exercises 17–26 show that the given function is of exponential order.

17. t^3 **18.** $t^{1/2}$

19. $t^2 e^{3t}$ **20.** $\sin t$

21. $\sinh t$ **22.** $t \sin 2t$

23. $\dfrac{1}{t} \sin kt$ **24.** t^n

25. $t^n e^{kt}$ **26.** $t^n \sin kt$

27. Show that all bounded functions are of exponential order.

28. Show that the Laplace transform of any polynomial exists.

29. Indicate which of the following functions are piecewise continuous and of exponential order on $t \ge 0$.

(a) $t^{1/2}$ (b) e^{t^2} (c) $\dfrac{1}{t - 2}$ (d) $\sin e^{t^2}$ (e) t^{-3}

30. The fact that a function has a Laplace transform does not mean that its derivative will have a Laplace transform. Give an example of a function that has a Laplace transform but whose derivative does not. (HINT: See the functions in Exercise 29.)

31. Show that if a function f is piecewise continuous and of exponential order, and $\mathcal{L}\{f(t)\} = F(s)$, then $\lim\limits_{s \to \infty} sF(s)$ is finite.

32. Using the result of Exercise 31, show that $\sqrt{\pi/s}$ is not the Laplace transform of a function that is piecewise continuous and of exponential order. (NOTE: $\sqrt{\pi/s}$ is the Laplace transform of $t^{-1/2}$.)

6–3   THE INVERSE LAPLACE TRANSFORM

In Section 6–1 we showed how to find the Laplace transform of a function $f(t)$. It is also important to be able to reverse this process and reconstruct a function $f(t)$ whose Laplace transform $F(s)$ is given.

DEFINITION

> If there exists a function $f(t)$ such that $\mathcal{L}\{f(t)\} = F(s)$, then $f(t)$ is called the **inverse Laplace transform** of $F(s)$. We designate the inverse Laplace transform by $\mathcal{L}^{-1}$ and write
>
> $$f(t) = \mathcal{L}^{-1}\{F(s)\}$$

The determination of an inverse Laplace transform is less direct than that of finding a Laplace transform, just as in calculus finding an antiderivative is less direct than finding a derivative. Basically, to find an inverse Laplace transform, you must be familiar with the formulas for finding the Laplace transform. See Table 6–1.

Example 1

Find $\mathcal{L}^{-1}\{1/(s + 2)\}$.

SOLUTION   Since we know that $\mathcal{L}\{e^{-2t}\} = 1/(s + 2)$, it follows that

$$\mathcal{L}^{-1}\left\{\frac{1}{s + 2}\right\} = e^{-2t} \qquad \blacksquare$$

Procedurally, to find an inverse Laplace transform, you must learn to use Table 6–1 in reverse. However, in most cases, the given transform will not be in a form that allows direct use of the table, so the given $F(s)$ will have to be algebraically manipulated into a form that can be found in the table. Of fundamental importance to this discussion is the fact that the inverse Laplace transform has the linearity property. Thus

$$\mathcal{L}^{-1}\{c_1 F_1(s) + c_2 F_2(s)\} = c_1 \mathcal{L}^{-1}\{F_1(s)\} + c_2 \mathcal{L}^{-1}\{F_2(s)\}$$

where c_1 and c_2 represent any constants. The validity of the property follows from the definition of $\mathcal{L}^{-1}\{F(s)\}$ and the corresponding linearity property for $F(s)$.

Example 2

Find $\mathcal{L}^{-1}\left\{\dfrac{2s + 1}{s^2 + 4}\right\}$.

SOLUTION   The function f, of which $(2s + 1)/(s^2 + 4)$ is its transform, is not directly obtainable from Table 6–1. Consequently, we first write

$$\mathcal{L}^{-1}\left\{\frac{2s + 1}{s^2 + 4}\right\} = 2\mathcal{L}^{-1}\left\{\frac{s}{s^2 + 4}\right\} + \mathcal{L}^{-1}\left\{\frac{1}{s^2 + 4}\right\}$$

Then, from Table 6–1,

$$\mathscr{L}^{-1}\left\{\frac{s}{s^2 + 4}\right\} = \cos 2t$$

and

$$\mathscr{L}^{-1}\left\{\frac{1}{s^2 + 4}\right\} = \tfrac{1}{2}\mathscr{L}^{-1}\left\{\frac{2}{s^2 + 4}\right\} = \tfrac{1}{2} \sin 2t$$

Hence

$$\mathscr{L}^{-1}\left\{\frac{2s + 1}{s^2 + 4}\right\} = 2 \cos 2t + \tfrac{1}{2} \sin 2t$$ ∎

In many cases the given transform is a proper rational function of s in which Table 6–1 can be used only after the given transform is decomposed into its partial fractions. Some additional techniques for manipulating $F(s)$ are covered in later sections, but for now we concentrate on the use of partial fractions.

Example 3

Evaluate $\mathscr{L}^{-1}\left\{\dfrac{s + 5}{s^2 - 2s - 3}\right\}$.

SOLUTION   The denominator factors into $(s - 3)(s + 1)$. Hence the partial fractions are

$$\frac{s + 5}{(s - 3)(s + 1)} = \frac{A}{s - 3} + \frac{B}{s + 1}$$

where A and B are constants to be determined. Clearing the fractions, we get

$$s + 5 = A(s + 1) + B(s - 3)$$

If $s = 3$, we have

$$8 = 4A \qquad \text{or} \qquad A = 2$$

Similarly, if $s = -1$,

$$4 = -4B \qquad \text{or} \qquad B = -1$$

Finally,

$$\mathscr{L}^{-1}\left\{\frac{s + 5}{s^2 - 2s - 3}\right\} = \mathscr{L}^{-1}\left\{\frac{2}{s - 3}\right\} + \mathscr{L}^{-1}\left\{\frac{-1}{s + 1}\right\}$$

$$= 2e^{3t} - e^{-t}$$ ∎

Example 4

Evaluate $\mathcal{L}^{-1}\left\{\dfrac{s^2}{(s+1)^3}\right\}$.

SOLUTION   The partial fractions are

$$\frac{s^2}{(s+1)^3} = \frac{A}{s+1} + \frac{B}{(s+1)^2} + \frac{C}{(s+1)^3}$$

Clearing the fractions,

$$s^2 = A(s+1)^2 + B(s+1) + C$$

Expanding,

$$s^2 = As^2 + 2As + A + Bs + B + C$$

Collecting like terms,

$$s^2 = As^2 + (2A + B)s + (A + B + C)$$

Equating the like powers of s yields the system of equations

$$A = 1$$
$$2A + B = 0$$
$$A + B + C = 0$$

The solution of this system is $A = 1$, $B = -2$, and $C = 1$. Making these substitutions, we have

$$\mathcal{L}^{-1}\left\{\frac{s^2}{(s+1)^3}\right\} = \mathcal{L}^{-1}\left\{\frac{1}{s+1}\right\} + \mathcal{L}^{-1}\left\{\frac{-2}{(s+1)^2}\right\} + \mathcal{L}^{-1}\left\{\frac{1}{(s+1)^3}\right\}$$

$$= e^{-t} - 2te^{-t} + \tfrac{1}{2}t^2 e^{-t}$$

$$= (1 - 2t + \tfrac{1}{2}t^2)e^{-t} \qquad \blacksquare$$

Example 5

Evaluate $\mathcal{L}^{-1}\left\{\dfrac{9s+14}{(s-2)(s^2+4)}\right\}$.

SOLUTION   Since the denominator is composed of a linear factor and an irreducible quadratic factor, the required partial fractions are

$$\frac{9s+14}{(s-2)(s^2+4)} = \frac{A}{s-2} + \frac{Bs+C}{s^2+4}$$

Clearing the fractions,

$$9s + 14 = A(s^2 + 4) + (Bs + C)(s - 2)$$

Expanding on the right and collecting like terms,

$$9s + 14 = (A + B)s^2 + (-2B + C)s + (4A - 2C)$$

Equating the coefficients of corresponding powers of s, we get

$$A + B = 0$$
$$-2B + C = 9$$
$$4A - 2C = 14$$

Solving this system yields $A = 4$, $B = -4$, and $C = 1$. Therefore

$$\mathcal{L}^{-1}\left\{\frac{9s + 14}{(s - 2)(s^2 + 4)}\right\} = \mathcal{L}^{-1}\left\{\frac{4}{s - 2}\right\} + \mathcal{L}^{-1}\left\{\frac{-4s + 1}{s^2 + 4}\right\}$$

$$= 4\mathcal{L}^{-1}\left\{\frac{1}{s - 2}\right\} - 4\mathcal{L}^{-1}\left\{\frac{s}{s^2 + 4}\right\}$$

$$+ \mathcal{L}^{-1}\left\{\frac{1}{s^2 + 4}\right\}$$

$$= 4e^{2t} - 4\cos 2t + \tfrac{1}{2}\sin 2t \qquad\blacksquare$$

EXERCISES FOR SECTION 6–3

Find the indicated inverse Laplace transforms.

1. $\mathcal{L}^{-1}\left\{\dfrac{1}{s - 3}\right\}$

2. $\mathcal{L}^{-1}\left\{\dfrac{s}{s^2 + 9}\right\}$

3. $\mathcal{L}^{-1}\left\{\dfrac{2}{s^2 + 9}\right\}$

4. $\mathcal{L}^{-1}\left\{\dfrac{s + 2}{s^2 + 1}\right\}$

5. $\mathcal{L}^{-1}\left\{\dfrac{1}{s^3}\right\}$

6. $\mathcal{L}^{-1}\left\{\dfrac{3}{s^5}\right\}$

7. $\mathcal{L}^{-1}\left\{\dfrac{2}{(s - 3)^5}\right\}$

8. $\mathcal{L}^{-1}\left\{\dfrac{1}{(s + 1)^3}\right\}$

9. $\mathcal{L}^{-1}\left\{\dfrac{s - 3}{s^2 - 16}\right\}$

10. $\mathcal{L}^{-1}\left\{\dfrac{1}{s} + \dfrac{s}{s^2 - 9}\right\}$

11. $\mathcal{L}^{-1}\left\{\dfrac{s + 1}{s^2 + 4s - 5}\right\}$

12. $\mathcal{L}^{-1}\left\{\dfrac{s + 2}{s^2 - s - 6}\right\}$

13. $\mathcal{L}^{-1}\left\{\dfrac{s + 12}{s^3 + s^2 - 6s}\right\}$

14. $\mathcal{L}^{-1}\left\{\dfrac{1}{(s - 2)^3}\right\}$

15. $\mathcal{L}^{-1}\left\{\dfrac{1}{(s + 4)^3}\right\}$

16. $\mathcal{L}^{-1}\left\{\dfrac{s^2 + 3s - 6}{s(s - 1)^2}\right\}$

17. $\mathcal{L}^{-1}\left\{\dfrac{s + 1}{s^3 - s^2}\right\}$

18. $\mathcal{L}^{-1}\left\{\dfrac{s^2 - 3s + 6}{(s^2 - 1)(3 - 2s)}\right\}$

19. $\mathcal{L}^{-1}\left\{\dfrac{2(s^2 + 1)}{s^3 + 2s}\right\}$ **20.** $\mathcal{L}^{-1}\left\{\dfrac{s^2 + s + 2}{s^3 + 2s}\right\}$

21. $\mathcal{L}^{-1}\left\{\dfrac{s^3 + s^2 - s + 1}{(s - 1)^2(s^2 + 1)}\right\}$ **22.** $\mathcal{L}^{-1}\left\{\dfrac{5 + 3s - s^2}{(s^2 + 2s - 3)(s^2 + 5)}\right\}$

6—4   THE FIRST SHIFTING THEOREM

One of the most useful properties of the Laplace transform is contained in the following theorem, which is one of the two "shifting" theorems in the theory of Laplace transforms. When we talk about shifting a function, we mean that its graph is moved to the left or right some specified number of units. For instance, the graph of $F(s)$ is shown in Figure 6–5(a), and the graph, shifted a units to the right, is shown in Figure 6–5(b). The shifted function is denoted by $F(s - a)$.

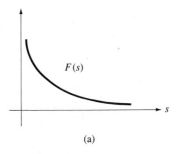

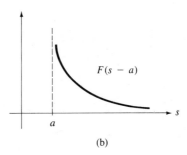

(a)   (b)

FIGURE 6—5

THEOREM 6—2   *The First Shifting Theorem*

Let $\mathcal{L}\{f(t)\} = F(s)$. Then

$$\mathcal{L}\{e^{at}f(t)\} = F(s - a)$$

PROOF   By definition of $\mathcal{L}\{e^{at}f(t)\}$, we write

$$\mathcal{L}\{e^{at}f(t)\} = \int_0^\infty e^{-st}[e^{at}f(t)]\, dt$$

$$= \int_0^\infty e^{-(s-a)t}f(t)\, dt = F(s - a)$$

Example 1

Use the first shifting theorem to obtain the Laplace transform of (a) $e^t t^2$ and (b) $e^{3t} \sin t$.

SOLUTION   (a) Since $\mathcal{L}\{t^2\} = \dfrac{2}{s^3}$, it follows that $\mathcal{L}\{e^t t^2\} = \dfrac{2}{(s-1)^3}$.

(b) Since $\mathcal{L}\{\sin t\} = \dfrac{1}{s^2 + 1}$, it follows that $\mathcal{L}\{e^{3t} \sin t\} = \dfrac{1}{(s-3)^2 + 1}$.   ■

Example 2

Use the first shifting theorem to find $\mathcal{L}\{e^{-t} g(t)\}$, where

$$g(t) = \begin{cases} 2t, & 0 \le t < 3 \\ -1, & t \ge 3 \end{cases}$$

SOLUTION   The Laplace transform of $g(t)$ was found in Example 3 of Section 6–2 to be

$$\mathcal{L}\{g(t)\} = \frac{2}{s^2} - \frac{2e^{-3s}}{s^2} - \frac{7e^{-3s}}{s}$$

Therefore by the first shifting theorem we have

$$\mathcal{L}\{e^{-t} g(t)\} = \frac{2}{(s+1)^2} - \frac{2e^{-3(s+1)}}{(s+1)^2} - \frac{7e^{-3(s+1)}}{s+1}$$ ■

The first shifting theorem is also valuable for finding inverse transforms. As a theorem for inverse transforms it takes the form

$$\mathcal{L}^{-1}\{F(s-a)\} = e^{at} f(t)$$

Example 3

Find $\mathcal{L}^{-1}\left\{ \dfrac{3}{(s-5)^2 + 9} \right\}$.

SOLUTION   The given transform is that of $\sin 3t$ with s replaced by $s - 5$. Hence the first shifting theorem yields

$$\mathcal{L}^{-1}\left\{ \frac{3}{(s-5)^2 + 9} \right\} = e^{5t} \mathcal{L}^{-1}\left\{ \frac{3}{s^2 + 9} \right\} = e^{5t} \sin 3t$$ ■

COMMENT: To use the first shifting theorem to find an inverse transform, first express the given Laplace transform as a function of $s - a$.

Example 4

Find $\mathcal{L}^{-1}\left\{\dfrac{1}{s^2 + 4s + 10}\right\}$.

SOLUTION   Completing the square on $s^2 + 4s + 10$, we get

$$\mathcal{L}^{-1}\left\{\frac{1}{s^2 + 4s + 10}\right\} = \mathcal{L}^{-1}\left\{\frac{1}{s^2 + 4s + 4 + 6}\right\}$$

$$= \mathcal{L}^{-1}\left\{\frac{1}{(s + 2)^2 + (\sqrt{6})^2}\right\}$$

$$= \frac{1}{\sqrt{6}}\, e^{-2t} \sin \sqrt{6}\, t \qquad \blacksquare$$

Example 5

Find $\mathcal{L}^{-1}\left\{\dfrac{s}{s^2 + 6s + 13}\right\}$.

SOLUTION   Completing the square on the quadratic function, we write

$$\mathcal{L}^{-1}\left\{\frac{s}{s^2 + 6s + 13}\right\} = \mathcal{L}^{-1}\left\{\frac{s}{(s + 3)^2 + 4}\right\}$$

To obtain a function of $s + 3$ in the numerator, we write $s = s + 3 - 3$, so that

$$\mathcal{L}^{-1}\left\{\frac{s}{s^2 + 6s + 13}\right\} = \mathcal{L}^{-1}\left\{\frac{(s + 3) - 3}{(s + 3)^2 + 4}\right\}$$

$$= \mathcal{L}^{-1}\left\{\frac{s + 3}{(s + 3)^2 + 4}\right\} - \mathcal{L}^{-1}\left\{\frac{3}{(s + 3)^2 + 4}\right\}$$

$$= e^{-3t} \cos 2t - \tfrac{3}{2}e^{-3t} \sin 2t \qquad \blacksquare$$

EXERCISES FOR SECTION 6–4

In Exercises 1–12 use the first shifting theorem to find the Laplace transform of the given function.

1. $t^2 e^{2t}$

2. $e^{-2t} \sin 5t$

3. $e^{2t} \sinh t$

4. $e^t(\cos 2t - 3 \sin 5t)$

5. $P(t)e^{at}$, where $P(t) = a_n t^n + a_{n-1}t^{n-1} + \cdots + a_2 t^2 + a_1 t + a_0$

6. $e^{2t} \sin 3t \cos 3t$

7. $e^{-t} \cos^2 2t$

8. $e^{-3t} (1 + \sin 4t)^2$

9. $e^t \cdot g(t)$, where $g(t) = \begin{cases} t, & 0 \le t < 2 \\ 2, & t \ge 2 \end{cases}$

10. $e^{2t} \cdot h(t)$, where $h(t) = \begin{cases} e^t, & 0 \le t < 2 \\ 0, & t \ge 2 \end{cases}$

11. $e^{-t} \cdot g(t)$, where $g(t) = \begin{cases} 2, & 0 \le t < 5 \\ 0, & t \ge 5 \end{cases}$

12. $e^{3t} \cdot f(t)$, where $f(t) = \begin{cases} t, & 0 \le t < 1 \\ 0, & t \ge 1 \end{cases}$

In Exercises 13–22 find the inverse transform.

13. $\dfrac{1}{s^2 + 3s + 3}$

14. $\dfrac{1}{s^2 + 3s + 1}$

15. $\dfrac{s + 5}{(s + 2)^3}$

16. $\dfrac{1}{s^2 - s + 1}$

17. $\dfrac{s}{s^2 + 2s + 5}$

18. $\dfrac{s}{s^2 + 6s + 9}$

19. $\dfrac{s + 1}{s^2 - 6s + 13}$

20. $\dfrac{2s}{s^2 + 10s + 34}$

21. $\dfrac{1}{s(s + 1)(s^2 + 4s + 5)}$

22. $\dfrac{s^2}{(s + 1)^4}$

23. Show that $\mathcal{L}\{f(at)\} = \dfrac{1}{a} F\left(\dfrac{s}{a}\right)$, where $F(s) = \mathcal{L}\{f(t)\}$.

6–5   TRANSFORMS OF DERIVATIVES AND INTEGRALS

The Laplace transform of the derivative function exists under certain conditions on f and f' and corresponds to the algebraic operation of multiplication of the Laplace transform of the function by s. It is this property that makes the Laplace transform so useful in solving differential equations.

If a function f is continuous and of exponential order for $t \ge 0$, and if f' is piecewise continuous for $t \ge 0$, then the Laplace transform of f' exists and is given by

$$\mathcal{L}\{f'(t)\} = \int_0^\infty e^{-st} f'(t)\, dt$$

Using integration by parts with

$u = e^{-st}$	$dv = f'(t)\, dt$
$du = -se^{-st}\, dt$	$v = f(t)$

we write $\mathcal{L}\{f'(t)\}$ as

$$\mathcal{L}\{f'(t)\} = f(t)e^{-st}\Big|_0^\infty + s \int_0^\infty e^{-st}f(t)\,dt$$

Since $f(t)$ is of exponential order, $f(t)e^{-st}$ approaches 0 as t increases without bound (assuming $s > 0$). Also, the integral on the right is the Laplace transform of $f(t)$, so $\mathcal{L}\{f'(t)\}$ is given by the formula

$$\mathcal{L}\{f'(t)\} = s\mathcal{L}\{f(t)\} - f(0) \tag{6-8}$$

A formula for $\mathcal{L}\{f''(t)\}$ can be found by letting $g(t) = f'(t)$. Then it follows that $g'(t) = f''(t)$, and

$$\begin{aligned}
\mathcal{L}\{f''(t)\} &= \mathcal{L}\{g'(t)\} = s\mathcal{L}\{g(t)\} - g(0) \\
&= s\mathcal{L}\{f'(t)\} - f'(0) \\
&= s[s\mathcal{L}\{f(t)\} - f(0)] - f'(0)
\end{aligned}$$

Expanding the last expression yields the formula

$$\mathcal{L}\{f''(t)\} = s^2\mathcal{L}\{f(t)\} - sf(0) - f'(0) \tag{6-9}$$

The general case for the Laplace transform of the nth derivative of a function is given in the following theorem:

THEOREM 6-3

If $f(t), f'(t), f''(t), \ldots, f^{(n-1)}(t)$ are continuous and of exponential order for $t \geq 0$, and if $f^{(n)}(t)$ is piecewise continuous for $t \geq 0$, then

$$\mathcal{L}\{f^{(n)}(t)\} = s^n\mathcal{L}\{f(t)\} - s^{n-1}f(0) - s^{n-2}f'(0) - \cdots - f^{(n-1)}(0)$$

Example 1

Find $\mathcal{L}\{\sin kt\}$ using the formula for the Laplace transform of f''.

SOLUTION We let $f(t) = \sin kt$ and note that $f'(t) = k\cos kt$ and $f''(t) = -k^2\sin kt$. Then $f(0) = 0$ and $f'(0) = k$, so by Equation 6-9 we have

$$\mathcal{L}\{f''(t)\} = s^2\mathcal{L}\{f(t)\} - k$$

Substituting in this equation for $f(t)$ and $f''(t)$, we get

$$\mathcal{L}\{-k^2\sin kt\} = s^2\mathcal{L}\{\sin kt\} - k$$

Finally, solving for $\mathcal{L}\{\sin kt\}$ yields

$$\mathcal{L}\{\sin kt\} = \frac{k}{s^2 + k^2}$$

which agrees with the transform given in Table 6-1. ■

Example 2

Find $\mathcal{L}\{t\}$, using the formula for the Laplace transform of f'.

SOLUTION   If $f(t) = t$, then $f'(t) = 1$. Since $f(0) = 0$, Formula 6–8 is

$$\mathcal{L}\{f'(t)\} = s\mathcal{L}\{f(t)\} - 0$$

Or, substituting for f and f',

$$\mathcal{L}\{1\} = s\mathcal{L}\{t\}$$

Since $\mathcal{L}\{1\} = 1/s$, we have

$$\frac{1}{s} = s\mathcal{L}\{t\}$$

which means that $\mathcal{L}\{t\} = 1/s^2$.   ∎

Laplace Transforms of Integrals

Theorem 6–3 may be used to generate a formula for the Laplace transform of the indefinite integral. That is, we can find a formula for

$$\mathcal{L}\left\{ \int_0^t f(x)\, dx \right\}$$

THEOREM 6–4

If f is piecewise continuous and of exponential order for $t \geq 0$, then

$$\mathcal{L}\left\{ \int_0^t f(x)\, dx \right\} = \frac{1}{s}\mathcal{L}\{f(t)\} = \frac{1}{s}F(s) \tag{6–10}$$

PROOF   Let $g(t) = \int_0^t f(x)\, dx$. Then $g'(t) = f(t)$ and $g(0) = 0$. Further, $g(t)$ may be shown to be of exponential order, and hence the result of Theorem 6–3 may be applied. Thus

$$\mathcal{L}\{g'(t)\} = s\mathcal{L}\{g(t)\} - g(0)$$

or

$$\mathcal{L}\{f(t)\} = s\mathcal{L}\left\{ \int_0^t f(x)\, dx \right\}$$

Rearranging, we write

$$\mathcal{L}\left\{\int_0^t f(x)\,dx\right\} = \frac{1}{s}\mathcal{L}\{f(t)\}$$

which completes the proof.

Notice that the Laplace transform of the indefinite integral corresponds to division of the transform of f by s. As corollary to Theorem 6–4 we have

$$\mathcal{L}^{-1}\left\{\frac{1}{s}F(s)\right\} = \int_0^t f(x)\,dx \tag{6-11}$$

Example 3

Given $\mathcal{L}\{f(t)\} = 1/s(s^2 + 4)$, find $f(t)$ by using Formula 6–11.

SOLUTION   We note that $\mathcal{L}\{\frac{1}{2}\sin 2t\} = 1/(s^2 + 4)$. Further,

$$\int_0^t \frac{\sin 2x}{2}\,dx = -\left.\frac{\cos 2x}{4}\right|_0^t = \frac{1}{4} - \frac{\cos 2t}{4}$$

Now writing $f(t)$ as an inverse transform, we have

$$f(t) = \mathcal{L}^{-1}\left\{\frac{1}{s}\frac{1}{s^2 + 4}\right\} = \int_0^t \frac{\sin 2x}{2}\,dx = \frac{1 - \cos 2t}{4} \qquad \blacksquare$$

EXERCISES FOR SECTION 6–5

1. Use Formula 6–8 and $\mathcal{L}\{\sin kt\} = k/(s^2 + k^2)$ to derive $\mathcal{L}\{\cos kt\}$.

2. Use Formula 6–8 to derive $\mathcal{L}\{e^{at}\}$.

3. Use Formula 6–9 to derive $\mathcal{L}\{\cos kt\}$.

4. Use Formula 6–9 and $\mathcal{L}\{1\} = 1/s$ to derive $\mathcal{L}\{t^2\}$.

5. Evaluate $\mathcal{L}\left\{\int_0^t \cos x\,dx\right\}$.   6. Evaluate $\mathcal{L}\left\{\int_0^t x\,dx\right\}$.

Use Formula 6–11 to find $f(t)$ for the given $F(s)$ in Exercises 7–12.

7. $1/s(s^2 + 9)$	8. $1/s(s - 2)$
9. $1/(s^2 + 2s)$	10. $1/(s^2 + 3s)$
11. $1/(s^3 - 9s)$	12. $2/s(s^2 + 2)$

6–6   INITIAL-VALUE PROBLEMS

The Laplace transforms of the derivatives of a function contain terms that require the values of the function and its derivatives at $t = 0$. Because of this property the Laplace transform is uniquely suited to initial-value problems involving linear differential equations with constant coefficients. The purpose of this section is to show how the Laplace transform is used to solve initial-value problems.

Example 1

Use the Laplace transform to solve the initial-value problem

$$y' + 2y = 0, \qquad y(0) = 1$$

SOLUTION   Taking the Laplace transform of both sides of the given equation yields

$$\mathcal{L}\{y' + 2y\} = \mathcal{L}\{0\}$$

Denoting $\mathcal{L}\{y(t)\}$ by $Y(s)$ and using Formula 6–8, we have

$$sY(s) - 1 + 2Y(s) = 0$$

Solving for $Y(s)$,

$$Y(s) = \frac{1}{s + 2}$$

The function $y(t)$ is then found by taking the inverse transform of this equation. Thus

$$y(t) = \mathcal{L}^{-1}\left\{\frac{1}{s + 2}\right\} = e^{-2t}$$

is the desired solution.   ∎

The Laplace transform technique described in Example 1 is typical. We summarize the Laplace transform procedure for solving linear differential equations as follows:

1. Take the Laplace transform of both sides of the given differential equation, making use of the linearity property of the transform.

2. Solve the transformed equation for the Laplace transform of the solution function.

3. Find the inverse transform of the expression $F(s)$ found in step 2.

Notice that when the Laplace transform method is used, the initial conditions at $t = 0$ are automatically included. Further, notice that the solution is a particular solution corresponding to the initial conditions. Thus one of the advantages of the Laplace transform method is that it shows the effect of the initial conditions on the solution function.

Example 2

Solve the initial-value problem

$$y' + 3y = 3, \qquad y(0) = 0$$

SOLUTION   Taking the Laplace transform of both sides and letting $\mathcal{L}\{y\} = Y(s)$, we get

$$[sY(s) - 0] + 3Y(s) = \frac{3}{s}$$

Solving for $Y(s)$ yields

$$Y(s) = \frac{3}{s(s + 3)}$$

The partial fractions expansion of the right-hand side allows us to write

$$Y(s) = \frac{1}{s} - \frac{1}{s + 3}$$

Therefore the solution is

$$y = \mathcal{L}^{-1}\left\{\frac{1}{s}\right\} - \mathcal{L}^{-1}\left\{\frac{1}{s + 3}\right\}$$
$$= 1 - e^{-3t}$$

∎

Example 3

Solve the initial-value problem

$$\ddot{x} + 4x = e^{-t}, x(0) = 2, \dot{x}(0) = 1$$

SOLUTION   Letting $\mathcal{L}\{x(t)\} = X(s)$, we get

$$[s^2X(s) - 2s - 1] + 4X(s) = \frac{1}{s + 1}$$

Solving for $X(s)$,

$$X(s) = \frac{2s^2 + 3s + 2}{(s + 1)(s^2 + 4)}$$

The partial fractions for this are

$$\frac{2s^2 + 3s + 2}{(s + 1)(s^2 + 4)} = \frac{A}{s + 1} + \frac{Bs + C}{s^2 + 4}$$

from which $A = \frac{1}{5}$, $B = \frac{9}{5}$, and $C = \frac{6}{5}$. Thus

$$X(s) = \frac{\frac{1}{5}}{s + 1} + \frac{\frac{9}{5}s + \frac{6}{5}}{s^2 + 4}$$

Taking the inverse transform yields

$$x(t) = \tfrac{1}{5}e^{-t} + \tfrac{9}{5}\cos 2t + \tfrac{3}{5}\sin 2t$$ ∎

Example 4

Solve the initial-value problem

$$y'' + 2y' + 5y = 10, \; y(0) = 1, \; y'(0) = 0$$

SOLUTION   Taking the Laplace transform of both sides of the given equation and letting $Y = \mathcal{L}\{y\}$, we have

$$\mathcal{L}\{y''\} + 2\mathcal{L}\{y'\} + 5\mathcal{L}\{y\} = \mathcal{L}\{10\}$$

$$[s^2Y - s] + 2[sY - 1] + 5Y = \frac{10}{s}$$

Solving for Y yields

$$Y = \frac{s^2 + 2s + 10}{s(s^2 + 2s + 5)}$$

The partial fractions for this function are

$$\frac{s^2 + 2s + 10}{s(s^2 + 2s + 5)} = \frac{A}{s} + \frac{Bs + C}{s^2 + 2s + 5}$$

$$s^2 + 2s + 10 = A(s^2 + 2s + 5) + s(Bs + C)$$

$$= (A + B)s^2 + (2A + C)s + 5A$$

Equating the powers of s, we get

$$1 = A + B \qquad 2 = 2A + C \qquad 10 = 5A$$

Solving this system of equations for A, B, and C yields $A = 2$, $B = -1$, and $C = -2$. Hence the Laplace transform of the solution function can be written

$$Y = \frac{2}{s} - \frac{s + 2}{s^2 + 2s + 5}$$

$$= \frac{2}{s} - \frac{(s + 1) + 1}{(s + 1)^2 + 4}$$

$$= \frac{2}{s} - \frac{s + 1}{(s + 1)^2 + 4} - \frac{1}{(s + 1)^2 + 4}$$

Finally, taking the inverse transform, the solution function is

$$y = 2 - e^{-t} \cos 2t - \tfrac{1}{2} e^{-t} \sin 2t \qquad\blacksquare$$

Example 5

Find the current in a series RC circuit if the driving voltage is $v = e^{-t}$ for $t \geq 0$, and the initial charge on the capacitor is zero. See Figure 6–6.

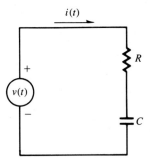

i(t)

R

+

v(t)

−

C

FIGURE 6–6

SOLUTION   The equation for the circuit is Kirchhoff's law in the form

$$Ri + \frac{1}{C}\int_0^t i(x)\, dx = e^{-t}, \qquad q(0) = 0$$

See Section 5–6. Taking the Laplace transform of both sides and letting the transform of the current be $I(s)$, we get

$$RI(s) + \frac{1}{Cs} I(s) = \frac{1}{s + 1}$$

Solving for $I(s)$,

$$I(s) = \frac{sC}{(RCs + 1)(s + 1)} = \frac{1}{R} \frac{s}{(s + 1/RC)(s + 1)}$$

Expressing the fraction on the right in terms of its partial fractions yields

$$I(s) = \frac{-1}{R(RC - 1)}\left[\frac{1}{s + 1/RC}\right] + \frac{C}{RC - 1}\left[\frac{1}{s + 1}\right]$$

Finally, the inverse transform yields

$$i(t) = \frac{-1}{R(RC - 1)} e^{-t/RC} + \frac{C}{RC - 1} e^{-t}$$ ∎

EXERCISES FOR SECTION 6–6

Use the Laplace transform method to solve each of the initial-value problems in Exercises 1–22.

1. $y' + 3y = 0, y(0) = 2$

2. $y' - 4y = 0, y(0) = -1$

3. $x''(t) + 9x(t) = 0, x(0) = 3, x'(0) = 0$

4. $y'' + 4y = 0, y(0) = 1, y'(0) = 1$

5. $y'' + y' = 0, y(0) = 0, y'(0) = 1$

6. $\ddot{x} - 2\dot{x} = 0, x(0) = 0, \dot{x}(0) = -2$

7. $y'' + 3y' + 2y = 0, y(0) = 0, y'(0) = -2$

8. $d^2x/dt^2 - dx/dt - 6x = 0, x(0) = 0, x'(0) = 1$

9. $\dot{x} + 2x = 4, x(0) = 0$

10. $y' - y = e^{2t}, y(0) = 2$

11. $d^2y/dt^2 + 4y = 1, y(0) = 0, y'(0) = 0$

12. $y'' - y = 10, y(0) = 0, y'(0) = -2$

13. $y'' - 2y' = -4, y(0) = 0, y'(0) = 0$

14. $d^2x/dt^2 + 4dx/dt = 2, x(0) = 0, \dot{x}(0) = -2$

15. $dx/dt - 3x = \sin 2t, x(0) = 2$

16. $y' + 4y = \cos 3t, y(0) = 0$

17. $d^2x/dt^2 + dx/dt - 2x = -4, x(0) = 2, \dot{x}(0) = 3$

18. $3\ddot{x} + 6\dot{x} + 3x = 9, x(0) = 0, \dot{x}(0) = 6$

19. $y'' - 6y' + 9y = te^{3t}, y(0) = 3, y'(0) = 2$

20. $y' + 3y = e^{-t} \sin t, y(0) = 1$

21. $y' + 2y = e^{-2t} \cos 2t, y(0) = 1$

22. $y'' + 2y' = e^{-t} \cos t, y(0) = 0, y'(0) = 0$

23. A 96-lb rocket is propelled from rest by a motor that generates a constant thrust of 1000 lb. The rocket experiences a drag force numerically equal to its velocity. Find the displacement of the rocket at $t = 5$ sec.

24. A spring is stretched 0.5 ft by an 8-lb weight. Assuming there is no damping force, find the equation for the displacement of the weight if it is pushed up 0.25 ft and released.

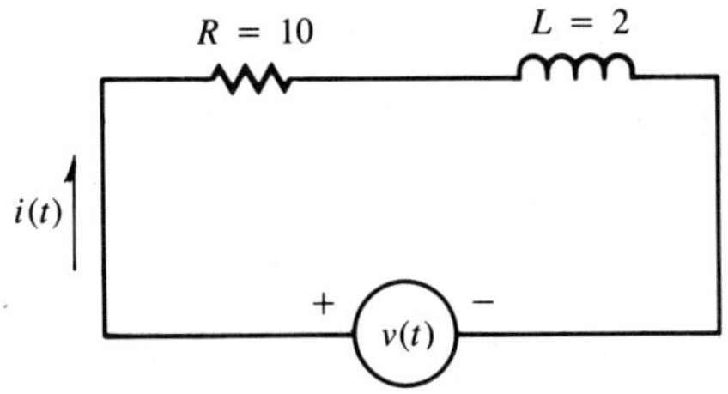

FIGURE 6–7

25. Solve the spring-mass problem in Exercise 24 if the weight is pushed up 0.25 ft then given a downward velocity of 4 ft/sec.

26. Determine the amplitude and period of the oscillation described in Exercise 25.

27. Find the current in the circuit shown in Figure 6–7 if the impressed voltage is 12 v and the initial current is zero.

28. Find the current in the circuit in Figure 6–7 if the impressed voltage varies with time according to $v = \sin 2t$. Assume $i(0) = 0$.

29. Find the current in a series RC circuit with $R = 5$, $C = 1$, and $v = 3t$. Assume $q(0) = 0$.

REVIEW EXERCISES FOR CHAPTER 6

In Exercises 1–10 evaluate the given Laplace transform.

1. $\mathcal{L}\{5 + 2e^{-3t}\}$

2. $\mathcal{L}\{t^5 - te^{2t}\}$

3. $\mathcal{L}\{3 \sin 4t + t^7 e^{-t}\}$

4. $\mathcal{L}\{5 \sinh 3t - 2 \cosh 2t\}$

5. $\mathcal{L}\left\{\dfrac{t^2}{e^{4t}}\right\}$

6. $\mathcal{L}\{(t + 5)e^t\}$

7. $\mathcal{L}\{(e^t + \sin 5t)e^{-3t}\}$

8. $\mathcal{L}\left\{\dfrac{5 \cos 2t}{e^{3t}}\right\}$

9. $\mathcal{L}\{10 \, e^{t/2} \sin 3t \cos 3t\}$

10. $\mathcal{L}\{4 \cos^2 \sqrt{2}t\}$

11. Show that the function

$$f(t) = \begin{cases} 0, & 0 \le t < 5 \\ 1, & t \ge 5 \end{cases}$$

is piecewise continuous on $t \ge 0$ and find its Laplace transform.

12. Show that the function

$$g(t) = \begin{cases} e^{-t}, & 0 \le t < 1 \\ 0, & t \ge 1 \end{cases}$$

is piecewise continuous on $t \ge 0$, and find its Laplace transform.

In Exercises 13–16 indicate which of the functions are piecewise continuous and of exponential order on t ≥ 0.

13. t^{-2} **14.** $3 + \cos t^2$ **15.** te^{2t} **16.** $\sqrt{t+3}$

In Exercises 17–26 find the indicated inverse Laplace transform.

17. $\mathcal{L}^{-1}\left\{\dfrac{3s+2}{s^2+3}\right\}$ **18.** $\mathcal{L}^{-1}\left\{\dfrac{1-s}{s^2+2}\right\}$

19. $\mathcal{L}^{-1}\left\{\dfrac{2s-9}{s(s+3)}\right\}$ **20.** $\mathcal{L}^{-1}\left\{\dfrac{15}{(s+1)(s-2)}\right\}$

21. $\mathcal{L}^{-1}\left\{\dfrac{3s^2+7s+4}{s^3+4s}\right\}$ **22.** $\mathcal{L}^{-1}\left\{\dfrac{2s^2-4s-10}{(s+3)(s^2+1)}\right\}$

23. $\mathcal{L}^{-1}\left\{\dfrac{3}{s^2+6s+11}\right\}$ **24.** $\mathcal{L}^{-1}\left\{\dfrac{2}{s^2-2s+5}\right\}$

25. $\mathcal{L}^{-1}\left\{\dfrac{2s-15}{s^2-18s+17}\right\}$ **26.** $\mathcal{L}^{-1}\left\{\dfrac{1-s}{s^2+2s+10}\right\}$

Use the Laplace transform to solve the initial-value problems in Exercises 27–30.

27. $\ddot{x} + 4x = 10, x(0) = 0, \dot{x}(0) = 0$

28. $y'' - 3y' = 1, y(0) = 2, y'(0) = 0$

29. $y'' + 2y' = 4e^{-2t}, y(0) = 0, y'(0) = -1$

30. $y'' - 6y' + 9y = e^{3t}, y(0) = 0, y'(0) = 0$

7

More Laplace Transforms

The suspension system of a car softens the effects of roadway irregularities on the passengers. In simplified cases the response x of the car to a bump in the road is described by the differential equation

$$m\frac{d^2x}{dt^2} + c\frac{dx}{dt} + kx = f(t)$$

where m, c, and k are constants of the suspension system and $f(t)$ is the force experienced by the car. Methods are developed in this chapter that allow the solution of this differential equation when $f(t)$ contains a finite number of finite discontinuities.

7–1   DERIVATIVES OF LAPLACE TRANSFORMS

If $f(t)$ is piecewise continuous and of exponential order for $t \geq 0$, it can be shown that the Laplace transform integral may be differentiated with respect to s by differentiating under the integral sign. That is, if

$$F(s) = \int_0^\infty e^{-st} f(t)\, dt$$

then

$$\frac{d}{ds} F(s) = \int_0^\infty \frac{d}{ds} e^{-st} f(t)\, dt$$

$$= \int_0^\infty -te^{-st} f(t)\, dt = \int_0^\infty e^{-st}[-tf(t)]\, dt$$

The last integral is $\mathcal{L}\{-tf(t)\}$. Therefore

$$\frac{d}{ds} F(s) = \mathcal{L}\{-tf(t)\}$$

Similarly,

$$\frac{d^2}{ds^2} F(s) = \frac{d}{ds} \int_0^\infty e^{-st}[(-t)f(t)]\, dt$$

$$= \int_0^\infty e^{-st}[(-t)^2 f(t)]\, dt$$

$$= \mathcal{L}\{(-t)^2 f(t)\}$$

Observing the pattern in these two cases, we state the general case in the following theorem:

THEOREM 7–1

If $F(s) = \mathcal{L}\{f(t)\}$ and n is a positive integer, then the nth derivative of $F(s)$ is given by

$$\frac{d^n}{ds^n} F(s) = \mathcal{L}\{(-t)^n f(t)\} \tag{7–1}$$

COMMENT: Differentiation of $F(s)$ with respect to s corresponds to multiplication of $f(t)$ by $-t$.

Example 1

Use Theorem 7–1 to find $\mathcal{L}\{t \sin kt\}$.

SOLUTION   We know that $\mathcal{L}\{\sin kt\} = k/(s^2 + k^2)$. Therefore from Equation 7–1 we know that

$$\frac{d}{ds}\left[\frac{k}{s^2 + k^2}\right] = \mathcal{L}\{(-t)\sin kt\}$$

or

$$\mathcal{L}\{t \sin kt\} = -\frac{d}{ds}\left[\frac{k}{s^2 + k^2}\right] = \frac{2ks}{(s^2 + k^2)^2}$$ ∎

Example 2

Find $\mathcal{L}\{t^2 \sin kt\}$.

SOLUTION   From Example 1 we know that $\mathcal{L}\{t \sin kt\} = 2ks/(s^2 + k^2)^2$. We proceed as follows:

$$\begin{aligned}
\mathcal{L}\{t^2 \sin kt\} &= \mathcal{L}\{(-t)(t \sin kt)\} \\
&= -\frac{d}{ds}\left[\frac{2ks}{(s^2 + k^2)^2}\right] \\
&= -\frac{(s^2 + k^2)^2 2k - 8ks^2(s^2 + k^2)}{(s^2 + k^2)^4} \\
&= \frac{2k(3s^2 - k^2)}{(s^2 + k^2)^3}
\end{aligned}$$ ∎

As a convenient method of finding certain inverse Laplace transforms, consider Theorem 7–1 with $n = 1$. That is,

$$\mathcal{L}\{-tf(t)\} = \frac{d}{ds}F(s)$$

Taking the inverse transform of both sides and solving for $f(t)$, we get

$$f(t) = -\frac{1}{t}\mathcal{L}^{-1}\left\{\frac{d}{ds}F(s)\right\}$$ (7–2)

Example 3

Evaluate $f(t) = \mathcal{L}^{-1}\{\ln(s - 2)/(s + 2)\}$.

SOLUTION   Using Equation 7–2, we have

$$\begin{aligned}
f(t) &= -\frac{1}{t}\mathcal{L}^{-1}\left\{\frac{d}{ds}\ln\frac{s - 2}{s + 2}\right\} \\
&= -\frac{1}{t}\mathcal{L}^{-1}\left\{\frac{4}{s^2 - 4}\right\} = -\frac{2}{t}\sinh 2t
\end{aligned}$$ ∎

Example 4

Evaluate $f(t) = \mathcal{L}^{-1}\{\pi/2 - \tan^{-1} s\}$.

SOLUTION   Using Equation 7–2, we have

$$
\begin{aligned}
f(t) &= -\frac{1}{t} \mathcal{L}^{-1}\left\{\frac{d}{ds}(\pi/2 - \tan^{-1} s)\right\} \\
&= -\frac{1}{t} \mathcal{L}^{-1}\left\{\frac{-1}{1 + s^2}\right\} \\
&= \frac{1}{t} \sin t
\end{aligned}
$$

■

EXERCISES FOR SECTION 7–1

In Exercises 1–12 use Theorem 7–1 to find the Laplace transform of the given function.

1. $t \cos kt$

2. $t^2 \cos kt$

3. $t^2 e^{2t}$

4. $t^4 e^{3t}$

5. $t^n e^{kt}$, n a positive integer

6. $t \sin t \cos t$

7. $\sin kt - t \cos kt$

8. $t \sin^2 t$

9. $t^2(\sin kt + \cos kt)$

10. $t^2 \sinh kt$

11. $t^2 e^{3t} \sin 5t$

12. $te^{3t} \cos^2 4t$

In Exercises 13–22 use Equation 7–2 to find $f(t)$ for the given Laplace transform.

13. $\ln \dfrac{s + 1}{s - 1}$

14. $\ln \dfrac{s - 2}{s - 1}$

15. $\ln\left(1 - \dfrac{5}{s}\right)$

16. $\ln\left[(s - 3)/(s - 1)\right]$

17. $\ln\left[s^2/(s^2 + 4)\right]$

18. $\ln\left[(s^2 + 4)/(s^2 + 1)\right]$

19. $\pi/2 - \tan^{-1}(s + 4)$

20. $\cot^{-1} s$

21. $\tan^{-1}(2/s)$

22. $\cot^{-1}(3s)$

7–2   PERIODIC FUNCTIONS

Periodic functions dominate the study of many technical subjects and, aside from the trigonometric functions, the techniques of Chapter 5 are difficult to apply when the driving function is periodic. The next theorem shows how to find the Laplace transform of any periodic function.

THEOREM 7–2

If $f(t)$ is piecewise continuous on any finite interval and periodic with period p for $t \geq 0$, then

$$\mathcal{L}\{f(t)\} = \frac{1}{1 - e^{-sp}} \int_0^p e^{-st} f(t) \, dt \tag{7–3}$$

PROOF   Since $f(t)$ is periodic with period p, we may write

$$f(t) = f(t + p) = f(t + 2p) = \cdots = f(t + np) = \cdots$$

By assumption, $f(t)$ is piecewise continuous, so the Laplace transform may be written

$$\mathcal{L}\{f(t)\} = \int_0^p e^{-st} f(t) \, dt + \int_p^\infty e^{-st} f(t) \, dt$$

We make the substitution $t = u + p$ in the second integral, observing that $du = dt$. Further, $u = 0$ when $t = p$, and u is infinite when t is infinite. Thus

$$\mathcal{L}\{f(t)\} = \int_0^p e^{-st} f(t) \, dt + \int_0^\infty e^{-s(u+p)} f(u + p) \, du$$

$$= \int_0^p e^{-st} f(t) \, dt + \int_0^\infty e^{-sp} e^{-su} f(u) \, du$$

$$= \int_0^p e^{-st} f(t) \, dt + e^{-sp} \int_0^\infty e^{-su} f(u) \, du$$

But $\int_0^\infty e^{-su} f(u) \, du = \mathcal{L}\{f(t)\}$, so

$$\mathcal{L}\{f(t)\} = \int_0^p e^{-st} f(t) \, dt + e^{-sp} \mathcal{L}\{f(t)\}$$

Solving for $\mathcal{L}\{f(t)\}$ yields

$$\mathcal{L}\{f(t)\} = \frac{1}{1 - e^{-sp}} \int_0^p e^{-st} f(t) \, dt$$

Example 1

Find the Laplace transform of the periodic triangular wave shown in Figure 7–1.

$$f(t) = 2t, \, 0 \leq t < 2 \qquad \text{and} \qquad f(t + 2) = f(t)$$

SOLUTION   The function $f(t)$ is periodic with period 2, so we use Theorem 7–2 to obtain

$$\mathcal{L}\{f(t)\} = \frac{1}{1 - e^{-2s}} \int_0^2 e^{-st} [2t] \, dt$$

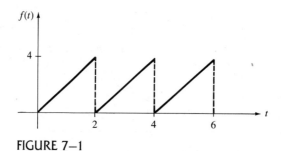

FIGURE 7–1

Using integration by parts with

$u = t$	$dv = e^{-st}\,dt$
$du = dt$	$v = -\dfrac{1}{s}e^{-st}$

we proceed as follows:

$$\mathcal{L}\{f(t)\} = \frac{2}{1 - e^{-2s}}\left[-\frac{t}{s}e^{-st}\Big|_0^2 + \int_0^2 \frac{1}{s}e^{-st}\,dt \right]$$

$$= \frac{2}{1 - e^{-2s}}\left[\frac{-te^{-st}}{s} - \frac{e^{-st}}{s^2} \right]_0^2$$

$$= \frac{2}{1 - e^{-2s}}\left[-\frac{2e^{-2s}}{s} - \frac{e^{-2s}}{s^2} + \frac{1}{s^2} \right]$$

$$= \frac{2}{1 - e^{-2s}}\left[\frac{1}{s^2}(1 - e^{-2s}) - \frac{2e^{-2s}}{s} \right]$$

$$= \frac{2}{s^2} - \frac{4e^{-2s}}{s(1 - e^{-2s})}$$ ∎

Example 2

Find the Laplace transform of the function whose graph is shown in Figure 7–2. The graph is called the "half-wave rectification of the sine wave."

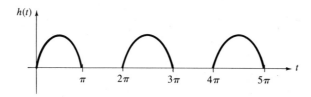

FIGURE 7–2

$$h(t) = \begin{cases} \sin t, & 0 \le t < \pi \\ 0, & \pi \le t < 2\pi \end{cases}$$

$$h(t) = h(t + 2\pi)$$

SOLUTION Using Theorem 7–2, we get

$$\mathcal{L}\{h(t)\} = \frac{1}{1 - e^{-2\pi s}} \int_0^\pi e^{-st} \sin t \, dt$$

$$= \frac{1}{1 - e^{-2\pi s}} \left[\frac{e^{-st}(-s \sin t - \cos t)}{s^2 + 1} \right]_0^\pi$$

$$= \frac{1}{1 - e^{-2\pi s}} \frac{1 + e^{-\pi s}}{s^2 + 1} = \frac{1}{(1 - e^{-\pi s})(s^2 + 1)} \qquad \blacksquare$$

Example 3

Find the Laplace transform of the rectangular waveform shown in Figure 7–3.

$$f(t) = \begin{cases} 1, & 0 \le t < 1 \\ -1, & 1 \le t < 2 \end{cases}$$

$$f(t) = f(t + 2)$$

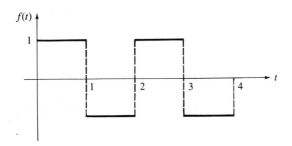

FIGURE 7–3

SOLUTION

$$\mathcal{L}\{f(t)\} = \frac{1}{1 - e^{-2s}} \left[\int_0^1 e^{-st}(1) \, dt + \int_1^2 e^{-st}(-1) \, dt \right]$$

$$= \frac{1}{1 - e^{-2s}} \frac{1 - 2e^{-s} + e^{-2s}}{s}$$

$$= \frac{(1 - e^{-s})^2}{s(1 - e^{-s})(1 + e^{-s})} = \frac{1 - e^{-s}}{s(1 + e^{-s})} \qquad \blacksquare$$

Example 4

Find the Laplace transform of the "staircase" function shown in Figure 7–4.

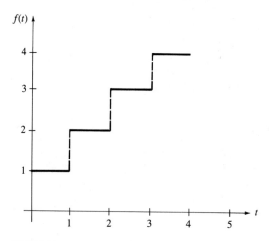

FIGURE 7–4

$$f(t) = \begin{cases} 1, & 0 \le t < 1 \\ 2, & 1 \le t < 2 \\ 3, & 2 \le t < 3 \\ \vdots & \\ n, & n - 1 \le t < n \\ \vdots & \end{cases}$$

SOLUTION   This function is not periodic; further, its Laplace transform could be found by direct application of the definition. However, by observing that the staircase function is equal to the difference of the two functions shown in Figure 7–5, then the transform can be found as follows:

$$\mathscr{L}\{\text{Staircase function}\} = \mathscr{L}\{t + 1\} - \mathscr{L}\{\text{Triangular wave function}\}$$

$$= \frac{1}{s^2} + \frac{1}{s} - \frac{1}{s^2} + \frac{e^{-s}}{s(1 - e^{-s})}$$

$$= \frac{1}{s} + \frac{e^{-s}}{s(1 - e^{-s})}$$

The Laplace transform of the triangular wave was found in a manner similar to Example 1. Simplifying the expression on the right yields

$$\mathscr{L}\{\text{Staircase function}\} = \frac{1}{s(1 - e^{-s})}$$ ■

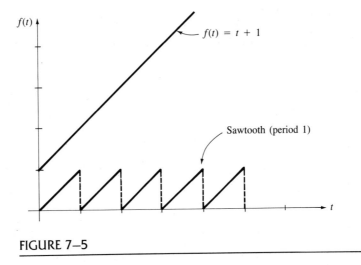

$f(t) = t + 1$

Sawtooth (period 1)

FIGURE 7–5

EXERCISES FOR SECTION 7–2

1. Use Theorem 7–2 to find $\mathscr{L}\{\sin t\}$.

2. Use Theorem 7–2 to find $\mathscr{L}\{\cos 3t\}$.

In Exercises 3–10 find the Laplace transform of the given function.

3. The triangular wave shown in Figure 7–6.

$$f(t) = \begin{cases} t, & 0 \le t < 1 \\ 2 - t, & 1 \le t < 2 \end{cases}$$
$$f(t) = f(t + 2)$$

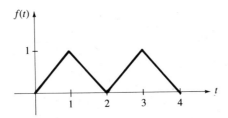

FIGURE 7–6

4. The full-wave rectification of the sine wave shown in Figure 7–7.

$$f(t) = \sin t, \, 0 \le t < \pi$$
$$f(t) = f(t + \pi)$$

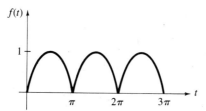

FIGURE 7–7

5. $f(t) = |\cos kt|$ **6.** $f(t) = \begin{cases} 1, & 0 \le t < \frac{1}{2}L \\ 0, & \frac{1}{2}L \le t < L \end{cases}$

$$f(t) = f(t + L)$$

7. $f(t) = \begin{cases} 2, & 0 < t < 1 \\ 0, & 1 < t < 2, \quad f(t) = f(t + 3) \\ -2, & 2 < t < 3 \end{cases}$

8. $f(t) = e^t \cdot \begin{pmatrix} \text{Rectangular wave} \\ \text{shown in Figure 7–3} \end{pmatrix}$ **9.** $f(t) = e^t \cdot \begin{pmatrix} \text{Staircase wave} \\ \text{shown in Figure 7–4} \end{pmatrix}$

10. The function similar to the rectangular wave in Figure 7–3, except that its amplitude is h, its period is p, and it is moved upward k units.

In Exercises 11–14 find the Laplace transform of the solution function of the initial-value problem $y'' + 4y' + 3y = f(t)$, $y(0) = y'(0) = 0$, for the given driving function.

11. $f(t) = $ the waveform in Figure 7–1.

12. $f(t) = $ the waveform in Figure 7–2.

13. $f(t) = $ the waveform in Figure 7–3.

14. $f(t) = $ the waveform in Figure 7–4.

7–3   THE CONVOLUTION THEOREM; INTEGRAL EQUATIONS

In Section 6–5 we derived the formula

$$\mathcal{L}\left\{ \int_0^t f(\tau) \, d\tau \right\} = \frac{1}{s} \mathcal{L}\{f(t)\}$$

for the Laplace transform of an indefinite integral (see Theorem 6–4). This theorem is a special case of the convolution theorem to be derived in this section. We first define the operation of convolution.

DEFINITION

Let f and g be piecewise continuous functions for $t \geq 0$. Then the **convolution** of f and g is denoted by $f*g$ and is defined by the integral

$$f(t)*g(t) = \int_0^t f(\tau)g(t - \tau) \, d\tau$$

COMMENT: In taking the convolution of two functions, the order of the two functions is immaterial. That is,

$$f(t)*g(t) = g(t)*f(t)$$

Example 1

Find the convolution of $\sin t$ and $\cos t$.

SOLUTION

$$\sin t * \cos t = \int_0^t \sin(t - \tau) \cos \tau \, d\tau \qquad = \int_o^t \cos \tau \, \sin(t - \tau) \, d\tau$$

$$= \tfrac{1}{2} \int_0^t [\sin t + \sin(t - 2\tau)] \, d\tau$$

$$= \left[\frac{\tau \sin t}{2} + \frac{\cos(t - 2\tau)}{4} \right]_0^t$$

$$= \tfrac{1}{2} t \sin t \qquad\qquad ■$$

THEOREM 7–3

If f and g are piecewise continuous and of exponential order for $t \geq 0$, then the Laplace transform of $f*g$ is given by the product of the transform of f and the transform of g. That is,

$$\mathcal{L}\{f*g\} = F(s)G(s)$$

PROOF

$$\mathcal{L}\{f*g\} = \int_0^\infty e^{-st} \left[\int_0^t f(t - \tau)g(\tau) \, d\tau \right] dt$$

$$= \lim_{B \to \infty} \int_0^B e^{-st} \left[\int_0^t f(t - \tau)g(\tau) \, d\tau \right] dt$$

The iterated integral is equal to the double integral of $e^{-st}f(t - \tau)g(\tau)$ over the region R in the $t\tau$-plane, between the t-axis and the line $\tau = t$. (See Figure 7–8.) The change of variable $u = t - \tau$, $v = \tau$, changes the region R into the region R^* in the uv-plane. Hence

$$\mathcal{L}\{f*g\} = \lim_{B \to \infty} \int_0^B \int_0^{B-v} e^{-s(u+v)}f(u)g(v) \, du \, dv$$

$$= \int_0^\infty \left[\int_0^\infty e^{-su}f(u) \, du \right] e^{-sv}g(v) \, dv$$

$$= F(s)G(s)$$

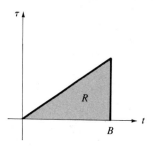
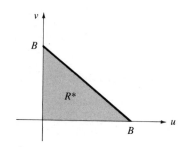

FIGURE 7–8

Example 2

Evaluate $\mathcal{L}\{f(t)*g(t)\}$ for $f(t) = e^{-t}$ and $g(t) = \sin 2t$.

SOLUTION   Using Theorem 7–3,

$$\mathcal{L}\{f(t)*g(t)\} = \mathcal{L}\{e^{-t}\}\mathcal{L}\{\sin 2t\}$$

$$= \frac{1}{s + 1} \cdot \frac{2}{s^2 + 4} = \frac{2}{(s + 1)(s^2 + 4)} \qquad \blacksquare$$

Example 3

Use the convolution theorem to find

$$\mathcal{L}^{-1}\left\{ \frac{1}{s(s^2 + 1)} \right\}$$

SOLUTION   Note that

$$\frac{1}{s(s^2 + 1)} = \frac{1}{s} \cdot \frac{1}{s^2 + 1} = F(s)G(s)$$

Since

$$\mathcal{L}^{-1}\left\{ \frac{1}{s} \right\} = 1 \qquad \text{and} \qquad \mathcal{L}^{-1}\left\{ \frac{1}{s^2 + 1} \right\} = \sin t$$

it follows that

$$\mathcal{L}^{-1}\left\{\frac{1}{s(s^2 + 1)}\right\} = 1*\sin t = \int_0^t [1] \sin \tau \, d\tau = 1 - \cos t \qquad \blacksquare$$

Example 4

Use the convolution theorem to find $\mathcal{L}^{-1}\left\{\dfrac{1}{(s + 1)(s - 2)}\right\}$.

SOLUTION   Since $\mathcal{L}^{-1}\{1/(s + 1)\} = e^{-t}$ and $\mathcal{L}^{-1}\{1/(s - 2)\} = e^{2t}$, we write

$$\mathcal{L}^{-1}\left\{\frac{1}{(s + 1)(s - 2)}\right\} = e^{-t}*e^{2t} = \int_0^t e^{-\tau} e^{2(t-\tau)} \, d\tau$$

$$= \int_0^t e^{2t} e^{-3\tau} \, d\tau$$

$$= e^{2t} \int_0^t e^{-3\tau} \, d\tau$$

$$= e^{2t}[-\tfrac{1}{3}e^{-3t} + \tfrac{1}{3}]$$

$$= \tfrac{1}{3}[e^{2t} - e^{-t}] \qquad \blacksquare$$

Example 5

Use the convolution theorem to find $\mathcal{L}^{-1}\{s^2/(s^2 + 1)^2\}$.

SOLUTION   Notice that the given transform is the square of $s/(s^2 + 1)$, whose inverse transform is $\cos t$. Hence

$$\mathcal{L}^{-1}\left\{\frac{s^2}{(s^2 + 1)^2}\right\} = \cos t * \cos t = \int_0^t \cos(t - \tau) \cos \tau \, d\tau$$

$$= \tfrac{1}{2}\int_0^t [\cos t + \cos(t - 2\tau)] \, d\tau = [\tfrac{1}{2}\tau \cos t - \tfrac{1}{4} \sin (t - 2\tau)]_0^t$$

$$= \tfrac{1}{2}t \cos t - \tfrac{1}{4} \sin (-t) + \tfrac{1}{4} \sin t = \tfrac{1}{2}[\sin t + t \cos t] \qquad \blacksquare$$

The convolution theorem has several important applications. Consider, for example, the initial-value problem

$$a_2 y'' + a_1 y' + a_0 y = f(t)$$
$$y(0) = y'(0) = 0$$

Taking the Laplace transform of both sides, we have

$$Y(s) \cdot (a_2 s^2 + a_1 s + a_0) = F(s)$$

Letting $K(s) = a_2 s^2 + a_1 s + a_0$ and solving for $Y(s)$, we get

$$Y(s) = \frac{1}{K(s)} F(s)$$

Then if $\mathcal{L}^{-1}\{1/K(s)\} = k(t)$, we have

$$y(t) = k(t) * f(t) \tag{7-4}$$

This manner of expressing the solution of an initial-value problem clearly relates the driving function with the solution.

Example 6

Use (7–4) to write the solution to the initial-value problem

$$y'' + y = f(t), \; y(0) = y'(0) = 0$$

Then write the solution for $f(t) = 1$.

SOLUTION   The solution for any driving function $f(t)$, using Equation 7–4, is given by

$$y(t) = k(t) * f(t)$$

where

$$k(t) = \mathcal{L}^{-1}\left\{\frac{1}{s^2 + 1}\right\} = \sin t$$

Thus, the solution for $y(t)$ is

$$y(t) = \sin t * f(t)$$

For $f(t) = 1$, $y(t)$ is

$$y(t) = (\sin t) * 1 = \int_0^t \sin \tau \, d\tau = 1 - \cos t \qquad \blacksquare$$

The convolution theorem may be used to great advantage in solving equations with an unknown function under an indefinite integral sign. Equations of this type are called **integral equations.** To use the Laplace transform, the indefinite integral must be a convolution integral or an indefinite integral of the unknown function.

Example 7

Find the function y if

$$y = t + \int_0^t y(\tau) \sin (t - \tau) \, d\tau$$

SOLUTION We take the Laplace transform of both sides and denote the transform of y by $Y(s)$. Thus

$$Y(s) = \frac{1}{s^2} + Y(s)\frac{1}{s^2 + 1}$$

Solving for $Y(s)$, we obtain

$$Y(s) = \frac{s^2 + 1}{s^4} = \frac{1}{s^2} + \frac{1}{s^4}$$

The inverse transform yields

$$y = t + \tfrac{1}{6}t^3 \qquad\blacksquare$$

Example 8

Solve the integral equation

$$y(t) = t + \int_0^t y(x)\, dx + \int_0^t (t - x)\, y(x)\, dx$$

SOLUTION Taking the Laplace transform of both sides, we obtain

$$Y(s) = \frac{1}{s^2} + \frac{Y(s)}{s} + \frac{Y(s)}{s^2}$$

Solving for $Y(s)$,

$$Y(s) = \frac{1}{s^2 - s - 1}$$

Letting $s^2 - s - 1 = s^2 - s + \tfrac{1}{4} - \tfrac{5}{4}$, the transform becomes

$$Y(s) = \frac{1}{(s^2 - s + \tfrac{1}{4}) - \tfrac{5}{4}}$$

$$= \frac{1}{(s - \tfrac{1}{2})^2 - (\sqrt{5}/2)^2}$$

Taking the inverse transform, the desired function is

$$y(t) = \frac{2}{\sqrt{5}}\, e^{t/2} \sinh \frac{\sqrt{5}}{2}t \qquad\blacksquare$$

EXERCISES FOR SECTION 7-3

In Exercises 1–6 find the Laplace transform of the given function.

1. $\displaystyle\int_0^t (t - u)\sin 2u\, du$

2. $\displaystyle\int_0^t e^{-(t-u)}\cos u\, du$

3. $2*e^{3t}$ **4.** $t*\cos t$

5. $t^3*\sin t$ **6.** e^t*t

In Exercises 7–16 use the convolution theorem to find the inverse transform of the given function of s.

7. $\dfrac{1}{s(s^2 + 4)}$ **8.** $\dfrac{1}{s^2(s - 1)}$

9. $\dfrac{3}{2s(s^2 + 9)}$ **10.** $\dfrac{1}{s(s - 2)}$

11. $\dfrac{1}{s^2(s + 3)}$ **12.** $\dfrac{3}{s(s^2 + 2)}$

13. $\dfrac{2}{s^2 + s - 6}$ **14.** $\dfrac{1}{s^2 + 3s + 2}$

15. $\dfrac{1}{(s^2 + 1)^2}$ **16.** $\dfrac{s}{(s^2 + 1)^2}$

Solve the initial-value problems in Exercises 17–20 in terms of a general input function, and then for the specific $f(t)$ that is given. In all cases $y(0) = y'(0) = 0$.

17. $y'' + 2y' + y = f(t); f(t) = 5$

18. $y'' + 3y' + 2y = f(t); f(t) = \sin t$

19. $y'' + y' - 2y = f(t); f(t) = t$

20. $y'' + y = f(t); f(t) = \sin t$

Solve each of the equations in Exercises 21–29.

21. $y(t) = 1 + \displaystyle\int_0^t y(u) \sin(t - u)\, du$

22. $y(t) = \sin t + \displaystyle\int_0^t \sin(t - u)\, y(u)\, du$

23. $y'(t) + 3y(t) + 2\displaystyle\int_0^t y(u)\, du = 1,\ y(0) = 0$

24. $y(t) = t^2 + \displaystyle\int_0^t y(u) \sin(t - u)\, du$

25. $y(t) = t + 4\displaystyle\int_0^t (u - t)^2\, y(u)\, du$

26. $y(t) = t^2 - 2\displaystyle\int_0^t y(t - u) \sinh 2u\, du$

27. $y(t) = t + \displaystyle\int_0^t y(t - u)\, e^{-u}\, du$

28. $y(t) = t + 2\displaystyle\int_0^t \cos(t - u)\, y(u)\, du$

29. Show that $f(t)*g(t) = g(t)*f(t)$.

7-4   A STEP FUNCTION

Piecewise continuous functions have been described in previous sections by giving the appropriate "rule" over each subinterval of the domain. For example, the function $f(t)$ whose value is 2 for $0 \le t < 4$, and t for $t \ge 4$, is written

$$f(t) = \begin{cases} 2, & 0 \le t < 4 \\ t, & t \ge 4 \end{cases}$$

Sometimes it is desirable to express a piecewise continuous function by a single rule for its domain. This may be done by using a function, called the unit step function, as a building block. This function is also called the Heaviside function in honor of English engineer Oliver Heaviside (1850–1925), who pioneered much of the application of the Laplace transform.

DEFINITION

The **unit step function** is denoted by $u_a(t)$ and defined by

$$u_a(t) = \begin{cases} 0, & t < a \\ 1, & t > a \end{cases}$$

The real constant a indicates the point at which the "step" occurs. See Figure 7–9.

The function $u_a(t)$ is sometimes called a "turn-on" function because, when multiplied by another function, its effect is to suppress the function for $t < a$ and turn it on for $t > a$. Thus $u_a(t)f(t)$ means

$$u_a(t)f(t) = \begin{cases} 0, & t < a \\ f(t), & t > a \end{cases}$$

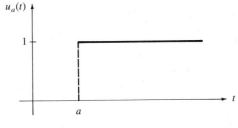

FIGURE 7–9

Combinations of unit step functions may be used to express other functions such as the unit pulse function, which is defined below and depicted in Figure 7–10.

DEFINITION

The **unit pulse function** is denoted by $P_{ab}(t)$ and defined by

$$P_{ab}(t) = \begin{cases} 1, & a < t < b \\ 0, & t \text{ not in } (a, b) \end{cases}$$

In terms of the unit step function, $P_{ab}(t)$ may be expressed as

$$P_{ab}(t) = u_a(t) - u_b(t)$$

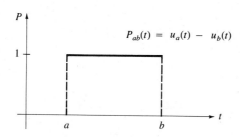

$$P_{ab}(t) = u_a(t) - u_b(t)$$

FIGURE 7–10

Notice that $u_a(t) - u_b(t)$ is zero for $t < a$ since both $u_a(t)$ and $u_b(t)$ are zero. For $a < t < b$, $u_a(t) = 1$ and $u_b(t) = 0$, so $u_a(t) - u_b(t) = 1$. For the interval $t > b$, $u_a(t) = 1$ and $u_b(t) = 1$, so $u_a(t) - u_b(t) = 0$.

Example 1

Express

$$f(t) = \begin{cases} 2, & 0 < t < 4 \\ t, & t > 4 \end{cases}$$

in terms of the unit step function.

SOLUTION   We begin by observing that $f(t) = 2$ for $0 < t < 4$ can be written as $2[P_{04}(t)] = 2[u_0(t) - u_4(t)]$. Also, $f(t) = t$ for $t > 4$ can be written as $tu_4(t)$. Adding the above expressions, we get

$$f(t) = 2[u_0(t) - u_4(t)] + tu_4(t)$$
$$= 2u_0(t) - 2u_4(t) + tu_4(t)$$

Since $u_0(t) = 1$ for $t \geq 0$, $f(t)$ can also be written

$$f(t) = 2 - 2u_4(t) + tu_4(t)$$ ■

Example 2

Express $g(t)$ in terms of the unit step function. See Figure 7–11.

$$g(t) = \begin{cases} 4 - t^2, & 0 < t < 1 \\ 0, & t > 1 \end{cases}$$

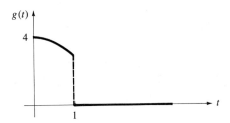

FIGURE 7–11

SOLUTION Since $4 - t^2$ is "on" for $0 < t < 1$ and is zero everywhere else, we can write

$$g(t) = P_{01}(t) \cdot (4 - t^2)$$
$$= (4 - t^2)[u_0(t) - u_1(t)]$$ ■

THEOREM 7–4

Let $u_a(t)$ be a unit step function for $t \geq 0$; then

$$\mathcal{L}\{u_a(t)\} = \frac{e^{-as}}{s}, \qquad s > 0 \tag{7–5}$$

PROOF $u_a(t)$ is piecewise continuous and of exponential order for $t \geq 0$, so its Laplace transform exists. By definition,

$$\mathcal{L}\{u_a(t)\} = \int_0^\infty e^{-st}[u_a(t)] \, dt$$

$$= \int_0^a e^{-st}[0] \, dt + \int_a^\infty e^{-st}[1] \, dt$$

$$= \lim_{B \to \infty} \frac{e^{-st}}{-s} \Big|_a^B = \frac{e^{-as}}{s}, \qquad s > 0$$

Notice that if $a = 0$, Formula 7–5 reduces to

$$\mathcal{L}\{u_0(t)\} = \mathcal{L}\{1\} = \frac{1}{s}$$

for $t > 0$. We also have, as an immediate result of Theorem 7–4, the inverse transform formula

$$\mathcal{L}^{-1}\left\{\frac{e^{-as}}{s}\right\} = u_a(t) \tag{7-6}$$

Example 3

Find the function $f(t)$ and sketch its graph if

$$f(t) = \mathcal{L}^{-1}\left\{\frac{1}{s}(1 - e^{-2s})^2\right\}$$

SOLUTION   Carrying out the indicated squaring operation, we get

$$f(t) = \mathcal{L}^{-1}\left\{\frac{1}{s}(1 - 2e^{-2s} + e^{-4s})\right\}$$

$$= \mathcal{L}^{-1}\left\{\frac{1}{s} - \frac{2e^{-2s}}{s} + \frac{e^{-4s}}{s}\right\}$$

$$= \mathcal{L}^{-1}\left\{\frac{1}{s}\right\} - 2\mathcal{L}^{-1}\left\{\frac{e^{-2s}}{s}\right\} + \mathcal{L}^{-1}\left\{\frac{e^{-4s}}{s}\right\}$$

$$= 1 - 2u_2(t) + u_4(t)$$

To write $f(t)$ without the step function, consider the interval $0 < t < 2$, for which $u_2(t)$ and $u_4(t)$ both equal zero. Then

$$f(t) = 1, \qquad 0 < t < 2$$

Also, on the interval $2 < t < 4$ we know that $u_2(t) = 1$ and $u_4(t) = 0$, so

$$f(t) = 1 - 2 = -1, \qquad 2 < t < 4$$

Finally, for $t > 4$ both $u_2(t)$ and $u_4(t)$ equal 1, so

$$f(t) = 1 - 2 + 1 = 0, \qquad t > 4$$

Combining these results, we have

$$f(t) = \begin{cases} 1, & 0 < t < 2 \\ -1, & 2 < t < 4 \\ 0, & t > 4 \end{cases}$$

Notice the economy of notation when the unit step function is used to represent $f(t)$. Three rules are required to describe this function if the unit step function is not used. The graph of $f(t)$ is shown in Figure 7–12.   ∎

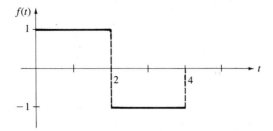

FIGURE 7–12

Example 4

Express the periodic square wave shown in Figure 7–13 in terms of unit step functions and then find its Laplace transform.

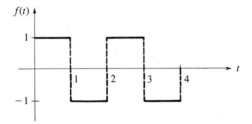

FIGURE 7–13

SOLUTION   The square wave may be thought of as a repeated pulse function alternating between 1 and -1. Denoting the function by f, we write

$$f(t) = P_{01}(t) - P_{12}(t) + P_{23}(t) - P_{34}(t) + \cdots$$
$$= [u_0(t) - u_1(t)] - [u_1(t) - u_2(t)] + [u_2(t) - u_3(t)] - \cdots$$
$$= u_0(t) - 2u_1(t) + 2u_2(t) - 2u_3(t) + \cdots$$
$$= u_0(t) + 2 \sum_{n=1}^{\infty} (-1)^n u_n(t)$$

The Laplace transform of this function may be found by using the formula for periodic functions developed in Theorem 7–2. Thus

$$\mathscr{L}\{\text{Square wave}\} = \frac{1}{1 - e^{-2s}} \left[\int_0^1 e^{-st}\, dt - \int_1^2 e^{-st}\, dt \right]$$

$$= \frac{1}{1 - e^{-2s}} \left[\frac{e^{-st}}{-s} \bigg|_0^1 - \frac{e^{-st}}{-s} \bigg|_1^2 \right]$$

$$= \frac{1}{1 - e^{-2s}} \left[\frac{1}{s} - \frac{2e^{-s}}{s} + \frac{e^{-2s}}{s} \right]$$

$$= \frac{(1 - e^{-s})^2}{s(1 - e^{-s})(1 + e^{-s})}$$

$$= \frac{1 - e^{-s}}{s(1 + e^{-s})}$$ ∎

EXERCISES FOR SECTION 7–4

In Exercises 1–12 sketch the graph of the given function for $t \geq 0$.

1. $u_3(t)$ **2.** $u_5(t) - u_7(t)$

3. $u_0(t) - u_3(t)$ **4.** $tu_2(t)$

5. $t(u_1(t) - u_2(t))$ **6.** $u_1(t) - 3u_4(t) - 4u_5(t)$

7. $(t^2 - 4)u_2(t)$ **8.** $u_\pi(t) \sin t$

9. $u_\pi(t) \sin (t - \pi)$ **10.** $t(u_0(t) - u_1(t)) + (u_1(t) - u_3(t)) +$
                                                          $(u_3(t) - u_4(t))(4 - t)$

11. $u_1(t) + u_2(t) + u_3(t) + u_4(t) + \cdots$ **12.** $u_1(t) - 2u_2(t) + 2u_3(t) - 2u_4(t) + \cdots$

In Exercises 13–20 represent the given functions in terms of unit step functions.

13. $f(t) = \begin{cases} 1, & 0 < t < 2 \\ t, & t > 2 \end{cases}$ **14.** $f(t) = \begin{cases} 5, & 0 < t < 3 \\ 2t - 1, & t > 3 \end{cases}$

15. $f(t) = \begin{cases} t^2, & 0 < t < 4 \\ 0, & t > 4 \end{cases}$ **16.** $f(t) = \begin{cases} e^{-t}, & 0 < t < 2 \\ 0, & t > 2 \end{cases}$

17. $f(t) = \begin{cases} 0, & 0 < t < 2 \\ 4, & 2 < t < 5 \\ 0, & t > 5 \end{cases}$ **18.** $f(t) = \begin{cases} \cos t, & 0 < t < \pi/2 \\ 0, & t > \pi/2 \end{cases}$

19. $f(t) = \begin{cases} 0, & 0 < t < 1 \\ t, & 1 < t < 2 \\ 0, & t > 2 \end{cases}$ **20.** $f(t) = \begin{cases} 4, & 0 < t < 1 \\ 2, & 1 < t < 2 \\ 0, & t > 2 \end{cases}$

In Exercises 21–24 find $f(t)$ and sketch its graph.

21. $f(t) = \mathcal{L}^{-1}\left\{\dfrac{1}{s} - \dfrac{e^{-2s}}{s}\right\}$ **22.** $f(t) = \mathcal{L}^{-1}\left\{\dfrac{1}{s^2} + \dfrac{2e^{-3s}}{s}\right\}$

23. $f(t) = \mathcal{L}^{-1}\left\{\dfrac{3e^{-s}}{s} - \dfrac{e^{-2s}}{s}\right\}$ **24.** $f(t) = \mathcal{L}^{-1}\left\{\dfrac{1}{s}(e^{-2s} + 3e^{-5s})\right\}$

7–5   THE SECOND SHIFTING THEOREM

Recall that the first shifting theorem specifies that the transform function is translated to the right or left if the function of t is multiplied by an exponential. The second shifting theorem considers the effect of multiplying the transform by an exponential. Such a multiplication results in a "shift" of the inverse transform function.

THEOREM 7–5   *The Second Shifting Theorem*

Let $F(s) = \mathcal{L}\{f(t)\}$. Then

$$\mathcal{L}\{u_a(t)f(t - a)\} = e^{-as}F(s), a > 0$$

or, which is the same thing,

$$\mathcal{L}^{-1}\{e^{-as}F(s)\} = u_a(t)f(t - a)$$

PROOF   By definition,

$$e^{-as}F(s) = \int_0^\infty e^{-s(x+a)}f(x)dx$$

In the integral we make the substitution $t = x + a$:

$$e^{-as}F(s) = \int_a^\infty e^{-st}f(t - a)dt$$

$$= \int_0^\infty e^{-st}u_a(t)f(t - a)dt$$

$$= \mathcal{L}\{u_a(t)f(t - a)\}$$

From Theorem 7–5 we make the following observations:

1.   When finding the transform of a function times a unit step function $u_a(t)$, the function must first be expressed in terms of $t - a$. Then the result of the theorem may be applied.

2.   The multiplication of the transform by the exponential e^{-as} causes a shift of the inverse function to the right and a "turn-on" at $t = a$.

Recall that the first shifting theorem showed how a shift of the transform function occurs when the function of t is multiplied by e^{at}. The first and second shifting theorems are analogous, but the "turn-on" is unique to the second theorem.

Example 1

Use the unit step function to find $\mathcal{L}\{f(t)\}$ if

$$f(t) = \begin{cases} 0, & 0 < t < 2 \\ t^2 - t + 5, & t > 2 \end{cases}$$

See Figure 7–14.

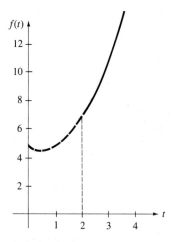

FIGURE 7–14

SOLUTION   The given function can be written

$$f(t) = (t^2 - t + 5)u_2(t)$$

To use Theorem 7–5, we must express $f(t)$ in terms of $(t - 2)$. By inspection we add and subtract $4t$ and 4 to the given quadratic to obtain the term $(t - 2)^2$. Thus

$$t^2 - t + 5 = t^2 - 4t + 4 - t + 5 + 4t - 4$$

Combining terms,

$$t^2 - t + 5 = (t - 2)^2 + 3t + 1$$

Finally, adding and subtracting 6, we get

$$t^2 - t + 5 = (t - 2)^2 + 3(t - 2) + 7$$

Therefore the Laplace transform is given by

$$\mathscr{L}\{(t^2 - t + 5)u_2(t)\} = \mathscr{L}\{(t - 2)^2 u_2(t)\} + 3\mathscr{L}\{(t - 2)u_2(t)\} + 7\mathscr{L}\{u_2(t)\}$$
$$= \frac{2}{s^3} e^{-2s} + \frac{3}{s^2} e^{-2s} + \frac{7}{s} e^{-2s} \qquad \blacksquare$$

COMMENT: The expansion in powers of $t - 2$ in the previous example was done by inspection. We could have proceeded more formally by using the Taylor expansion of a quadratic function about $t = a$. Thus

$$f(t) = f(a) + f'(a)(t - a) + \frac{f''(a)}{2}(t - a)^2$$

For $f(t) = t^2 - t + 5$ we have $f(2) = 7$, $f'(2) = 3$, and $f''(2) = 2$. Substituting these values and $a = 2$ into the Taylor expansion yields

$$f(t) = 7 + 3(t - 2) + (t - 2)^2$$

which is the same result obtained in Example 1.

Example 2

Find the inverse of the transform function

$$e^{-4s}/s^2$$

SOLUTION   Since $\mathcal{L}^{-1}\{1/s^2\} = t$, the factor e^{-4s} causes a shift to $t - 4$ and a "turn-on" at that point:

$$\mathcal{L}^{-1}\left\{\frac{e^{-4s}}{s^2}\right\} = u_4(t)(t - 4)$$ ∎

Example 3

Find $\mathcal{L}^{-1}\left\{\dfrac{e^{-s}}{(s + 2)^2 + 9}\right\}$

SOLUTION   From the first shifting theorem and the table of Laplace transforms we have

$$\mathcal{L}^{-1}\left\{\frac{1}{(s + 2)^2 + 9}\right\} = \tfrac{1}{3} e^{-2t} \sin 3t$$

Multiplication by the exponential factor e^{-s} causes a translation and multiplication by a unit step function. Therefore

$$\mathcal{L}^{-1}\left\{\frac{e^{-s}}{(s + 2)^2 + 9}\right\} = [\tfrac{1}{3} e^{-2(t-1)} \sin 3(t - 1)]u_1(t)$$ ∎

In some applications it will be necessary to express transforms of the form $\dfrac{1}{1 - u}$ as a geometric series $1 + u + u^2 + u^3 + \cdots$ before taking the inverse transform. The following two examples are typical and, while seeming similar, lead to significantly different results.

Example 4

Find $f(t)$ if $\mathcal{L}\{f(t)\} = \dfrac{1}{s(1 - e^{-2s})}$.

SOLUTION   We expand $\dfrac{1}{1 - e^{-2s}}$ in a geometric series as follows:

$$\frac{1}{1 - e^{-2s}} = 1 + e^{-2s} + e^{-4s} + e^{-6s} + \cdots$$

Hence

$$\mathcal{L}^{-1}\left\{\frac{1}{s(1 - e^{-2s})}\right\} = \mathcal{L}^{-1}\left\{\frac{1}{s}\right\} + \mathcal{L}^{-1}\left\{\frac{e^{-2s}}{s}\right\} + \mathcal{L}^{-1}\left\{\frac{e^{-4s}}{s}\right\} + \cdots$$

$$= u_0(t) + u_2(t) + u_4(t) + \cdots$$

(Do you recognize this function? See Figure 7–4.) ■

Example 5

Find and sketch $\mathcal{L}^{-1}\left\{\dfrac{1}{(s + 1)(1 - e^{-s})}\right\}$.

SOLUTION

$$\frac{1}{(s + 1)(1 - e^{-s})} = \frac{1}{s + 1}(1 + e^{-s} + e^{-2s} + \cdots + e^{-ns} + \cdots)$$

$$= \frac{1}{s + 1} + \frac{e^{-s}}{s + 1} + \frac{e^{-2s}}{s + 1} + \cdots + \frac{e^{-ns}}{s + 1} + \cdots$$

Taking the inverse transform term by term, we obtain

$$\mathcal{L}^{-1}\left\{\frac{1}{(s + 1)(1 - e^{-s})}\right\} = e^{-t} + e^{-(t-1)}u_1(t) + e^{-(t-2)}u_2(t)$$

$$+ \cdots + e^{-(t-n)}u_n(t) + \cdots$$

Thus

$$f(t) = \begin{cases} e^{-t}, & 0 < t < 1 \\ e^{-t}(1 + e), & 1 < t < 2 \\ e^{-t}(1 + e + e^2), & 2 < t < 3 \\ \quad\vdots \\ e^{-t}(1 + e + e^2 + \cdots + e^n) = \dfrac{e^{n+1} - 1}{e - 1}e^{-t} \\ \quad = \dfrac{e^{-(t-n-1)}}{e - 1} - \dfrac{e^{-t}}{e - 1}, \qquad n < t < n + 1 \\ \quad\vdots \end{cases}$$

The expression for $f(t)$ over the general interval $(n, n + 1)$ contains two terms. The first term, $e^{-(t-n-1)}/(e - 1)$, is a periodic function that decreases exponentially from $e/(e - 1)$ to $1/(e - 1)$ on the interval. It has a jump discontinuity of magnitude

$$\frac{e}{e - 1} - \frac{1}{e - 1} = 1$$

at every integer value of t. The second term, $e^{-t}/(e - 1)$, is a transient that dies away rapidly with increasing t. Figure 7–15 shows the contribution that each term makes.

COMMENT: When t is relatively large, the second term has little effect on the value of f(t), and f(t) is approximately the periodic function given in the first term.

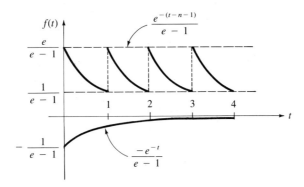

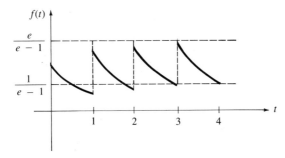

FIGURE 7–15

∎

EXERCISES FOR SECTION 7–5

In Exercises 1–10 represent the functions in terms of unit step functions and find their Laplace transforms.

1. $f(t) = \begin{cases} 2, & 0 < t < 1 \\ t, & 1 < t \end{cases}$

2. $f(t) = \begin{cases} 3, & 0 < t < 2 \\ t + 1, & 2 < t \end{cases}$

3. $f(t) = \begin{cases} e^{-t}, & 0 < t < 3 \\ 0, & 3 < t \end{cases}$

4. $f(t) = \begin{cases} \sin 3t, & 0 < t < \pi \\ 0, & \pi < t \end{cases}$

5. $f(t) = \begin{cases} t^2, & 0 < t < 3 \\ 9, & 3 < t < 5 \\ 0, & 5 < t \end{cases}$

6.

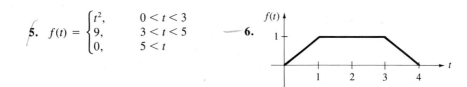

7. $f(t)$

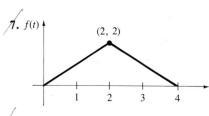

8. $f(t)$

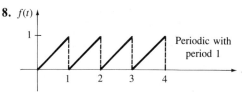

Periodic with period 1

9. $f(t)$

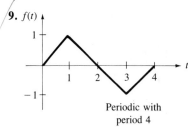

Periodic with period 4

10. $f(t)$

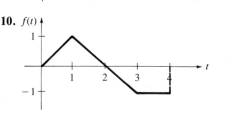

In Exercises 11–20 find the inverse Laplace transform of the given function. Sketch the graph of the function of t.

11. $\dfrac{e^{-3s}}{s^2}$

12. $\dfrac{e^{-4s} - e^{-s}}{s^3}$

13. $\dfrac{e^{-s}}{s^2 + 4}$

14. $\dfrac{e^{-s}}{(s - 1)(s - 2)}$

15. $\dfrac{se^{-2s}}{s^2 + 4}$

16. $\dfrac{1}{s(s + 1)(1 + e^{-s})}$

17. $\dfrac{1}{(1 + e^{-s})s^2}$

18. $\dfrac{1}{s(1 - e^{-s})}$

19. $\dfrac{1}{(s + a)(1 + e^{-ks})}$

20. $\dfrac{1}{(s + 1)^2(1 - e^{-s})}$

In Exercises 21–25 find the Laplace transform of the given function.

21. $tu_2(t)$

22. $t^2u_3(t)$

23. $\cos t\, u_\pi(t)$

24. $(t - 3)^2\, u_2(t)$

25. $e^t u_3(t)$

7–6   MORE INITIAL-VALUE PROBLEMS

In Chapter 6 we solved second-order differential equations having elementary driving functions. With the techniques you have acquired in this chapter you can now solve equations having discontinuous driving functions. The transform method yields a particular solution of an initial-value problem. If initial conditions are not given, we may assume arbitrary values, in which case the transform method gives the general solution.

Example 1

Using the method of Laplace transforms, solve the initial-value problem

$$x'(t) + 2x(t) = f(t), \qquad x(0) = 0$$

where

$$f(t) = \begin{cases} 2, & 0 < t < 2 \\ t, & t > 2 \end{cases}$$

SOLUTION   Writing $f(t)$ in terms of a unit step function, we get

$$f(t) = 2 - 2u_2(t) + tu_2(t)$$
$$= 2 + (t - 2)u_2(t)$$

The Laplace transform of $f(t)$ is then

$$\mathcal{L}\{f(t)\} = \frac{2}{s} + \frac{1}{s^2}e^{-2s}$$

Using this result and letting $\mathcal{L}\{x(t)\} = X(s)$, the given differential equation transforms into

$$sX(s) - 0 + 2X(s) = \frac{2}{s} + \frac{1}{s^2}e^{-2s}$$

from which

$$X(s) = \frac{2}{s(s + 2)} + \frac{1}{s^2(s + 2)}e^{-2s} \qquad\qquad (7\text{--}7)$$

By partial fractions expansions we find that

$$\frac{2}{s(s + 2)} = \frac{1}{s} - \frac{1}{s + 2}$$

and

$$\frac{1}{s^2(s + 2)} = -\frac{\frac{1}{4}}{s} + \frac{\frac{1}{2}}{s^2} + \frac{\frac{1}{4}}{s + 2}$$

Hence Equation 7–7 becomes

$$X(s) = \frac{1}{s} - \frac{1}{s + 2} - \left(\frac{\frac{1}{4}}{s} - \frac{\frac{1}{2}}{s^2} - \frac{\frac{1}{4}}{s + 2}\right)e^{-2s}$$

Since $x(t) = \mathcal{L}^{-1}\{X(s)\}$, we have

$$x(t) = 1 - e^{-2t} - [\tfrac{1}{4} - \tfrac{1}{2}(t - 2) - \tfrac{1}{4}e^{-2(t-2)}]\,u_2(t) \qquad\blacksquare$$

Example 2

Using the method of Laplace transforms, solve the initial-value problem

$$y''(t) + 2y'(t) = \phi(t), \, y(0) = 0, \, y'(0) = 3$$

where

$$\phi(t) = \begin{cases} 4, & 0 < t < 1 \\ 0, & t > 1 \end{cases}$$

SOLUTION   We can write $\phi(t)$ in terms of a step function as

$$\phi(t) = 4 - 4u_1(t)$$

The transform of $\phi(t)$ is

$$\mathcal{L}\{\phi(t)\} = \frac{4}{s} - \frac{4}{s}e^{-s}$$

Therefore the transform of the given equation is

$$s^2Y(s) - 0 - 3 + 2[sY(s) - 0] = \frac{4}{s} - \frac{4}{s}e^{-s}$$

Solving for $Y(s)$,

$$Y(s) = \frac{3}{s(s+2)} + \frac{4}{s^2(s+2)} - \frac{4}{s^2(s+2)}e^{-s} \qquad (7\text{-}8)$$

Using partial fractions expansions, we note that

$$\frac{3}{s(s+2)} = \frac{\frac{3}{2}}{s} - \frac{\frac{3}{2}}{s+2}$$

and

$$\frac{4}{s^2(s+2)} = -\frac{1}{s} + \frac{2}{s^2} + \frac{1}{s+2}$$

Making these substitutions in Equation 7–8,

$$Y(s) = \frac{\frac{3}{2}}{s} - \frac{\frac{3}{2}}{s+2} - \frac{1}{s} + \frac{2}{s^2} + \frac{1}{s+2} + \left(\frac{1}{s} - \frac{2}{s^2} - \frac{1}{s+2}\right)e^{-s}$$

Taking the inverse transform, the solution function is

$$y(t) = \tfrac{1}{2} + 2t - \tfrac{1}{2}e^{-2t} + [1 - 2(t-1) - e^{-2(t-1)}]u_1(t) \qquad \blacksquare$$

Example 3

A spring is attached to a 16-lb block resting on a frictionless plane. A horizontal force of 4 lb is applied to the block through the spring for 3 sec and then released. Describe the resulting motion if the block is initially at rest and the spring constant is equal to 2.

$$m\frac{d^2x}{dt^2} + c\frac{dx}{dt} + Kx = 0$$

→ damping constant

SOLUTION   The differential equation of the system is

$$\frac{16}{32}\ddot{x}(t) + 2x(t) = F(t),\ x(0) = 0,\ \dot{x}(0) = 0 \tag{7–9}$$

where the applied force $F(t)$ is given by

$$F(t) = \begin{cases} 4, & 0 < t < 3 \\ 0, & t > 3 \end{cases}$$

Using the unit step function, we can write $F(t)$ in the form

$$F(t) = 4 - 4u_3(t)$$

Therefore Equation 7–9 can be written

$$\frac{1}{2}\ddot{x}(t) + 2x(t) = 4 - 4u_3(t)$$

or

$$\ddot{x}(t) + 4x(t) = 8 - 8u_3(t)$$

Letting $\mathcal{L}\{x(t)\} = X(s)$, we have

$$s^2 X(s) + 4X(s) = \frac{8}{s} - \frac{8}{s}e^{-3s}$$

and therefore

$$X(s) = \frac{8}{s(s^2 + 4)} - \frac{8}{s(s^2 + 4)}e^{-3s}$$

Using partial fractions expansions,

$$X(s) = \frac{2}{s} - \frac{2s}{s^2 + 4} - \frac{2}{s}e^{-3s} + \frac{2s}{s^2 + 4}e^{-3s}$$

The appropriate inverse Laplace transforms yield

$$x(t) = 2 - 2\cos 2t - 2u_3(t) + 2\cos 2(t - 3)\,u_3(t)$$
$$= 2(1 - \cos 2t) - 2[1 - \cos 2(t - 3)]\,u_3(t)$$

Writing $x(t)$ as a two-rule function

$$x(t) = \begin{cases} 2(1 - \cos 2t) = 4\sin^2 t, & 0 < t < 3 \\ 2[\cos 2(t - 3) - \cos 2t], & t > 3 \end{cases} \tag{7–10}$$

The graph of this function is shown in Figure 7–16. Notice that the maximum deviation from the equilibrium position is 4 ft and occurs at $t = 1.57$ sec. For $t > 3$ we use the identity

$$\cos A - \cos B = -2\sin\frac{1}{2}(A + B)\sin\frac{1}{2}(A - B)$$

in Equation 7–10 to obtain

$$x(t) = 4 \sin\frac{1}{2}(2t + 2t - 6) \sin\frac{1}{2}(2t - 2t + 6)$$

$$= (4 \sin 3) \sin (2t - 3)$$

$$= 0.5645 \sin (2t - 3)$$

■

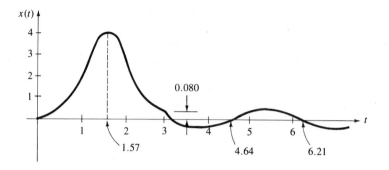

FIGURE 7–16

Example 4

Find the response of an RL circuit to a single square wave, $u_a(t) - u_b(t)$. Assume that the initial current is zero.

SOLUTION   The differential equation of the circuit is

$$L\frac{di}{dt} + Ri = u_a(t) - u_b(t)$$

Taking the Laplace transform of both sides and letting $\mathcal{L}\{i(t)\} = I(s)$,

$$LsI(s) + RI(s) = \frac{e^{-as} - e^{-bs}}{s}$$

from which

$$I(s) = \frac{e^{-as}}{Ls(s + R/L)} - \frac{e^{-bs}}{Ls(s + R/L)}$$

$$= \frac{e^{-as}}{R}\left(\frac{1}{s} - \frac{1}{s + R/L}\right) - \frac{e^{-bs}}{R}\left(\frac{1}{s} - \frac{1}{s + R/L}\right)$$

Hence

$$i(t) = \frac{1}{R}[1 - e^{-R(t-a)/L}]u_a(t) - \frac{1}{R}[1 - e^{-R(t-b)/L}]u_b(t)$$

The graph of the current is shown in Figure 7–17.

■

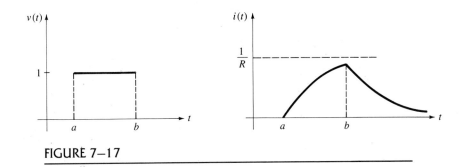

FIGURE 7–17

EXERCISES FOR SECTION 7–6

Use Laplace transform methods to solve each of the initial-value problems in Exercises 1–10.

1. $y'(t) + 2y(t) = 5u_1(t); y(0) = 5$

2. $y'(t) + y(t) = u_1(t) - u_2(t); y(0) = 0$

3. $x'(t) + x(t) = (t - 4)^2 u_4(t); x(0) = 0$

4. $y' + 3y = 4(t - 2)u_2(t); y(0) = 0$

5. $y'' + 4y' + 4y = u_1(t); y(0) = y'(0) = 0$

6. $y'' + 2y' + y = u_0(t); y(0) = 1, y'(0) = 0$

7. $\ddot{x} + x = (t - 2)u_2(t); x(0) = \dot{x}(0) = 0$

8. $\dfrac{d^2x}{dt^2} + 4\dfrac{dx}{dt} - 5x = 3(t - 1)u_1(t); x(0) = 0, \dot{x}(0) = 0$

9. $y''(t) + y(t) = \phi(t); y(0) = y'(0) = 0$, where

$$\phi(t) = \begin{cases} t, & 0 < t < 2 \\ 2, & t > 2 \end{cases}$$

10. $x''(t) - x(t) = f(t), x(0) = x'(0) = 0$, where

$$f(t) = \begin{cases} 0, & 0 < t < 4 \\ t - 4, & t > 4 \end{cases}$$

11. A spring is stretched 2 feet by an 8-lb force. An 8-lb weight is attached to the spring and allowed to come to rest. Describe the motion of the weight if a 20-lb force is applied to the spring for 1 second and then released.

12. In Exercise 11 what is the maximum displacement of the weight from its equilibrium position?

13. A 32-lb weight is attached to a spring ($k = 1$) and allowed to come to rest. Describe the motion of the weight if the driving function is given by $f(t) = (t - 2)u_2(t)$. Assume $x(0) = \dot{x}(0) = 0$.

14. Solve Exercise 13 if there is a damping force equal to twice the instantaneous velocity.

15. A mass of 1 slug is attached to a spring having a spring constant of 9 and a damping force equal to 6 times its instantaneous velocity. Determine the equation of motion of the mass if the forcing function is given by

$$f(t) = \begin{cases} 3, & 0 < t < 1 \\ 6, & t > 1 \end{cases}$$

Assume $x(0) = \dot{x}(0) = 0$.

16. Compute the position of the mass in Exercise 15 at $t = 1$ sec and $t = 2$ sec.

17. An RL series circuit having $R = 18$ ohms and $L = 2$ henrys has an applied voltage of $v(t) = 6$ volts. Find the current in the circuit for $t > 0$ if the voltage is removed after 2 seconds. Assume the current is initially zero.

18. Find the current in the circuit in Exercise 17 if the applied voltage is given by $v(t) = 2(t - 1)u_1(t)$.

19. Find the charge/time equation in a series RC circuit if $R\dfrac{dq}{dt} + \dfrac{1}{C}q = v(t)$ and $v(t)$ is given by

$$v(t) = \begin{cases} E, & 0 < t < T \\ 0, & t > T \end{cases}$$

Assume $q(0) = 0$.

20. Find the response of an RC series circuit to a single square wave voltage of height h between $t = a$ and $t = b$. Assume zero initial charge.

REVIEW EXERCISES FOR CHAPTER 7

In Exercises 1–12 find the Laplace transform of the indicated function.

1. $f(t) = t \cosh 3t$ (Use Theorem 7–1)

2. $g(t) = 3t \cos^2 5t$ (Use Theorem 7–1)

3. $g(t) = \begin{cases} 2, & 0 < t < 2 \\ 1, & 2 < t < 4 \end{cases}$

   $g(t) = g(t + 4)$

4. $h(t) = \begin{cases} t, & 0 < t < 1 \\ 1, & 1 < t < 2 \end{cases}$

   $h(t) = h(t + 2)$

5. $f(t) = e^{3t}$(Triangular wave in Figure 7–1)

6. $f(t) = e^{-2t}$(Wave in Figure 7–2)

7. $p(t) = t*\sin 2t$

8. $m(t) = t^2*e^{3t}$

9. $g(t) = \begin{cases} t, & 0 < t < 1 \\ 1, & 1 < t < 2 \\ 0, & t > 2 \end{cases}$ (Use Theorem 7–5)

10. $h(t) = \begin{cases} e^{t/2}, & 0 < t < 2 \\ 0, & t > 2 \end{cases}$ (Use Theorem 7–5)

11. $f(t) = (t^2 + 3t)u_2(t)$

12. $F(t) = e^{2t}u_1(t)$

In Exercises 13–20 find the inverse Laplace transform.

13. $\ln\left[\dfrac{s+3}{s-3}\right]$

14. $\cot^{-1}(2s)$

15. $\dfrac{7}{s^2 - 7s - 8}$ (Use Theorem 7–3)

16. $\dfrac{1}{3s(s^2 + 4)}$ (Use Theorem 7–3)

17. $\dfrac{2e^{-s}}{s} - \dfrac{e^{-4s}}{s^3}$

18. $\dfrac{e^{-2s}}{s^2 + 16}$

19. $\dfrac{3e^{-s} + 2e^{-2s}}{s^2 + 9}$

20. $\dfrac{8}{3s(1 - e^{-2s})}$

21. Solve the initial-value problem

$$y'' + 4y = u_2(t) - u_3(t),\ y(0) = 0,\ y'(0) = 0$$

22. Solve the initial-value problem

$$y'' - y' = t[u_0(t) - u_2(t)],\ y(0) = 0,\ y'(0) = 0$$

23. Solve the initial-value problem

$$y'' - y = f(t);\ y(0) = 0;\ y'(0) = 0$$

in terms of a convolution for a general function $f(t)$ and then for the driving function $f(t) = e^{2t}$.

8

Infinite Series Methods

How high can a vertical column of uniform cross section be extended upward until it will buckle under its own weight? The answer to this question is contained in the solution of the differential equation

$$EI \frac{d^2\theta}{dx^2} + gkx\theta = 0$$

where E, I, g, and k are constants of the column, x is the distance below the top of the column, and θ is the deflection of the column from vertical at x. (See Figure 8–1.) Methods discussed in this chapter are used to solve this problem.

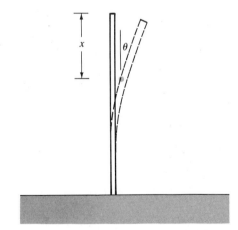

FIGURE 8–1

In Example 1 of Section 5–3 we showed that the general solution of

$$y'' + y = \sec x$$

is expressible in the form

$$y = c_1 \cos x + c_2 \sin x + \ln|\cos x|\cos x + x \sin x$$

The solution of $y'' + y = \sec x$ is typical of all previous solutions in that it is expressed as a linear combination of elementary functions. Unfortunately, many important differential equations exist for which this is not the case. For instance, the solution of

$$x^2 y'' + xy' + (x^2 - m^2)y = 0$$

cannot be expressed in terms of rational algebraic, trigonometric, exponential, and logarithmic functions. Differential equations like this one are usually solved by assuming that the form of the solution is an infinite series. Before discussing methods of solution, we review some basic facts about infinite series.

Power Series: A Brief Review

A power series about a point x_o is an infinite series of the form

$$\sum_{n=0}^{\infty} a_n(x - x_o)^n$$

where the numbers $a_0, a_1, a_2, \ldots, a_n, \ldots$ are called the coefficients of the power series. We say the power series **converges** if

$$\lim_{k \to \infty} \sum_{n=0}^{k} a_n(x - x_o)^n$$

exists. The value of the limit is called the sum of the power series at the point. If the limit does not exist, the power series is said to **diverge.** The interval of values of x for which a power series converges is called the **interval of convergence** and denoted $|x - x_o| < R$, where x_o is called the center of the power series and R is called the **radius of convergence.** The value of R can be found using either of the following formulas.

$$R = \lim_{n \to \infty} \left| \frac{a_n}{a_{n+1}} \right|$$

or

$$R = \lim_{n \to \infty} |a_n|^{-1/n}$$

if either limit exists. The endpoints of the interval of convergence are $-R + x_o$ and $R + x_o$. The power series must be checked for convergence at each of the endpoints. If $R = 0$, the series converges only at x_o; if $R = \infty$, the series converges for all values of x.

Within a common interval of convergence two power series may be added term by term. That is,

$$\sum_{n=0}^{\infty} a_n x^n + \sum_{n=0}^{\infty} b_n x^n = \sum_{n=0}^{\infty} (a_n + b_n) x^n$$

Further, within the interval of convergence the power series represents a function whose derivative and integral may be found from term-by-term differentiation and integration. That is, if

$$f(x) = \sum_{n=0}^{\infty} a_n (x - x_o)^n$$

then

$$f'(x) = \sum_{n=1}^{\infty} n a_n (x - x_o)^{n-1}$$

(NOTE: *The first term of the power series here begins with* $n = 1$.)

and

$$\int_{x_0}^{x} f(t)\, dt = \sum_{n=0}^{\infty} a_n \frac{(x - x_o)^{n+1}}{n + 1}$$

A power series expansion of $f(x)$ around $x = 0$ is called a **Maclaurin series.** The Maclaurin series of $f(x)$ is given by

$$f(x) = \sum_{n=0}^{\infty} \frac{f^{(n)}(0)}{n!} x^n$$

where $f^{(n)}(0)$ means the value of the nth derivative of f at $x = 0$. Recall that $f^{(0)} = f$ and $0! = 1$. Four well-known Maclaurin series are

$$\frac{1}{1 - x} = 1 + x + x^2 + x^3 + \cdots + x^n \cdots = \sum_{n=0}^{\infty} x^n \qquad (|x| < 1)$$

$$e^x = 1 + x + \frac{x^2}{2!} + \frac{x^3}{3!} + \cdots + \frac{x^n}{n!} + \cdots = \sum_{n=0}^{\infty} \frac{x^n}{n!}$$

$$\sin x = x - \frac{x^3}{3!} + \frac{x^5}{5!} - \cdots + \frac{(-1)^n}{(2n + 1)!} x^{2n+1} + \cdots = \sum_{n=0}^{\infty} \frac{(-1)^n}{(2n + 1)!} x^{2n+1}$$

$$\cos x = 1 - \frac{x^2}{2!} + \frac{x^4}{4!} - \frac{x^6}{6!} + \cdots + \frac{(-1)^n}{(2n)!} x^{2n} + \cdots = \sum_{n=0}^{\infty} \frac{(-1)^n}{(2n)!} x^{2n}$$

The **Taylor series** of $f(x)$ around $x = x_o$ is given by

$$f(x) = \sum_{n=0}^{\infty} \frac{f^{(n)}(x_o)}{n!} (x - x_o)^n$$

8–1 TAYLOR SERIES SOLUTIONS

When using the method of Taylor series to solve differential equations, we assume the solution to be of the form

$$y(x) = \sum_{n=0}^{\infty} \frac{y^{(n)}(x_o)}{n!} (x - x_o)^n$$

Hence to write the solution, we need to know the values of the unknown function and its derivatives at $x = x_o$. The first few terms yield an **approximating polynomial** to the infinite series solution. Later we will address the rather important question of how we know when an assumption of this type is valid. For now, if the assumption turns out to be invalid, it will become obvious in the procedure of the technique. (See Exercise 22.)

The values of the required derivatives are found directly from the initial conditions and the differential equation. Determining the various derivatives can be time consuming and tedious and is the major limitation to this method. However, the method is general enough to apply to all types of differential equations, both linear and nonlinear.

> *COMMENT: The form of the solution series depends upon where the values of the function and its derivatives are given. If the values are given at $x = 0$, the series is of the form $\sum_{n=0}^{\infty} \frac{y^{(n)}(0)}{n!} x^n$; if they are given at $x = 1$, the series is of the form $\sum_{n=0}^{\infty} \frac{y^{(n)}(1)}{n!} (x - 1)^n$; and so on.*

Example 1

Find the first five nonzero terms of the Taylor series expression for the solution to the initial-value problem

$$y'' = y^2 + x, \ y(1) = 2, \ y'(1) = -1$$

SOLUTION   The value of the solution function and its first derivative are given at $x = 1$. Consequently, we want to write a Taylor series for $y(x)$ in powers of $(x - 1)$. The value of $y''(1)$ is determined from the differential equation to be

$$y''(1) = [y(1)]^2 + 1 = 4 + 1 = 5$$

Differentiating both sides of the differential equation leads to the following expressions for $y'''(x)$ and $y^{(4)}(x)$:

$$y'''(x) = 2yy' + 1$$
$$y^{(4)}(x) = 2(y')^2 + 2yy''$$

from which

$$y'''(1) = 2(2)(-1) + 1 = -3$$
$$y^{(4)}(1) = 2(-1)^2 + 2(2)(5) = 22$$

The first five terms of the Taylor series expression for the solution $y(x)$ are

$$y(x) = 2 - (x - 1) + \frac{5}{2}(x - 1)^2 - \frac{1}{2}(x - 1)^3 + \frac{11}{12}(x - 1)^4 + \cdots \qquad \blacksquare$$

Example 2

Find the Taylor series solution about $x = 0$ of the initial-value problem $y'' + y = 0$, $y(0) = a_0$, $y'(0) = a_1$. Initial conditions on the solution and its derivative are to be left arbitrary.

SOLUTION   The solution of the given problem is assumed to be of the form

$$y(x) = \sum_{n=0}^{\infty} \frac{y^{(n)}(0)}{n!} x^n$$

The given differential equation can be arranged in the form $y'' = -y$. Consequently, the derivatives of y, starting with y'', are $y'' = -y$, $y''' = -y'$, $y^{(4)} = -y''$, ..., $y^{(n)} = -y^{(n-2)}$. To determine $y^{(n)}(0)$, we use $y(0) = a_0$ and $y'(0) = a_1$. Thus

$$y''(0) = -y(0) = -a_0, \; y'''(0) = -y'(0) = -a_1, \; y^{(4)}(0) = -y''(0) = a_0,$$
$$y^{(5)}(0) = a_1, \; y^{(6)}(0) = -a_0, \; \ldots, \; y^{(n)}(0) = -y^{(n-2)}(0)$$

Notice that

$$y''(0) = -a_0, \; y^{(4)}(0) = a_0, \; y^{(6)}(0) = -a_0, \; \ldots$$

and

$$y'''(0) = -a_1, \; y^{(5)}(0) = a_1, \; y^{(7)}(0) = -a_1, \; \ldots$$

Thus for n a positive integer

$$y^{(2n)}(0) = (-1)^n a_0 \qquad \text{and} \qquad y^{(2n+1)}(0) = (-1)^n a_1$$

and hence the solution can be written in the form

$$y = a_0 \sum_{n=0}^{\infty} \frac{(-1)^n}{(2n)!} x^{2n} + a_1 \sum_{n=0}^{\infty} \frac{(-1)^n}{(2n+1)!} x^{2n+1}$$

We recognize the two infinite series as being the Maclaurin series for the elementary functions $\cos x$ and $\sin x$. Hence the solution is

$$y = a_0 \cos x + a_1 \sin x$$

Note that this solution agrees with the result obtained by using the methods of Chapter 4 for solving homogeneous linear differential equations with constant coefficients. $\blacksquare$

Example 3

Find three terms of the Taylor series expansion about $x = 2$ of the solution to the initial-value problem

$$y'' + y' - \ln y = e^x, \quad y(2) = a_0, \quad y'(2) = a_1$$

SOLUTION   Since $y'' = e^x + \ln y - y'$, we use the initial conditions $y(2) = a_0$, $y'(2) = a_1$ to obtain $y''(2) = e^2 + \ln a_0 - a_1$. Hence the first three terms of the Taylor series are

$$y(x) = a_0 + a_1(x - 2) + \frac{e^2 + \ln a_0 - a_1}{2}(x - 2)^2 + \cdots \qquad \blacksquare$$

Example 4

Solve the initial-value problem $y' - y + x^2 + x + 1 = 0$, $y(0) = 1$ by assuming a Taylor series solution valid near $x = 0$.

SOLUTION   Let

$$y = \sum_{n=0}^{\infty} \frac{y^{(n)}(0)}{n!} x^n$$

The differential equation gives

$$\begin{aligned}
y' &= y - x^2 - x - 1 \\
y'' &= y' - 2x - 1 \\
y''' &= y'' - 2 \\
y^{(4)} &= y'''
\end{aligned}$$

and for $n \geq 4$,

$$y^{(n)} = y^{(n-1)}$$

Therefore $y(0) = 1$, $y'(0) = 0$, $y''(0) = -1$, $y'''(0) = -3$, and for $n \geq 3$, $y^{(n)}(0) = -3$. Consequently, the solution is

$$\begin{aligned}
y &= 1 - \frac{x^2}{2} - 3 \sum_{n=3}^{\infty} \frac{x^n}{n!} \\
&= 1 - \frac{x^2}{2} - 3 \sum_{n=0}^{\infty} \frac{x^n}{n!} + 3 \sum_{n=0}^{2} \frac{x^n}{n!} \\
&= 1 - \frac{x^2}{2} - 3 \sum_{n=0}^{\infty} \frac{x^n}{n!} + 3 + 3x + \frac{3x^2}{2}
\end{aligned}$$

The infinite series in this last expression can be recognized as the Maclaurin series for e^x. Hence

$$y = 4 + 3x + x^2 - 3e^x$$

This solution is also obtainable by any of the previous methods for solving non-homogeneous linear differential equations. $\qquad \blacksquare$

Comments on the Form of a Series Solution

From a practical standpoint it makes little difference whether the solution is expressed in terms of elementary functions or in terms of an infinite power series. In either case we can evaluate the solution to any specified accuracy.

When an infinite power series form of a solution is assumed, we must decide how to state the answer. There are four basic levels that can be reached, each of which yields a satisfactory method of stating a solution under certain circumstances.

Level 1 The solution may be represented by a few nonzero terms of the power series. This is the minimum, in the way of an answer, that should be given. (See Examples 1 and 3.)

Level 2 A recursion formula may be obtainable that expresses the nth coefficient (or nth derivative) in terms of previous coefficients (or derivatives). For example, we might know that $y^{(n)}(0) = -y^{(n-1)}(0)$ or that $a_n = a_{n-1} + a_{n-2}$. Recursion formulas permit writing as many terms of the solution series as desired.

Level 3 If the recursion formula is simple enough, a general formula may be found for the coefficient of x^n. (See Examples 2 and 4. Also, see Example 5 below.)

Level 4 The infinite series may be expressible in terms of known elementary functions. In Example 2 we were able to express the solution in terms of trigonometric functions. In Example 4 the solution was expressed in terms of e^x and a polynomial.

Example 5

Write the power series centered at $x = 0$ whose coefficients are related by the formula

$$a_0 = 1, a_1 = 0, a_n = -\frac{2}{n}a_{n-2} \text{ for } n \geq 2$$

SOLUTION Such a power series has the form $\sum\limits_{n=0}^{\infty} a_n x^n$. Because the recursion formula for a_n is in terms of the coefficient a_{n-2}, we know that if n is even, a_n will be in terms of a_0, and if n is odd, a_n will be in terms of a_1. Such a series is often written in rearranged form as

$$\sum_{n=0}^{\infty} a_n x^n = \sum_{k=0}^{\infty} a_{2k}x^{2k} + \sum_{k=0}^{\infty} a_{2k+1}x^{2k+1}$$

The first series on the right is a series of even terms and the second is a series of odd terms. Since a_1 is zero, so are all the odd coefficients, and the second series vanishes. The recursion formula for the even coefficients is

$$a_{2k} = -\frac{1}{k}a_{2k-2}, \; k \ge 1$$

Repeatedly applying this formula to $a_{2k-2}, a_{2k-4}, \ldots, a_8, a_6, a_4,$ and $a_2,$

$$a_{2k} = \left(-\frac{1}{k}\right)\left(-\frac{1}{k-1}\right)\left(-\frac{1}{k-2}\right) \cdots \left(-\frac{1}{4}\right)\left(-\frac{1}{3}\right)\left(-\frac{1}{2}\right)\left(-\frac{1}{1}\right)a_0$$

$$= \frac{(-1)^k}{k!} a_0$$

Hence the power series is

$$\sum_{k=0}^{\infty} \frac{(-1)^k}{k!} x^{2k} = \sum_{k=0}^{\infty} \frac{(-x^2)^k}{k!} = e^{-x^2}$$ ∎

EXERCISES FOR SECTION 8–1

Use the method of Taylor series to solve the differential equations in Exercises 1–10. Write at least five nonzero terms of the series.

1. $y' = x + y$, $y(0) = 1$

2. $y' = y$, $y(0) = 1$

3. $y'' = y$, $y(0) = 1$, $y'(0) = 1$

4. $y' + 2y = x^2$, $y(0) = 1$

5. $y'' + 4y' + 3y = 0$,
   $y(0) = 1$, $y'(0) = 1$

6. $y' = y^2$, $y(0) = 2$

7. $y'' = xy$, $y(1) = 2$, $y'(1) = -2$

8. $y'' = y^3$, $y(1) = 1$, $y'(1) = -1$

9. $y'' = \ln y$, $y(1) = 1$, $y'(1) = 1$

10. $y'' = e^y$, $y(0) = 1$, $y'(0) = 1$

Exercises 11–16: Use Exercises 1–6 to obtain the general coefficient and write the solution in terms of elementary functions.

In Exercises 17–21 a recursion formula is given for the power series solution $\sum_{n=0}^{\infty} a_n x^n$ of a differential equation. Find the general term and write the solution.

17. $a_n = -\dfrac{1}{n-2}a_{n-1}$, $n \ge 4$, $a_0 = 1$, $a_1 = 0$, $a_2 = 1$, $a_3 = -1$

18. $a_n = \dfrac{n+2}{n}a_{n-2}$, $n \ge 2$, $a_0 = a_1 = 1$

19. $a_n = \dfrac{(n-5)(n-6)}{n^2}a_{n-2}$, $n \ge 2$, $a_0 = a_1 = 1$

20. $a_n = \dfrac{n-5}{n(n-1)}a_{n-2}$, $n \ge 2$, $a_0 = a_1 = 1$

21. $a_n = \dfrac{3}{4n} a_{n-2}, n \geq 2, a_0 = a_1 = 1$

22. Show that the method of Taylor series cannot be used to solve the differential equation

$$(x^2 - 3x + 2) y'' + 3y' - y = 0, y(1) = y'(1) = 1$$

(HINT: Try it!)

23. A bag of sand is attached to a spring with spring constant $k = 2$ lb/ft, as shown in Figure 8–2. The weight of the sand is initially 32 lb. The bag has a slight hole in it, so the sand is lost from the bag at a linear rate of 2 lb/min. The bag of sand is pulled down 1 ft from equilibrium and released. Determine the motion of the bag of sand as a function of time by writing the first three nonzero terms of the Taylor expansion which represents the motion.

Loss of sand

FIGURE 8–2

8–2   ORDINARY AND SINGULAR POINTS

In this section we develop criteria for determining when we may assume that a power series solution of a differential equation exists. For simplicity the discussion is limited to second-order linear differential equations with polynomial coefficients with no common factors.

DEFINITION

A linear differential equation

$$p_2(x)y'' + p_1(x)y' + p_0(x)y = f(x)$$

with polynomial coefficients p_0, p_1, and p_2 has an **ordinary point** at $x = x_o$ if $p_2(x_o) \neq 0$. Otherwise, the point is called a **singular point**. (NOTE: Singular points may be complex numbers.)

Example 1

(a) $x^2y'' + xy' - 2y = 0$ has a singular point at $x = 0$. All other points are ordinary.

(b) $(1 - x^2)y'' - 2xy' + 6y = 0$ has singular points at 1 and -1.

(c) $(x^2 + 1)y'' + 6xy' - 2y = 0$ has singular points at i and $-i$.   ■

The following theorem is fundamental to power series solutions near ordinary points.

THEOREM 8–1

If $x = x_o$ is an ordinary point of

$$p_2(x)y'' + p_1(x)y' + p_0(x)y = f(x) \tag{8-1}$$

then Equation 8–1 has a solution of the form

$$y = \sum_{n=0}^{\infty} a_n(x - x_o)^n$$

with a_0 and a_1 arbitrary. This series converges at least for the interval $|x - x_o| < R$, where R is the distance in the complex plane from x_o to the closest singular point of the differential equation. (NOTE: In some cases convergence can extend beyond the nearest singularity although we shall not discuss any such problems.)

COMMENT: By "arbitrary" we mean that these coefficients are established by auxiliary conditions to the given differential equation, such as initial conditions. The simplest conditions are $y(x_o) = a_0$ and $y'(x_o) = a_1$.

Example 2

(a) $(x - 5)y' + y = x$ has a power series solution around 0. Since $x = 5$ is a singular point, the series converges for at least $|x| < 5$.

(b) $(x^2 + 4)y'' + y' + y = e^x$ has a power series solution around $x = 0$. Since $2i$ and $-2i$ are singular points, the series converges at least for $-2 < x < 2$.

(c) $x^2y'' + xy' - 2y = 0$ has a power series solution around $x = 1$. Since the differential equation has a singular point at 0, the series solution converges for at least $0 < x < 2$.

(d) Theorem 8–1 does not guarantee a power series solution around $x = 0$ for $x^2y'' + xy' - 2y = 0$ since $x = 0$ is a singular point of the differential equation.   ■

EXERCISES FOR SECTION 8–2 *all*

In Exercises 1–10 determine if the differential equation has a power series solution about the given value of x. If it does, give the interval of convergence predictable from Theorem 8–1.

1. $(x^2 - 1)y' + y = 0,$ $x_o = 0$

2. $(x^2 - 1)y' + y = 0,$ $x_o = 5$

3. $(x^2 + 1)y' + y = 0,$ $x_o = 4$

4. $(x^2 + 1)y' + y = 0,$ $x_o = 0$

5. $x^2y'' - xy' + y = 0,$ $x_o = 1$

6. $x^2y'' - xy' + y = 0,$ $x_o = 0$

7. $(x - 1)y'' + y = x,$ $x_o = 1$

8. $(x - 1)y'' + y = x,$ $x_o = 0$

9. $(x - 1)(x^2 - 2)y'' + y = 0,$ $x_o = 1.5$

10. $(x^2 - 3x + 2)y'' + 3y' - y = 0,$ $x_o = 0$

11. Explain why the method of Taylor series did not work in Exercise 22 in Section 8–1.

8–3   POWER SERIES SOLUTIONS

Theorem 8–1 states that if x_o is an ordinary point of the differential equation

$$p_2(x)y'' + p_1(x)y' + p_0(x)y = f(x) \qquad (8\text{–}2)$$

we may assume the form of the solution to be

$$y = \sum_{n=0}^{\infty} a_n(x - x_o)^n \qquad (8\text{–}3)$$

To solve Equation 8–2, we substitute Equation 8–3 and its derivatives into the differential equation, and at the same time we write the driving function $f(x)$ in its Taylor (or Maclaurin) expansion. Values for a_n are then obtained by equating the coefficients of like powers of x. When a_0 and a_1 are arbitrary, the problem reduces to finding a_n for $n \geq 2$. The basic technique is described in Example 1 for a first-order differential equation.

Example 1

Use a power series to solve the differential equation $y' + xy = x^2 - 2x$ about $x = 0$. Obtain a recursion formula for the coefficients and write the first few terms of a power series for the solution function.

SOLUTION   The given differential equation has no singular points, so there is a power series solution of the form

$$y = \sum_{n=0}^{\infty} a_n x^n$$

with a_0 arbitrary, that will converge for all x. Substituting the power series for y and y' into $y' + xy = x^2 - 2x$, we obtain

$$\sum_{n=1}^{\infty} n a_n x^{n-1} + x \sum_{n=0}^{\infty} a_n x^n = x^2 - 2x$$

or

$$\sum_{n=1}^{\infty} n a_n x^{n-1} + \sum_{n=0}^{\infty} a_n x^{n+1} = x^2 - 2x$$

To combine the two series on the left-hand side of the equation, the exponent of x in both series must be the same. If we replace n with $n - 2$ in the second series, the equation becomes

$$\sum_{n=1}^{\infty} n a_n x^{n-1} + \sum_{n=2}^{\infty} a_{n-2} x^{n-1} = x^2 - 2x$$

These two series can be combined into one series if the beginning index in each series is the same, so we write the first term of the first series separately and then write the series as

$$a_1 x^0 + \sum_{n=2}^{\infty} (n a_n + a_{n-2}) x^{n-1} = x^2 - 2x$$

The coefficient a_0 is left as arbitrary. To find the other coefficients, we proceed by equating coefficients of like powers of x on both sides of this equation.

Coefficients of x^0: $a_1 = 0$
Coefficients of x^1: $2a_2 + a_0 = -2$
Coefficients of x^2: $3a_3 + a_1 = 1$
Coefficients of x^3: $4a_4 + a_2 = 0$
Coefficients of x^4: $5a_5 + a_3 = 0$
$\quad\quad\quad \vdots \quad\quad\quad\quad\quad \vdots$
Coefficients of x^{n-1}: $n a_n + a_{n-2} = 0, n \geq 4$

From this the required coefficients are

a_0 arbitrary

$a_1 = 0$

$$a_2 = -\frac{a_0 + 2}{2}$$

$$a_3 = \frac{1}{3}$$

For $n \geq 4$, a_n is given by the recursion formula

$$a_n = -\frac{1}{n} a_{n-2} \qquad \text{for} \qquad n \geq 4$$

Iteration of this formula for $n \geq 4$ yields

$$a_4 = -\frac{1}{4} a_2 = -\frac{1}{4}\left(-\frac{a_0 + 2}{2} \right) = \frac{a_0 + 2}{4 \cdot 2}$$

$$a_5 = -\frac{1}{5} a_3 = -\frac{1}{5}\left(\frac{1}{3} \right) = -\frac{1}{5 \cdot 3}$$

$$a_6 = -\frac{1}{6} a_4 = -\frac{a_0 + 2}{6 \cdot 4 \cdot 2}$$

$$a_7 = -\frac{1}{7} a_5 = \frac{1}{7 \cdot 5 \cdot 3}$$

$$a_8 = -\frac{1}{8} a_6 = \frac{a_0 + 2}{8 \cdot 6 \cdot 4 \cdot 2}$$

$$a_9 = -\frac{1}{9} a_7 = -\frac{1}{9 \cdot 7 \cdot 5 \cdot 3}$$

$$\vdots$$

The solution can then be written

$$y = \sum_{n=0}^{\infty} a_n x^n$$

$$= a_0 - \frac{a_0 + 2}{2} x^2 + \frac{1}{3} x^3 + \frac{a_0 + 2}{4 \cdot 2} x^4 - \frac{1}{5 \cdot 3} x^5 - \frac{a_0 + 2}{6 \cdot 4 \cdot 2} x^6$$

$$+ \frac{1}{7 \cdot 5 \cdot 3} x^7 + \cdots$$

$$= a_0 - (a_0 + 2)\left(\frac{1}{2} x^2 - \frac{1}{4 \cdot 2} x^4 + \frac{1}{6 \cdot 4 \cdot 2} x^6 + \cdots \right)$$

$$+ \left(\frac{1}{3} x^3 - \frac{1}{5 \cdot 3} x^5 + \frac{1}{7 \cdot 5 \cdot 3} x^7 - \cdots \right) \qquad \blacksquare$$

Example 2

Solve the differential equation $y'' + y = e^x$ by assuming a power series solution about the origin. Obtain the recursion formula and write the first few terms of the solution.

SOLUTION   Letting $y = \sum\limits_{n=0}^{\infty} a_n x^n$, then

$$y' = \sum_{n=1}^{\infty} a_n n x^{n-1}$$

and

$$y'' = \sum_{n=2}^{\infty} a_n n(n-1) x^{n-2}$$

The Maclaurin series expansion for e^x is

$$e^x = \sum_{n=0}^{\infty} \frac{x^n}{n!}$$

Substituting into the differential equation, we obtain

$$\sum_{n=2}^{\infty} n(n-1) a_n x^{n-2} + \sum_{n=0}^{\infty} a_n x^n = \sum_{n=0}^{\infty} \frac{x^n}{n!}$$

Rearranging the terms,

$$\sum_{n=2}^{\infty} n(n-1) a_n x^{n-2} + \sum_{n=0}^{\infty} \left(a_n - \frac{1}{n!} \right) x^n = 0$$

Since the right-hand side of the equality is identically zero, we equate each coefficient of every power of x on the left-hand side to zero. To evaluate a_n, the exponents in the two series must be made the same by shifting the index. Here we choose to replace n with $n - 2$ in the second series. Thus

$$\sum_{n=2}^{\infty} n(n-1) a_n x^{n-2} + \sum_{n=2}^{\infty} \left(a_{n-2} - \frac{1}{(n-2)!} \right) x^{n-2} = 0$$

We combine the two series to obtain

$$\sum_{n=2}^{\infty} x^{n-2} \left[n(n-1) a_n + a_{n-2} - \frac{1}{(n-2)!} \right] = 0$$

The coefficients of x^{n-2} for $n \geq 2$ must equal zero. Thus

$$n(n-1) a_n + a_{n-2} - \frac{1}{(n-2)!} = 0$$

Solving for a_n, we obtain the recursion formula

$$a_n = \frac{1}{n!} - \frac{1}{n(n-1)} a_{n-2}, \quad n = 2, 3, 4, \ldots$$

Iteration of this formula yields

$$a_2 = \left(\frac{1}{2!} - \frac{1}{2 \cdot 1}\right) a_0 = \frac{1}{2!} - \frac{1}{2!} a_0$$

$$a_3 = \frac{1}{3!} - \frac{1}{3 \cdot 2} a_1 = \frac{1}{3!} - \frac{1}{3!} a_1$$

$$a_4 = \frac{1}{4!} - \frac{1}{4 \cdot 3} a_2 = \frac{1}{4!} a_0$$

$$a_5 = \frac{1}{5!} - \frac{1}{5 \cdot 4} a_3 = \frac{1}{5!} a_1$$

$$\vdots$$

Since $y = \sum\limits_{n=0}^{\infty} a_n x^n$ and a_0 and a_1 are arbitrary, we get

$$y = a_0 + a_1 x + \sum_{n=2}^{\infty} a_n x^n$$

$$= a_0 + a_1 x + \left(\frac{1}{2!} - \frac{1}{2!} a_0\right) x^2 + \left(\frac{1}{3!} - \frac{1}{3!} a_1\right) x^3 + \frac{1}{4!} a_0 x^4$$

$$+ \frac{1}{5!} a_1 x^5 + \cdots$$

$$= a_0 \left(1 - \frac{1}{2!} x^2 + \frac{1}{4!} x^4 + \cdots\right) + a_1 \left(x - \frac{1}{3!} x^3 + \frac{1}{5!} x^5 + \cdots\right)$$

$$+ \frac{1}{2!} x^2 + \frac{1}{3!} x^3 + \cdots$$

$$= a_0 y_1 + a_1 y_2 + y_p$$

where y_1 and y_2 are, respectively, infinite series of even and odd terms, and $y_p =$
$\frac{1}{2!} x^2 + \frac{1}{3!} x^3 + \frac{1}{6!} x^6 + \frac{1}{7!} x^7 + \frac{1}{10!} x^{10} + \frac{1}{11!} x^{11} + \cdots$ ■

COMMENT: Since the differential equation in Example 2 is linear and of the second order, the form of the solution is $y = y_c + y_p$, where y_c consists of a linear combination of two linearly independent functions and y_p is a particular solution to the given nonhomogeneous equation. By using the methods of Chapters 4 and 5, which apply only to constant coefficient differential equations, you would obtain the solution in terms of elementary functions:

$$y = a_0^* \cos x + a_1^* \sin x + \frac{1}{2} e^x$$

The two different forms may be reconciled by letting $a_0^* = a_0 - \frac{1}{2}$ and $a_1^* = a_1 - \frac{1}{2}$ but in general there will be no need to make such a reconciliation.

Example 3

Solve the differential equation

$$(1 - x^2)y'' - 2xy' + 12y = 0$$

by assuming a solution in power series form valid near the origin. This differential equation is called **Legendre's** equation of order three.

SOLUTION   Assuming $y = \sum_{n=0}^{\infty} a_n x^n$, then

$$y' = \sum_{n=1}^{\infty} a_n n x^{n-1}$$

$$y'' = \sum_{n=2}^{\infty} a_n n(n - 1)x^{n-2}$$

Substituting into the given differential equation

$$(1 - x^2) \sum_{n=2}^{\infty} a_n n(n - 1)x^{n-2} - 2x \sum_{n=1}^{\infty} a_n n x^{n-1} + 12 \sum_{n=0}^{\infty} a_n x^n = 0$$

Carrying out the indicated multiplications,

$$\sum_{n=2}^{\infty} a_n n(n - 1)x^{n-2} - \sum_{n=2}^{\infty} a_n n(n - 1)x^n - 2 \sum_{n=1}^{\infty} a_n n x^n + 12 \sum_{n=0}^{\infty} a_n x^n = 0$$

We may change the indices of the second and third series to start at $n = 0$ since $n(n - 1)x^n$ is zero for $n = 0$ and $n = 1$ while nx^n is zero for $n = 0$. The last three series may then be combined to yield

$$\sum_{n=2}^{\infty} n(n - 1)a_n x^{n-2} - \sum_{n=0}^{\infty} a_n(n^2 - n + 2n - 12)x^n = 0$$

or

$$\sum_{n=2}^{\infty} n(n - 1)a_n x^{n-2} - \sum_{n=0}^{\infty} a_n(n + 4)(n - 3)x^n = 0$$

Shifting the index in the second of these summations so the exponent is $n - 2$, we get

$$\sum_{n=2}^{\infty} n(n - 1)a_n x^{n-2} - \sum_{n=2}^{\infty} a_{n-2}(n + 2)(n - 5)x^{n-2} = 0$$

This equation implies that a_0 and a_1 are both arbitrary and that for $n \geq 2$,

$$a_n = \frac{(n + 2)(n - 5)}{n(n - 1)} a_{n-2}$$

Iteration of this formula for $n \geq 2$ yields

$$a_2 = \frac{4(-3)}{2 \cdot 1} a_0 = -6a_0$$

$$a_3 = \frac{5(-2)}{3 \cdot 2} a_1 = -\frac{5}{3}a_1$$

$$a_4 = \frac{6(-1)}{4 \cdot 3} a_2 = \frac{6(-1)}{4 \cdot 3}[-6a_0] = 3a_0$$

$$a_5 = 0$$

$$a_6 = \frac{8 \cdot 1}{6 \cdot 5} a_4 = \frac{8 \cdot 1}{6 \cdot 5}(3a_0) = \frac{4}{5}a_0$$

$$a_7 = 0$$

$$\vdots$$

Noting that $a_{2k+1} = 0$ for $k \geq 2$, the solution of Legendre's differential equation of order three is

$$y = a_0 + a_1 x - 6a_0 x^2 - \frac{5}{3}a_1 x^3 + 3a_0 x^4 + \frac{4}{5}a_0 x^6 + \cdots$$

$$= a_1\left(x - \frac{5}{3}x^3\right) + a_0\left(1 - 6x^2 + 3x^4 + \frac{4}{5}x^6 + \cdots\right)$$

There are two linearly independent solutions here—namely, the third-degree polynomial

$$y_1 = a_1\left(x - \frac{5}{3}x^3\right)$$

called a Legendre polynomial, and the infinite series

$$y_2 = a_0\left(1 - 6x^2 + 3x^4 + \frac{4}{5}x^6 + \cdots\right)$$

Since the differential equation has a singular point at $x = 1$, this series converges for $|x| < 1$. This can also be shown by using the ratio test. We will examine the Legendre differential equation in more detail in Section 8–6.   ∎

Power series may also be used to find solutions centered at points other than $x = 0$. In the following example we assume a power series solution about $x = 1$. The example also exhibits how **multiterm recursion formulas** arise.

Example 4

Find a solution in power series form, centered at $x = 1$, to the differential equation $xy'' + y' + xy = 0$. Obtain a recursion formula and write the first few terms.

SOLUTION   The given differential equation may be written in the form

$$(x - 1)y'' + y'' + y' + (x - 1)y + y = 0$$

From Theorem 8–1 we may assume a solution of the form $y = \sum\limits_{n=0}^{\infty} a_n(x - 1)^n$ that will converge for $0 < x < 2$. Then

$$y' = \sum_{n=1}^{\infty} a_n n(x - 1)^{n-1} \qquad \text{and} \qquad y'' = \sum_{n=2}^{\infty} a_n n(n - 1)(x - 1)^{n-2}$$

Substituting into the differential equation,

$$(x - 1) \sum_{n=2}^{\infty} a_n n(n - 1)(x - 1)^{n-2} + \sum_{n=2}^{\infty} a_n n(n - 1)(x - 1)^{n-2}$$

$$+ \sum_{n=1}^{\infty} a_n n(x - 1)^{n-1} + (x - 1) \sum_{n=0}^{\infty} a_n(x - 1)^n + \sum_{n=0}^{\infty} a_n(x - 1)^n = 0$$

Simplifying,

$$\sum_{n=2}^{\infty} a_n n(n - 1)(x - 1)^{n-1} + \sum_{n=2}^{\infty} a_n n(n - 1)(x - 1)^{n-2}$$

$$+ \sum_{n=1}^{\infty} a_n n(x - 1)^{n-1} + \sum_{n=0}^{\infty} a_n(x - 1)^{n+1} + \sum_{n=0}^{\infty} a_n(x - 1)^n = 0$$

Combining terms with like exponents,

$$\sum_{n=1}^{\infty} a_n n^2(x - 1)^{n-1} + \sum_{n=2}^{\infty} a_n n(n - 1)(x - 1)^{n-2}$$

$$+ \sum_{n=0}^{\infty} a_n(x - 1)^{n+1} + \sum_{n=0}^{\infty} a_n(x - 1)^n = 0$$

Shifting indices in the various series so that the exponents are all $n - 2$, we get

$$\sum_{n=2}^{\infty} a_{n-1}(n - 1)^2(x - 1)^{n-2} + \sum_{n=2}^{\infty} a_n n(n - 1)(x - 1)^{n-2}$$

$$+ \sum_{n=3}^{\infty} a_{n-3}(x - 1)^{n-2} + \sum_{n=2}^{\infty} a_{n-2}(x - 1)^{n-2} = 0$$

We write the terms for $n = 2$ separately and then combine the series to obtain

$$(2a_2 + a_1 + a_0)x^0 + \sum_{n=3}^{\infty} \{a_n n(n - 1) + (n - 1)^2 a_{n-1}$$

$$+ a_{n-2} + a_{n-3}\}(x - 1)^{n-2} = 0$$

The following coefficient information is obtained from this equation with a_0 and a_1 arbitrary:

$$a_2 = \frac{- a_0 - a_1}{2}$$

and for $n \geq 3$, we have the recursion formula

$$a_n = -\frac{(n-1)^2 a_{n-1} + a_{n-2} + a_{n-3}}{n(n-1)}$$

This recursion formula is too complicated to expect a simple expression for the general coefficients. To simplify the solution, we let $a_0 = a_1 = 1$ and evaluate several coefficients. Thus the first few coefficients are $a_2 = -1$, $a_3 = \frac{1}{3}$, $a_4 = -\frac{1}{4}$, $a_5 = \frac{7}{30}$. The solution is then written

$$y = 1 + (x-1) - (x-1)^2 + \frac{(x-1)^3}{3}$$
$$-\frac{(x-1)^4}{4} + \frac{7(x-1)^5}{30} \cdots \quad \blacksquare$$

> COMMENT: The differential equation of the previous example is called **Bessel's equation of order zero.** The technique of solution used in the example is not standard. A solution valid near the origin is much more popular but requires a slight variation on the technique of assuming a power series solution. Section 8–6 contains a more detailed discussion of Bessel's equation.

Example 5

Find a power series solution of $y'' + xy' + (1-x)y = 0$ about $x = 0$.

SOLUTION   Letting $y = \sum\limits_{n=0}^{\infty} a_n x^n$ and substituting into the given differential equation, we obtain

$$\sum_{n=2}^{\infty} a_n n(n-1)x^{n-2} + x\sum_{n=1}^{\infty} a_n n x^{n-1} + (1-x)\sum_{n=0}^{\infty} a_n x^n = 0$$

Expanding,

$$\sum_{n=2}^{\infty} a_n n(n-1)x^{n-2} + \sum_{n=1}^{\infty} a_n n x^n + \sum_{n=0}^{\infty} a_n x^n - \sum_{n=0}^{\infty} a_n x^{n+1} = 0$$

Shifting indices in the last three series to make the exponent of x equal to $n - 2$, we get

$$\sum_{n=2}^{\infty} a_n n(n-1)x^{n-2} + \sum_{n=3}^{\infty} a_{n-2}(n-2)x^{n-2}$$
$$+ \sum_{n=2}^{\infty} a_{n-2}x^{n-2} - \sum_{n=3}^{\infty} a_{n-3}x^{n-2} = 0$$

Writing the terms for $n = 2$ separately and combining the terms of the resulting series into one series yields

$$(2a_2 + a_0) + \sum_{n=3}^{\infty} [n(n-1)a_n + (n-1)a_{n-2} - a_{n-3}] x^{n-2} = 0$$

From this result we obtain

$$a_2 = -\frac{a_0}{2}$$

and for $n \geq 3$, we get the recursion formula

$$a_n = \frac{a_{n-3}}{n(n-1)} - \frac{a_{n-2}}{n}$$

Iteration of this formula for $n \geq 3$ yields

$$a_3 = \frac{a_0}{6} - \frac{a_1}{3}$$

$$a_4 = \frac{a_1}{12} - \frac{a_2}{4} = \frac{a_1}{12} + \frac{a_0}{8}$$

$$a_5 = \frac{a_2}{20} - \frac{a_3}{5} = -\frac{a_0}{40} - \frac{1}{5}\left(\frac{a_0}{6} - \frac{a_1}{3}\right) = -\frac{7}{120}a_0 + \frac{1}{15}a_1$$

$$\vdots$$

Substituting these coefficients into $y = \sum\limits_{n=0}^{\infty} a_n x^n$, the solution may be written as

$$y = a_0\left(1 - \frac{1}{2}x^2 + \frac{1}{6}x^3 + \frac{1}{8}x^4 - \frac{7}{120}x^5 + \cdots\right)$$

$$+ a_1\left(x - \frac{1}{3}x^3 + \frac{1}{12}x^4 + \frac{1}{15}x^5 + \cdots\right) \qquad \blacksquare$$

EXERCISES FOR SECTION 8–3

In Exercises 1–16 solve the given differential equation by assuming an infinite power series solution around 0 and using the method of undetermined coefficients. (Let $a_0 = y(0)$ and $a_1 = y'(0)$.) Obtain a recursion formula for the coefficients and write the first few terms of the solution.

1. $y' + y = 0$
2. $y' + y = e^x$
3. $y' + y = e^{-x}$
4. $y'' + y = 0$
5. $y'' + y = x^2$
6. $y'' + y = e^{-x}$
7. $y'' + y' = x$
8. $y' + x^2 y = 0$
9. $y' + xy = 1$
10. $y'' + xy = 0$
11. $y'' + x^3 y = 0$
12. $(1 - x^2)y'' + 3xy' + 5y = 0$
13. $(1 - x^2)y'' + xy' - y = 0$
14. $(x^2 + 1)y'' - 2xy' - 10y = 0$
15. $y'' + xy' - 4y = 0$
16. $(x^2 + 1)y'' + 3xy' - 3y = 0$

17. Show that the series solution to Exercise 1 is $y = a_0 e^{-x}$.

18. Show that the solution to Exercise 2 can be written $y = a_0^* e^{-x} + \frac{1}{2} e^x$, where $a_0^* = a_0 - \frac{1}{2}$.

19. Show that the solution to Exercise 3 can be written $y = a_0 e^{-x} + x e^{-x}$.

20. Show that the solution to Exercise 7 can be written

$$y = a_0^* + a_1^* e^{-x} + \frac{1}{2} x^2 - x$$

where $a_0^* = a_0 + a_1 + 1$; $a_1^* = -a_1 - 1$.

21. Show that the solution to Example 1 can be written in the form

$$\sum_{k=1}^{\infty} \frac{(-1)^k}{k! \, 2^{k-1}} x^{2k} + \sum_{k=1}^{\infty} \frac{(-1)^{k+1} \, 2^k k!}{(2k+1)!} x^{2k+1} + a_0 \sum_{k=0}^{\infty} \frac{(-1)^k}{k! \, 2^k} x^{2k}$$

22. Solve **Hermite's equation** for the case $n = 2$ by assuming an infinite series solution around 0 and using the method of undetermined coefficients.

$$y'' - 2xy' + 2ny = 0$$

Obtain a recursion formula for the coefficients and write the solution up to a ninth-degree polynomial approximation.

23. Solve **Chebyshev's equation** of order two:

$$(1 - x^2)y'' - xy' + 4y = 0$$

Follow the same directions as in Exercise 22.

To find a power series solution about points other than $x = 0$, it is usually best to make the substitution $X = x - x_0$. Solve the equations in Exercises 24–27 by the power series method about the indicated points.

24. $y'' + (x - 2)y = 0$, $x = 2$
25. $y'' + (x - 2)y = 0$, $x = 1$
26. $xy'' - y = 0$, $x = 1$
27. $y'' + (x - 1)y' = 0$, $x = 1$

8–4   THE METHOD OF FROBENIUS

As a consequence of Theorem 8–1 we do not expect a power series solution around a singular point. However, near certain kinds of singular points a modified power series approach can be used to obtain a solution.

Consider a second-order differential equation with polynomial coefficients

$$p_2(x)y'' + p_1(x)y' + p_0(x)y = 0$$

Dividing by p_2, we get

$$y'' + b_1(x)y' + b_0(x) y = 0$$

where $b_1(x)$ and $b_0(x)$ are rational functions.

DEFINITION

A singular point $x = x_o$ of the differential equation

$$y'' + b_1(x)y' + b_0(x)y = 0$$

is said to be **regular** if both

$$\lim_{x \to x_o} (x - x_o)b_1(x) \qquad \text{and} \qquad \lim_{x \to x_o} (x - x_o)^2 b_0(x)$$

exist at $x = x_o$. If a singular point is not regular, it is **irregular**.

Example 1

Locate and classify the singular points of the differential equation

$$(x - 1)^3 x^2 (x - 2)y'' + 3x^3(x - 1)y' + (x - 1)y = 0$$

SOLUTION   Since the coefficients are all polynomials, the singular points are those values of x for which the coefficient of y'' is 0—that is, $x = 0, 1,$ and 2. Dividing by the coefficient of y'', we obtain

$$b_1(x) = \frac{3x}{(x - 1)^2(x - 2)}, \quad b_0(x) = \frac{1}{(x - 1)^2 x^2 (x - 2)}$$

Since

$$\lim_{x \to 1} (x - 1)b_1(x) = \lim_{x \to 1} \frac{3x}{(x - 1)(x - 2)}$$

is undefined as $x \to 1$, the singular point at $x = 1$ is irregular. The other two singular points are regular. ∎

Example 2

Locate and classify the singular points of

$$(x^2 + 4)x^3(x^2 - 1)^2 y'' + (x - 1)x^2 y' + y = 0$$

SOLUTION   The singular points are $x = 0, \pm 1,$ and $\pm 2i$. Notice that

$$b_1 = \frac{1}{(x^2 + 4)x(x - 1)(x + 1)^2}$$

and

$$b_0 = \frac{1}{(x^2 + 4)x^3(x - 1)^2(x + 1)^2}$$

Since $\lim\limits_{x\to0} x^2b_0$ and $\lim\limits_{x\to-1} (x + 1)b_1$ do not exist, there are irregular singular points at 0 and -1. The singular point $x = 1$ is regular since $\lim\limits_{x\to1} (x - 1)^2b_0$ and $\lim\limits_{x\to1} (x - 1)b_1$ exist. The singular point at $2i$ is regular since both $\lim\limits_{x\to2i} (x - 2i)^2b_0$ and $\lim\limits_{x\to2i} (x - 2i)b_1$ exist. Similarly, there is a regular singular point at $-2i$. ∎

To find a solution around a regular singular point, we use an infinite series whose form is a power of $x - x_o$ times a power series.

DEFINITION

A series of the type

$$F_r(x) = (x - x_o)^r \sum_{n=0}^{\infty} a_n(x - x_o)^n = \sum_{n=0}^{\infty} a_n(x - x_o)^{n+r} \qquad (8\text{–}4)$$

is called a **Frobenius series with index r.** The index r need not be an integer.

The fundamental theorem of this section relates the concept of a Frobenius series to solutions near regular singular points.

THEOREM 8–2

If $x = x_o$ is a regular singular point of

$$p_2(x)y'' + p_1(x)y' + p_0(x)y = 0$$

where p_2, p_1, and p_0 are polynomials, then there exists at least one solution of the form of a Frobenius series that converges at least for $0 < x - x_o < R$, where R is the distance from x_o to the next singular point of the differential equation.

To find a solution around a regular singular point, we must find the coefficients of the Frobenius series and the index r. Theorem 8–2 motivates the following procedure for solving $p_2(x)y'' + p_1(x)y' + p_0(x)y = 0$.

1. Assume a solution of the form of a Frobenius series.

2. Substitute this series into the differential equation and simplify.

3. Equate to 0 the coefficient of the lowest power of $(x - x_o)$. This equation is called the **indicial equation** and determines the value(s) of the index r. The value(s) of r permit a_0 to be chosen arbitrarily.

4. Using the value(s) of r computed in step 3, proceed as with power series to determine relationships between the coefficients of powers of x.

Example 3

Use the Frobenius series near the origin to solve the differential equation $2xy'' + 3y' - y = 0$.

SOLUTION   Since $x = 0$ is a regular singular point, we assume the solution to be of the form

$$\sum_{n=0}^{\infty} a_n x^{n+r}$$

Differentiating twice,

$$y' = \sum_{n=0}^{\infty} a_n(n + r)x^{n+r-1}$$

and

$$y'' = \sum_{n=0}^{\infty} a_n(n + r)(n + r - 1)x^{n+r-2}$$

Substituting these series into $2xy'' + 3y' - y = 0$, we get

$$2\sum_{n=0}^{\infty} a_n(n + r)(n + r - 1)x^{n+r-1} + 3\sum_{n=0}^{\infty} a_n(n + r)x^{n+r-1} - \sum_{n=0}^{\infty} a_n x^{n+r} = 0$$

Shifting the index in the third series and combining the first two yields

$$\sum_{n=0}^{\infty} a_n(n + r)(2n + 2r + 1)x^{n+r-1} - \sum_{n=1}^{\infty} a_{n-1}x^{n+r-1} = 0$$

Writing the term corresponding to $n = 0$ and combining the terms for $n \geq 1$ into one series,

$$a_0 r(2r - 1)x^{r-1} + \sum_{n=1}^{\infty} [a_n(n + r)(2n + 2r + 1) - a_{n-1}]x^{n+r-1} = 0$$

Equating the coefficient of x^{r-1} to zero yields the indicial equation

$$a_0 r(2r + 1) = 0$$

The values of the index r for which this coefficient is zero are $r = 0$ or $r = -\frac{1}{2}$. Hence two linearly independent solutions of the given differential equation have the form

$$y_1 = F_0(x) = \sum_{n=0}^{\infty} a_n x^n \qquad \text{and} \qquad y_2 = F_{-1/2}(x) = x^{-1/2} \sum_{n=0}^{\infty} a_n^* x^n$$

Since $a_n(n + r)(2n + 2r + 1) - a_{n-1} = 0$ for all $n \geq 1$, we have the following information on the coefficients for the two series:

1. a_0 is arbitrary, and for $n \geq 1$, $a_n = \dfrac{1}{n(2n + 1)} a_{n-1}$.

2. a_0^* is arbitrary, and for $n \geq 1$, $a_n^* = \dfrac{1}{n(2n - 1)} a_{n-1}^*$.

Iteration of the formula for a_n yields

$$n = 1, a_1 = \frac{1}{1 \cdot 3} a_0 = \frac{2}{1 \cdot 2 \cdot 3} a_0 = \frac{2a_0}{3!}$$

$$n = 2, a_2 = \frac{1}{2 \cdot 5} a_1 = \frac{1}{2 \cdot 3 \cdot 5} a_0 = \frac{2^2 a_0}{5!}$$

$$n = 3, a_3 = \frac{1}{3 \cdot 7} a_2 = \frac{1}{3 \cdot 7} \frac{2^2 a_0}{5!} = \frac{2^3 a_0}{7!}$$

$$\vdots$$

Notice that each term of a_n was multiplied by $\frac{2}{2}$ to make the denominator $(2n + 1)!$. The general form of a_n is then

$$a_n = \frac{2^n a_0}{(2n + 1)!}$$

Similarly, the general form of a_n^* is found to be

$$a_n^* = \frac{2^n a_0^*}{(2n)!}$$

The two solutions are

$$y_1 = a_0 \sum_{n=0}^{\infty} \frac{2^n}{(2n + 1)!} x^n, \quad y_2 = a_0^* x^{-1/2} \sum_{n=0}^{\infty} \frac{2^n}{(2n)!} x^n$$

Notice that the second of these is not a power series. ∎

Indicial Equations for Regular Singular Points at the Origin

If a differential equation of the form

$$p_2(x)y'' + p_1(x)y' + p_0(x)y = 0$$

has a regular singular point at $x = 0$, then it can be put into the form

$$x^2y'' + xb_1(x)y' + b_0(x)y = 0$$

where b_1 and b_0 are rational functions of x that exist at $x = 0$. The indicial equation for this important case can be obtained by substituting $y = F_r(x)$ from Equation 8–4 into the differential equation to obtain

$$\sum_{n=0}^{\infty} a_n(n + r)(n + r - 1)x^{n+r} + \sum_{n=0}^{\infty} b_1(x)a_n(n + r)x^{n+r}$$

$$+ \sum_{n=0}^{\infty} a_n b_0(x)x^{n+r} = 0$$

The indicial equation is obtained from the term $n = 0$. Thus

$$a_0[r(r - 1) + b_1(0) r + b_0(0)] = 0$$

Since a_0 is to be arbitrary, the indicial equation is

$$r^2 + [b_1(0) - 1] r + b_0(0) = 0 \qquad (8\text{–}5)$$

Example 4

Find the form of two Frobenius series solutions around the origin for the differential equation

$$2xy'' - 3y' - \left(\frac{3 + x}{x}\right)y = 0$$

SOLUTION   There is a regular singular point at the origin. The differential equation can be put into the form

$$x^2y'' - \frac{3}{2}xy' + \left(\frac{3 + x}{2}\right)y = 0$$

In this form we recognize $b_1(x) = -\frac{3}{2}$ and $b_0(x) = (3 + x)/2$, from which $b_1(0) = -\frac{3}{2}$ and $b_0(0) = \frac{3}{2}$. Hence by Equation 8–5 the indicial equation is

$$r^2 - \frac{5}{2}r + \frac{3}{2} = 0$$

which can be written

$$(r - 1)\left(r - \frac{3}{2}\right) = 0$$

Thus the two roots of the indicial equation are $r = 1$ and $r = \frac{3}{2}$, and by Equation 8–4 the forms of the two Frobenius series solutions are

$$F_1 = \sum_{n=0}^{\infty} a_n x^{n+1}$$

and

$$F_{3/2} = \sum_{n=0}^{\infty} a_n^* x^{n+3/2}$$

Since the indicial equation is a quadratic equation, two Frobenius series are always possible, but they *might not be linearly independent*. The most obvious case is when the two roots are equal.   ∎

Example 5

Use the method of Frobenius to determine the form of a solution valid near the origin. Determine the recursion formula and write the general term.

$$4x^2 y'' + (3x + 1)y = 0$$

SOLUTION   Since the origin is a regular singular point, we may assume a Frobenius series solution

$$y = \sum_{n=0}^{\infty} a_n x^{n+r}$$

Differentiating twice and substituting into the given equation,

$$4x^2 \sum_{n=0}^{\infty} a_n(n + r)(n + r - 1)x^{n+r-2} + (3x + 1) \sum_{n=0}^{\infty} a_n x^{n+r} = 0$$

Simplifying and expanding,

$$\sum_{n=0}^{\infty} 4a_n(n + r)(n + r - 1)x^{n+r} + \sum_{n=0}^{\infty} 3a_n x^{n+r+1} + \sum_{n=0}^{\infty} a_n x^{n+r} = 0$$

Shifting the index in the second series and combining the first and the third series,

$$\sum_{n=0}^{\infty} a_n[4(n + r)(n + r - 1) + 1]x^{n+r} + \sum_{n=1}^{\infty} 3a_{n-1}x^{n+r} = 0$$

Writing the first term of the first series and then combining,

$$a_0 x^r[4r(r - 1) + 1] + \sum_{n=1}^{\infty} [a_n(2n + 2r - 1)^2 + 3a_{n-1}]x^{n+r} = 0$$

The indicial equation is obtained from the coefficient of x^r:

$$4r(r - 1) + 1 = 0 \qquad \text{or} \qquad r = \frac{1}{2}$$

Hence the form of one solution is

$$y = x^{1/2} \sum_{n=0}^{\infty} a_n x^n$$

where the coefficient information in recursive form is

$$a_0 \text{ is arbitrary, and } a_n = -\frac{3}{4n^2} a_{n-1} \qquad \text{for} \qquad n \geq 1$$

The general coefficient is found to be

$$a_n = \frac{(-1)^n 3^n}{4^n} \cdot \frac{1}{(n!)^2}$$

Hence the solution is

$$y = \sum_{n=0}^{\infty} \frac{(-1)^n 3^n}{4^n (n!)^2} x^{n+1/2}$$

∎

The roots of the indicial equation can be used to give some idea of the nature of the solution.

- If the roots are distinct and do not differ by an integer, then there are two Frobenius series solutions.

- If the roots are equal, there is one Frobenius series solution.

- If the roots are distinct but differ by an integer, then there is a *possibility* of two linearly independent Frobenius series solutions, but the recursion relationship must be carefully examined.

Example 6

Use the method of Frobenius to obtain the form of at least one solution near the origin and the recursion formula between the coefficients for the differential equation

$$xy'' - (x + 5)y' + 3y = 0$$

SOLUTION   Since $x = 0$ is a regular singular point, a Frobenius-type solution may be assumed. Letting

$$y = \sum_{n=0}^{\infty} a_n x^{n+r}$$

and substituting into the differential equation, we obtain

$$\sum_{n=0}^{\infty} a_n(n + r)(n + r - 1)x^{n+r-1} - \sum_{n=0}^{\infty} a_n(n + r)x^{n+r}$$

$$- 5 \sum_{n=0}^{\infty} a_n(n + r)x^{n+r-1} + 3 \sum_{n=0}^{\infty} a_n x^{n+r} = 0$$

Shifting the index in the second and fourth of these series yields

$$\sum_{n=0}^{\infty} a_n(n + r)(n + r - 1)x^{n+r-1} - \sum_{n=1}^{\infty} a_{n-1}(n + r - 1)x^{n+r-1}$$

$$- 5 \sum_{n=0}^{\infty} a_n(n + r)x^{n+r-1} + 3 \sum_{n=1}^{\infty} a_{n-1}x^{n+r-1} = 0$$

Setting the coefficient of x^{r-1} equal to 0 ($n = 0$), gives the indicial equation

$$a_0 r(r - 1) - 5a_0 r = 0$$

Since a_0 is to remain arbitrary, we have

$$r(r - 6) - 0$$

from which $r = 0$ and 6. We thus have two possible Frobenius series solutions, one with index 0 and one with index 6. But are they, in fact, linearly independent? We must examine the coefficients. The necessary coefficient information is obtained by equating the coefficient of x^{n+r-1} to zero.

$$a_n(n + r)(n + r - 1) - a_{n-1}(n + r - 1) - 5a_n(n + r) + 3a_{n-1} = 0$$

Simplifying,

$$a_n(n + r)(n + r - 6) = a_{n-1}(n + r - 4) \tag{8–6}$$

We begin with the *lower* index $r = 0$ in hopes of obtaining two linearly independent solutions from this one index. (There would be no hope of picking up two solutions starting with the higher index.) If $r = 0$, the recursion relation of Equation 8–6 becomes, with a_0 arbitrary,

$$a_n n(n - 6) = a_{n-1}(n - 4) \tag{8–7}$$

For $n = 1$, $a_1(1)(-5) = a_0(-3)$, so that $a_1 = 3a_0/5$
For $n = 2$, $a_2(2)(-4) = a_1(-2)$, so that $a_2 = a_1/4 = 3a_0/20$
For $n = 3$, $a_3(3)(-3) = a_2(-1)$, so that $a_3 = a_2/9 = a_0/60$
For $n = 4$, $a_4(4)(-2) = a_3(0)$, so that $a_4 = 0$
For $n = 5$, $a_5(5)(-1) = a_4(1) = 0$, so that $a_5 = 0$
For $n = 6$, $a_6(6)(0) = a_5(2) = 0$, so that a_6 is arbitrary

For $n > 6$, $a_n = \dfrac{n - 4}{n(n - 6)} a_{n-1}$ with a_6 arbitrary

Thus we have the two linearly independent solutions:

$$y_1 = F_0(x) = a_0\left(1 + \frac{3}{5}x + \frac{3}{20}x^2 + \frac{1}{60}x^3\right)$$

$$y_2 = F_6(x) = a_6 x^6\left(1 + \frac{3}{7}x + \frac{3}{28}x^2 + \frac{5}{252}x^3 \cdots\right)$$

The solution y_2 is the same solution that would be obtained by using the larger index first. Hence as a general rule we start with the smaller index.   ■

In Example 6 we were able to obtain two linearly independent Frobenius series solutions because of the peculiar nature of the recursion formula. The differential equation in the next example has only one Frobenius solution.

Example 7

Find the form of a Frobenius series solution to the differential equation

$$xy'' + (x - 5)y' + 3y = 0$$

and determine a recursion formula.

SOLUTION   Substituting as in Example 6 into the differential equation, we obtain, after a shift of index in the second and fourth series,

$$\sum_{n=0}^{\infty} a_n(n + r)(n + r - 1)x^{n+r-1} + \sum_{n=1}^{\infty} a_{n-1}(n + r - 1)x^{n+r-1}$$

$$- 5 \sum_{n=0}^{\infty} a_n(n + r)x^{n+r-1} + \sum_{n=1}^{\infty} 3a_{n-1}x^{n+r-1} = 0$$

The term $n = 0$ gives the indicial equation

$$r(r - 1) - 5r = 0$$

from which $r = 0$ and 6. The recursion formula is obtained by setting the coefficient of x^{n+r-1} equal to 0:

$$a_n(n + r)(n + r - 1) + a_{n-1}(n + r - 1) - 5a_n(n + r) + 3a_{n-1} = 0$$

Simplifying,

$$a_n(n + r)(n + r - 6) = -a_{n-1}(n + r + 2) \tag{8–8}$$

For the index $r = 0$ this recursion relation takes the form

$$a_n(n)(n - 6) = -a_{n-1}(n + 2) \tag{8–9}$$

For $n = 1$, $a_1(1)(-5) = -a_0(3)$, so that $a_1 = 3a_0/5$
$n = 2$, $a_2(2)(-4) = -a_1(4)$, so that $a_2 = 3a_0/10$
$n = 3$, $a_3(3)(-3) = -a_2(5)$, so that $a_3 = a_0/6$
$n = 4$, $a_4(4)(-2) = -a_3(6)$, so that $a_4 = a_0/8$
$n = 5$, $a_5(5)(-1) = -a_4(7)$, so that $a_5 = 7a_0/40$
$n = 6$, $a_6(6)(0) = -a_5(8)$, so that $0 \cdot a_6 = 7a_0/5$

The only way this last condition can be satisfied is for a_0 to be 0. Hence a_0 is *not* arbitrary; moreover, $a_1 = a_2 = a_3 = a_4 = a_5 = 0$. This leaves a_6 arbitrary, and we have a Frobenius solution of the form

$$y = x^6 \sum_{n=0}^{\infty} a_n^* x^n = \sum_{n=0}^{\infty} a_{n+6} x^{n+6} = \sum_{n=6}^{\infty} a_n x^n$$

where the a_0^* of the first series is the same as the a_6 of the Frobenius series of index 0. The coefficients a_n are obtained from Equation 8–9, which may be written as

$$a_n = -\frac{n+2}{n(n-6)}a_{n-1}$$

∎

In summary, the method of Frobenius always gives at least one solution near a regular singular point but might not give a second linearly independent series. In the next section we will see how to obtain a second solution in the cases where the method of Frobenius yields only one solution.

EXERCISES FOR SECTION 8–4

In Exercises 1–5 determine which points are not ordinary points for the given equation. Of the singular points, which are regular and which are irregular?

1. $x^2y'' + xy' + y = e^x$
2. $(x^2 - 1)^2 xy'' + (x + 1)y' + xy = 0$

3. $x(x - 1)^3 y'' + xy' + (x - 1)^2y = 0$
4. $xy'' + \dfrac{1}{x}y' + y = 7$

5. $(2 - x)y'' + xy' + \dfrac{y}{(x-2)^2} = 0$

In Exercises 6–16 show that the origin is a regular singular point. Find the roots of the indicial equation. Use the method of Frobenius to find at least one solution.

6. $4xy'' + 2y' + y = 0$
7. $x(1 - x)y'' - 3y' + 2y = 0$

8. $4x^2y'' - 2x(x + 2)y' + (x + 3)y = 0$
9. $x(x + 1)y'' + (5x + 1)y' + 3y = 0$

10. $(x^2 + x)y'' + (x + 1)y' - y = 0$
11. $x^2y'' + xy' + (x^2 - 1)y = 0$

12. $x^2y'' + xy' + (x^2 - 4)y = 0$
13. $2x^2y'' + x(1 - x)y' - y = 0$

14. $xy'' + y' + xy = 0$
15. $xy'' + y = 0$

16. $4x^2y'' - 2x(x - 2)y' - (3x + 1)y = 0$

For each equation in Exercises 17–20 find the form of at least one solution about $x = 0$. Express the coefficients in terms of a recurrence relation.

17. $x(2x + 1)y'' + 3y' - xy = 0$
18. $3x^2y'' + (5x - x^2)y' + (2x^2 - 1)y = 0$

19. $2xy'' - (1 + x^3)y' + y = 0$
20. $x^2(3 + x^2)y'' + 5xy' - (1 + x)y = 0$

8–5   OBTAINING A SECOND SOLUTION

Reduction of Order

Recall from Section 4–3 that if one solution y_1 of a second-order differential equation

$$p_2y'' + p_1y' + p_0y = 0$$

is known, then a second solution y_2 is given in Equation 4–16 as

$$y_2 = y_1 \cdot \int \frac{e^{-\int (p_1/p_2)dx}}{y_1^2} dx$$

Example 1 shows how to use this method to obtain a second solution when one series solution is known.

Example 1

Obtain a second solution of the differential equation of Example 5 of the previous section.

SOLUTION   The differential equation of that example

$$4x^2y'' + (3x + 1)y = 0$$

has the solution

$$y_1 = x^{1/2} - \frac{3}{4}x^{3/2} + \frac{9}{64}x^{5/2} - \frac{3}{256}x^{7/2} \cdots$$

In this case $p_2(x) = 4x^2$, and $p_1(x) = 0$, so that

$$y_2 = y_1(x) \int \frac{dx}{y_1^2(x)}$$

$$= y_1 \int \frac{dx}{\left(x^{1/2} - \frac{3}{4}x^{3/2} + \frac{9}{64}x^{5/2} - \frac{3}{256}x^{7/2} \cdots \right)^2}$$

$$= y_1 \int \frac{dx}{x - \frac{3}{2}x^2 + \frac{27}{32}x^3 - \frac{15}{64}x^4 + \cdots} \qquad (squaring)$$

$$= y_1 \int \left(\frac{1}{x} + \frac{3}{2} + \frac{45}{32}x - \frac{69}{64}x^2 + \cdots \right) dx \qquad (dividing)$$

$$= y_1 \left(\ln x + \frac{3}{2}x + \frac{45}{64}x^2 + \frac{23}{64}x^3 + \cdots \right) \qquad (integrating)$$

$$= y_1 \ln x + y_1 \left(\frac{3}{2}x + \frac{45}{64}x^2 + \frac{23}{64}x^3 + \cdots \right)$$

$$= y_1 \ln x + \left(x^{1/2} - \frac{3}{4}x^{3/2} + \frac{9}{64}x^{5/2} - \cdots \right) \left(\frac{3}{2}x + \frac{45}{64}x^2 + \frac{23}{64}x^3 + \cdots \right)$$

$$= y_1 \ln x + \frac{3}{2}x^{3/2} - \frac{27}{64}x^{5/2} + \frac{11}{256}x^{7/2} \cdots \qquad \blacksquare$$

The presence of the logarithm term in the solution is characteristic of the cases where the indicial equation has equal roots and in some cases where the roots differ by an integer. We have the following theorem:

THEOREM 8–3

(a)   *If the indicial equation has two equal roots, then a second solution of $p_2(x)y'' + p_1(x)y' + p_0(x)y = 0$ is of the form*

$$y_2 = y_1(x) \ln x + \sum_{n=1}^{\infty} A_n x^{n+r} \qquad\qquad (8\text{–}10)$$

where y_1 is a Frobenius series solution of the differential equation.

(b)   *If the indicial equation has roots r_1 and r_2, where $r_1 - r_2$ is equal to a positive integer, then there is one Frobenius solution $F_{r_1}(x)$ and a second linearly independent solution may be found of the form*

$$y_2 = cy_1(x) \ln x + \sum_{n=0}^{\infty} A_n x^{n+r_2}$$

COMMENT: • As in the previous section the given functions are valid for $x > 0$ out to the nearest singularity. For solutions valid for $x < 0$, substitute $-x$ for x.
 • Note that in (b) the value of c is to be determined through substitution into the given differential equation. If $c = 0$, then y_2 is just $F_{r_2}(x)$, that is, the form of y_2 is also a Frobenius series. (See Example 6 of Section 8–4.)

Using the Logarithmic Form

The result of Theorem 8–3 can be used directly to obtain another procedure for finding a second solution to the differential equation when the roots of the indicial equation are equal.

Example 2

The second solution of the differential equation

$$4x^2y'' + (3x + 1)y = 0$$

was found by the method of reduction of order in Example 1. Using Theorem 8–3, we may instead assume that y_2 has the form

$$y_2 = y_1 \ln x + \sum_{n=1}^{\infty} A_n x^{n+1/2}$$

and find the A_n by the method of undetermined coefficients. Differentiating twice,

$$y_2' = y_1' \ln x + \frac{y_1}{x} + \sum_{n=1}^{\infty} \left(n + \frac{1}{2} \right) A_n x^{(n-1/2)}$$

and

$$y_2'' = y_1'' \ln x + 2\frac{y_1'}{x} - \frac{y_1}{x^2} + \sum_{n=1}^{\infty} \left(n^2 - \frac{1}{4}\right) A_n x^{(n-3/2)}$$

Substituting into the differential equation,

$$4x^2 y_1'' \ln x + 8xy_1' - 4y_1 + \sum_{n=1}^{\infty} (4n^2 - 1)A_n x^{(n+1/2)}$$

$$+ (3x + 1) y_1 \ln x + 3 \sum_{n=1}^{\infty} A_n x^{n+3/2} + \sum_{n=1}^{\infty} A_n x^{n+1/2} = 0 \quad (8\text{–}11)$$

Using the fact that y_1 is a solution of $4x^2 y'' + (3x + 1)y = 0$, the coefficient of the $\ln x$ term in (8–11) vanishes. The form of y_1 for this differential equation was found in Example 5, Section 8–4 to be

$$y_1 = \sum_{n=0}^{\infty} \frac{(-1)^n 3^n}{4^n (n!)^2} x^{n+1/2}$$

from which

$$y_1' = \sum_{n=0}^{\infty} \frac{\left(n + \frac{1}{2}\right)(-1)^n 3^n}{4^n (n!)^2} x^{n-1/2}$$

Substituting these expressions into (8–11) and shifting the index in the second-to-last series in the equation, we get

$$8\sum_{n=0}^{\infty} \frac{\left(n + \frac{1}{2}\right)(-1)^n 3^n}{4^n (n!)^2} x^{n+1/2} - 4\sum_{n=0}^{\infty} \frac{(-1)^n 3^n}{4^n (n!)^2} x^{n+1/2}$$

$$+ \sum_{n=1}^{\infty} (4n^2 - 1)A_n x^{n+1/2} + 3\sum_{n=2}^{\infty} A_{n-1} x^{n+1/2} + \sum_{n=1}^{\infty} A_n x^{n+1/2} = 0$$

The coefficient of $x^{1/2}$ is found to be zero (check for $n = 0$). The coefficient of $x^{3/2}$ is equated to zero (use $n = 1$) to obtain

$$-8\left(\frac{3}{2}\right) \cdot \left(\frac{3}{4}\right) + 4\left(\frac{3}{4}\right) + 3A_1 + A_1 = 0$$

from which $A_1 = \frac{3}{2}$.

For $n \geq 2$, we obtain, by setting the coefficient of $x^{n+1/2}$ equal to 0,

$$\frac{(8n + 4)(-1)^n 3^n}{4^n (n!)^2} - \frac{4(-1)^n 3^n}{4^n (n!)^2} + (4n^2 - 1)A_n + 3A_{n-1} + A_n = 0$$

Simplifying, we obtain

$$A_n(4n^2) = -3A_{n-1} - \frac{(-1)^n (4)3^n (2n)}{4^n (n!)^2} = -3A_{n-1} - \frac{(-1)^n (2n)3^n}{4^{n-1}(n!)^2}$$

This gives $A_2 = -\frac{27}{64}$, $A_3 = \frac{11}{256}$, which is in agreement with Example 1. ∎

Example 3

Find two linearly independent solutions of

$$x^2 y'' - (x + 2)y = 0$$

valid near 0.

SOLUTION Because $x = 0$ is a regular singular point, a Frobenius type series of the form

$$y = \sum_{n=0}^{\infty} a_n x^{n+r}$$

may be substituted into the differential equation to obtain

$$\sum_{n=0}^{\infty} a_n(n + r)(n + r - 1)x^{n+r} - \sum_{n=0}^{\infty} a_n x^{n+r+1} - 2\sum_{n=0}^{\infty} a_n x^{n+r} = 0$$

Shifting the index in the second of these series and combining terms, we have

$$a_0(r^2 - r - 2)x^r + \sum_{n=1}^{\infty} \{a_n[(n + r)(n + r - 1) - 2] - a_{n-1}\} x^{n+r} = 0$$

Equating the coefficient of x^r to zero and (at least temporarily) keeping a_0 arbitrary we obtain the indicial equation

$$r^2 - r - 2 = 0$$

whose roots are -1 and 2. Equating the coefficients of x^{n+r} to zero for $n \geq 1$, we have

$$a_n[(n + r)(n + r - 1) - 2] - a_{n-1} = 0 \qquad (8\text{–}12)$$

Thus we have two possible Frobenius solutions:

$$F_{-1} = \sum_{n=0}^{\infty} a_n x^{n-1} \quad \text{and} \quad F_2 = \sum_{n=0}^{\infty} b_n x^{n+2}$$

where coefficients in F_{-1} and F_2 both satisfy Equation 8–12.

We first seek the coefficients of F_{-1} in the hope that the two Frobenius series will be linearly independent. The coefficient information for $r = -1$ becomes

$$a_n[(n-1)(n-2) - 2] - a_{n-1} = 0$$

which simplifies to

$$a_n(n^2 - 3n) - a_{n-1} = 0$$

We have $a_1 = -\frac{1}{2}a_0$, $a_2 = \frac{1}{4}a_0$ and $a_3 \cdot 0 = a_2$. Thus, $a_2 = 0$, from which $a_1 = a_0 = 0$ and a_3 is arbitrary.

Continuing with a_3 arbitrary,

$$a_n = \frac{1}{n(n-3)} a_{n-1} \text{ for } n > 4$$

Computing the coefficients for $n = 4, 5$, and 6, we find (with $a_3 = 1$)

$$a_4 = \frac{1}{4}, \ a_5 = \frac{1}{40}, \ a_6 = \frac{1}{720}$$

from which

$$y_1 = F_{-1} = x^2 + \frac{1}{4}x^3 + \frac{1}{40}x^4 + \frac{1}{720}x^5 \cdots$$

If we now try to obtain y_2 from F_2, we get the following coefficient information:

$$b_n = \frac{1}{n(n+3)} b_{n-1} \text{ for } n > 0$$

We notice that the coefficient information for b_n is exactly the same as for a_n for $n > 4$. Hence, our attempt to obtain a second linearly independent solution from F_2 is doomed to failure.

Instead, we use Theorem 8–3(b) and assume a second solution of the form

$$y_2 = cy_1 \ln x + \sum_{n=0}^{\infty} A_n x^{n-1}$$

Substituting this into the given differential equation, we get

$$cx^2\left(y_1'' \ln x + \frac{2y_1'}{x} - \frac{y_1}{x^2}\right) + x^2 \sum_{n=0}^{\infty} A_n(n-1)(n-2)x^{n-3} - (x+2) cy_1 \ln x$$

$$- \sum_{n=0}^{\infty} A_n x^n - 2 \sum_{n=0}^{\infty} A_n x^{n-1} = 0$$

The coefficient of $c \ln x$ is $x^2 y_1'' - (x + 2)y_1$, which is zero since y_1 is a solution of the given differential equation. What remains is

$$2cxy_1' - cy_1 - (A_0 + 2A_1) + \sum_{n=2}^{\infty} [(n^2 - 3n)A_n - A_{n-1}] x^{n-1} = 0$$

Here we use $x^2 + \frac{1}{4}x^3 + \frac{1}{40}x^4 \ldots$ to approximate y_1.

$$2cx\left(2x + \frac{3}{4}x^2 + \frac{3}{40}x^3 \ldots\right) - c\left(x^2 + \frac{1}{4}x^3 + \frac{1}{40}x^4 \ldots\right)$$
$$- (A_0 + 2A_1) + x(-A_1 - 2A_2) + (-A_2)x^2$$
$$+ (-A_3 + 4A_4)x^3 \ldots = 0$$

Collecting like terms

$$(-A_0 - 2A_1) + x(-2A_2 - A_1) + (3c - A_2)x^2$$
$$+ \left(\frac{5}{4}c + 4A_4 - A_3\right)x^3 + \ldots = 0$$

Let $A_0 = 1$, then $A_1 = -\frac{1}{2}$, $A_2 = \frac{1}{4}$, $c = \frac{1}{12}$, $A_4 = \frac{1}{4}A_3 - \frac{5}{16}c = \frac{1}{4}A_3 - \frac{5}{192}$.

We note that the coefficient of x^2 is A_3 and is therefore arbitrary since this number multiplies y_2 and any multiple of y_2 is again a solution. We are thus justified in letting $A_3 = 0$. Then $A_4 = -\frac{5}{192}$ and the remainder of the coefficients may be obtained recursively as were the first few. Combining the above, we have

$$y_2 = \frac{1}{12}y_1 \ln x + \frac{1}{x} - \frac{1}{2} + \frac{1}{4}x - \frac{5}{192}x^3 + \cdots$$

and the general solution is

$$y = c_1 y_1 + c_2 y_2$$

∎

EXERCISES FOR SECTION 8–5

For the differential equations in Exercises 1–8 use the method of Frobenius to show that y_1 is a solution. Find the first few terms of a second linearly independent solution.

1. $x(x + 1)y'' + (5x + 1)y' + 3y = 0$; $y_1 = \sum_{n=0}^{\infty} (-1)^n (n + 1)(n + 2)x^n$

2. $x(x + 1)^2 + (1 - x^2)y' + (x - 1)y = 0$; $y_1 = 1 + x$

3. $xy'' + y' + 2y = 0$; $y_1 = \sum_{n=0}^{\infty} \frac{(-1)^n (2x)^n}{(n!)^2}$

4. $xy'' - 2y' + 2y = 0$; $y_1 = \sum_{n=0}^{\infty} \frac{(-1)^n (3)2^{n+1}}{(n + 3)! \, n!} x^{n+3}$

5. $xy'' + y' + xy = 0;\ y_1 = \displaystyle\sum_{n=0}^{\infty} \frac{(-1)^n x^{2n}}{2^{2n}(n!)^2}$

6. $xy'' + y = 0;\ y_1 = \displaystyle\sum_{n=0}^{\infty} \frac{(-1)^n x^{n+1}}{n!\,(n+1)!}$

7. $4x^2 y'' - 2x(x-2)y' - (3x+1)y = 0;\ y_1 = \displaystyle\sum_{n=0}^{\infty} \frac{x^{n+1/2}}{2^n(n!)}$

8. $x^2 y'' - x(x+1)y' + y = 0;\ y_1 = \displaystyle\sum_{n=0}^{\infty} \frac{x^{n+1}}{n!} = xe^x$

8–6 LEGENDRE POLYNOMIALS AND BESSEL FUNCTIONS

This section could have been titled "Functions Defined by Infinite Series." Earlier in this chapter we showed that differential equations may be solved by assuming that the solution function may be written as an infinite power series or a Frobenius series about some point. The differential equation essentially defines a function since the coefficients of the power series are determined by forcing the power series to be a solution.

While solutions in infinite series form are usually "nonelementary," this does not curtail their usefulness. Tables of values are constructed so that the functions become (in time and with use) as familiar as, for instance, the trigonometric functions. In fact, the sine and cosine functions may be essentially defined by the differential equation $y'' + m^2 y = 0$.

The name given to the solution function is the same as the name of the differential equation from which it arises. Thus the second-order differential equation $y'' + m^2 y = 0$ is called the "trigonometric" differential equation; the differential equation $y'' - m^2 y = 0$ might be called the "hyperbolic" differential equation, and so forth.

In this section we direct our attention to equations that result in two famous classes of functions. They are both nonelementary but nonetheless important to the applied scientist.

Legendre's Differential Equation

An equation that arises in numerous important physical applications, particularly in problems involving spherical symmetry, is the equation

$$(1 - x^2)y'' - 2xy' + m(m+1)y = 0 \tag{8–12}$$

which is called **Legendre's differential equation of order m.** The word *order* is used here in a different sense than used previously to label the highest derivative in the differential equation.

Note that the differential equation (8–12) has regular singular points at ± 1 and that 0 is an ordinary point. Thus we may assume a power series solution around the origin and then use the method of undetermined coefficients to particularize the series. Let

$$y = \sum_{n=0}^{\infty} a_n x^n$$

then

$$y' = \sum_{n=1}^{\infty} a_n n x^{n-1}$$

and

$$y'' = \sum_{n=2}^{\infty} a_n n(n - 1)x^{n-2}$$

Substituting into the Legendre differential equation,

$$(1 - x^2) \sum_{n=2}^{\infty} a_n n(n - 1)x^{n-2} - 2x \sum_{n=1}^{\infty} a_n n x^{n-1} + m(m + 1) \sum_{n=0}^{\infty} a_n x^n = 0$$

Carrying out the indicated multiplications,

$$\sum_{n=2}^{\infty} a_n n(n - 1)x^{n-2} - \sum_{n=2}^{\infty} a_n n(n - 1)x^n$$

$$- 2 \sum_{n=0}^{\infty} a_n n x^n + m(m + 1) \sum_{n=0}^{\infty} a_n x^n = 0$$

Combining the last three series,

$$\sum_{n=2}^{\infty} a_n n(n - 1)x^{n-2} - \sum_{n=0}^{\infty} a_n [n^2 - n + 2n - m(m + 1)]x^n = 0$$

The coefficient in the last series may be written

$$n^2 - m^2 + n - m = (n - m)(n + m) + (n - m)$$
$$= (n - m)(n + m + 1)$$

Thus the differential equation with the assumed power series solution substituted into it becomes

$$\sum_{n=2}^{\infty} a_n n(n - 1)x^{n-2} - \sum_{n=0}^{\infty} a_n (n - m)(n + m + 1)x^n = 0$$

Shifting the index in the second of these series,

$$\sum_{n=2}^{\infty} n(n - 1)a_n x^{n-2} - \sum_{n=2}^{\infty} (n - m - 2)(n + m - 1)a_{n-2}x^{n-2} = 0$$

This equation implies a_0 and a_1 are both arbitrary, and for $n \geq 2$,

$$a_n = \frac{(n - m - 2)(n + m - 1)}{n(n - 1)} a_{n-2}$$

We note that there are two linearly independent solutions to Legendre's differential equation determined by a_0, a_1, and the above recursion formula. The solution may be written

$$y(x) = y_0(x) + y_1(x)$$

where

$$y_0(x) = a_0\left(1 - \frac{m(m + 1)}{2!} x^2 + \frac{(m - 2)(m)(m + 1)(m + 3)}{4!} x^4 - \cdots\right)$$

$$y_1(x) = a_1\left(x - \frac{(m - 1)(m + 2)}{3!} x^3 + \frac{(m - 3)(m - 1)(m + 2)(m + 4)}{5!} x^5 + \cdots\right)$$

If m is a nonnegative integer (as it is for some important applications), then when $n = m + 2$, $a_{n+2} = 0$, and thus because of the nature of the recursion formula, one of the "infinite" series consists of a finite number of terms—which is to say it is a polynomial. This polynomial, of degree m, is called **Legendre's polynomial** and is denoted by $P_m(x)$. *The arbitrary coefficient is usually determined by requiring that the graph of each of Legendre's polynomials passes through* (1, 1).

Example 1

Derive the first four Legendre polynomials.

SOLUTION For $m = 0$, a_0 is arbitrary, $a_{2n} = 0$ for $n > 0$, and hence $P_0(x) = a_1$. Since $P_0(1) = 1$, it follows that $P_0(x) = 1$.

For $m = 1$, the value of a_1 is arbitrary, $a_{2n+1} = 0$ for $n > 0$, and hence $P_1(x) = a_1x$. Since $P_1(1) = 1$ and $a_1 = 1$, it follows that $P_1(x) = x$.

For $m = 2$, a_0 is arbitrary, $a_2 = [(-2)(3)/2] a_0 = -3a_0$, from which $P_2(x) = a_0 - 3a_0x^2$. Using the fact that $P_2(1) = 1$, $a_0 = -\frac{1}{2}$. Thus $P_2(x) = \frac{1}{2}(3x^2 - 1)$.

For $m = 3$, a_1 is arbitrary, $a_3 = [(-2)(5)/3 \cdot 2] a_1 = -\frac{5}{3} a_1$, and $a_{2n+1} = 0$ for $n > 1$. From the auxiliary condition, we get $a_1 = -\frac{3}{2}$. Thus $P_3 = \frac{1}{2}(5x^3 - 3x)$. (See Figure 8–3.) ∎

The polynomial solutions rather than the infinite series solutions are what tend to highlight the Legendre differential equation. These polynomials play an important role in certain applied problems analogous to the trigonometric functions. They are basic in the sense that rather general functions may be approximated arbitrarily closely by linear combinations of Legendre polynomials.

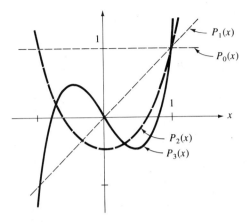

FIGURE 8–3

Bessel's Differential Equation

Another differential equation that arises in advanced studies in applied mathematics and physics is

$$x^2y'' + xy' + (x^2 - m^2)y = 0 \tag{8–13}$$

called **Bessel's equation of order _m_**. Note, as in the Legendre case, that the word _order_ does not refer to the differential equation itself.

Since $x = 0$ is a regular singular point of Bessel's equation, it is necessary to assume that the solution near the origin is of the form of a Frobenius series rather than a power series about the origin. Letting

$$y = x^r \sum_{n=0}^{\infty} a_n x^n$$

and substituting into the differential equation, we obtain

$$\sum_{n=0}^{\infty} [(n + r)(n + r - 1)a_n + (n + r)a_n - m^2 a_n]x^{n+r} + \sum_{n=0}^{\infty} a_n x^{n+r+2} = 0$$

or, by writing the terms for $n = 0$ and $n = 1$ separately, we have

$$a_0(r^2 - m^2)x^r + a_1[(r + 1)^2 - m^2]x^{r+1}$$
$$+ \sum_{n=2}^{\infty} \{[(n + r)^2 - m^2]a_n + a_{n-2}\}x^{n+r} = 0$$

To determine the coefficients a_n, we equate each of the coefficients of powers of x in this equation to zero. Thus equating the coefficient of x^r to zero, we get

$$a_0(r^2 - m^2) = 0$$

which gives the indicial equation $r^2 - m^2 = 0$. Thus $r = \pm m$ with a_0 "arbitrary."

Equating the coefficient of x^{r+1} to zero, we get

$$a_1[(r + 1)^2 - m^2] = 0$$

which implies that $a_1 = 0$ (since $r \neq m - 1$).

Finally, equating the coefficient of x^{n+r} to zero yields

$$[(n + r)^2 - m^2] a_n + a_{n-2} = 0$$

for $n \geq 2$, which means

$$a_n = \frac{-1}{(n + r)^2 - m^2} a_{n-2}$$

a recursion relationship valid for $n \geq 2$; hence all coefficients are determined. Although a_0 is arbitrary, it is customary to require

$$a_0 = \frac{1}{2^m m!}$$

from which the general coefficient formula may be written, after some effort, to be

$$a_{2n} = \frac{(-1)^n}{2^{2n+m} n! (m + n)!}$$

$$a_{2n+1} = 0$$

Therefore one solution of Bessel's equation, called **Bessel's function of the first kind,** is

$$J_m(x) = \sum_{n=0}^{\infty} \frac{(-1)^n}{2^{2n+m} n! (m + n)!} x^{2n+m}$$

Actually, any solution of this type, regardless of the choice of a_0, is called a Bessel function of the first kind.

Since Bessel's equation is a second-order differential equation, we would naturally expect another solution that is not a constant multiple of the first. Further, it would seem only natural that the "other solution" be found by letting $r = -m$ and using the same recursion relationship as before. But as can be shown, the solution J_{-m} is linearly dependent on J_m.

A second solution of Bessel's equation, which is independent of the first, may be found by the method of reduction of order and will include a term involving the logarithm of x. (See Section 8–5.) Consequently, the "second" solution of Bessel's equation when m is an integer is unbounded near the origin and is useful in applications only for $x \neq 0$.

More generally, any solution of Bessel's equation that is continuous for all values of x is called a **Bessel function of the first kind.** A solution of Bessel's equation that is continuous for all x except $x = 0$ and that becomes unbounded for x near 0 is known as a **Bessel function of the second kind.** Any linear combination of a Bessel function of the first kind with one of the second kind is a solution of Bessel's equation.

Some Bessel functions have at least a remote similarity to the sine and cosine functions. We can verify this by observing how much the first few terms of the functions J_0 and J_1 resemble the first few terms of the Maclaurin series for the cosine and the sine.

$$J_0(x) = 1 - \frac{x^2}{2^2(1!)^2} + \frac{x^4}{2^4(2!)^2} - \frac{x^6}{2^6(3!)^2} + \frac{x^8}{2^8(4!)^2} - + \cdots$$

$$\left(\cos x = 1 - \frac{x^2}{2!} + \frac{x^4}{4!} - \cdots \right)$$

$$J_1(x) = \frac{x}{2} - \frac{x^3}{2^3 1! 2!} + \frac{x^5}{2^5 3! 2!} - \frac{x^7}{2^7 4! 3!} + \frac{x^9}{2^9 5! 4!} - \cdots$$

$$\left(\sin x = x - \frac{x^3}{3!} + \frac{x^5}{5!} - \cdots \right)$$

Figure 8–4 shows the graphs of $J_0(x)$ and $J_1(x)$.

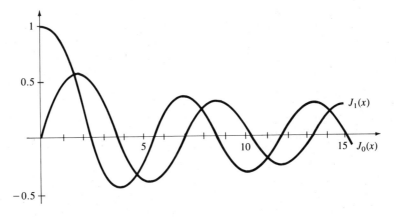

FIGURE 8–4

EXERCISES FOR SECTION 8–6

1.   Derive and sketch the graph of $P_4(x)$ and $P_5(x)$.

2.   Verify directly that $P_0(x)$, $P_1(x)$, $P_2(x)$, and $P_3(x)$, and $P_4(x)$ are solutions to Legendre's differential equation.

3.   Show that $P_m(x) = (-1)^m P_m(-x)$. What does this mean relevant to the graphs of the Legendre polynomials?

4.   Show that any polynomial of degree two or less can be written in terms of $P_0(x)$, $P_1(x)$, and $P_2(x)$.

5.   Write $1 + 2x - 3x^2$ in terms of Legendre polynomials.

6. Show that $J_0(0) = 1$ and $J_1(0) = 0$.

7. Show that $J_m(x) = (-1)^m J_m(-x)$.

By inspection, write one solution of each differential equation in Exercises 8–11.

8. $(1 - x^2)y'' - 2xy' + 20y = 0$

9. $(1 - x^2)y'' - 2xy' + 2y = 0$

10. $x^2 y'' + xy' + (x^2 - 4)y = 0$

11. $x^2 y'' + xy' + \frac{1}{4}(x^2 - 1)y = 0$

12. Show that, for large values of x, Bessel's differential equation of order m can be put into the form $xy'' + y' + xy = 0$. Now let $y = u/\sqrt{x}$ to obtain the equation $u'' + [(1/4x^2) + 1]u = 0$, which for large x becomes $u'' + u = 0$. Thus the approximate solution to Bessel's equation for large values of x is $y = (c_1 \cos x + c_2 \sin x)/\sqrt{x}$.

REVIEW EXERCISES FOR CHAPTER 8

In Exercises 1–5 use the method of Taylor series to obtain the first four nonzero terms of the solution of the given initial-value problem.

1. $y'' + xy = 0, y(0) = y'(0) = 1$

2. $y'' + xy = 1, y(1) = y'(1) = 1$

3. $y'' = \sin y, y(1) = y'(1) = 1$

4. $y' = \ln y + x^2, y(1) = 3$

5. $y'' + xy' + x^2 y = 0, y(0) = y'(0) = 1$

In Exercises 6–13 find a recursion formula between the coefficients of the power series solutions of each of the given differential equations. Assume $a_0 = y(0)$, and $a_1 = y'(0)$ are arbitrary.

6. $y'' + xy' + y = 0$

7. $y'' + xy' = 0$

8. $(1 - x^2)y'' + xy' + y = 0$

9. $(1 - x)y'' + 2y' = 0$

10. $y'' - xy' + (1 + x)y = 0$

11. $(1 - x^3)y'' - x^2 y' + xy = 0$

12. $y'' + (x + 1)y' - y = 0$

13. $y'' - xy' + (1 - x)y = 0$

In Exercises 14 and 15 find power series solutions in powers of $x - 1$.

14. $xy'' + (x - 1)y' = 0$

15. $y'' - (x - 1)y = 0$

Locate and classify the singular points in Exercises 16 and 17.

16. $(x^2 + 2x)y'' + (x - 3)y' + y = 0$

17. $(x^4 + x^2)y'' + (x^3 + 2x^2)y' + x^2 y = 0$

In Exercises 18–22 use Formulas 8–5 and 8–10 to obtain the form of two linearly independent solutions of the given differential equation.

18. $x^2y'' + xy' + (4x^2 - 3)y = 0$ **19.** $xy'' - 2y' + xy = 0$

20. $x^2y'' + xy' + (x^2 - \frac{1}{9}) = 0$ **21.** $xy'' + y' + 2y = 0$

22. $xy'' + y' + xy = 0$

In Exercises 23–25 obtain two linearly independent solutions of the given differential equation, one a Frobenius series and the other of the form of Formula 8–10.

23. $xy'' + (1 - x)y' - y = 0$ **24.** $x(1 - x)y'' + 2(1 - x)y' + 2y = 0$

25. $x^2y'' + (x^2 - 3x)y' + 3y = 0$

In Exercises 26 and 27 obtain two linearly independent Frobenius series solutions of the given differential equation.

26. $xy'' - (x + 3)y' + 2y = 0$ **27.** $x^2y'' - xy' - (x^2 + \frac{5}{4})y = 0$

9

Linear systems

In this chapter we shall discuss and analyze problems involving systems of differential equations. Physical problems that are modeled by systems of differential equations include electrical circuits like the one shown in Figure 9–1(a) and spring-mass systems like that shown in Figure 9–1(b). Both of these problems are described by a system of two linear differential equations. The rationale for these models is discussed in Section 9–8.

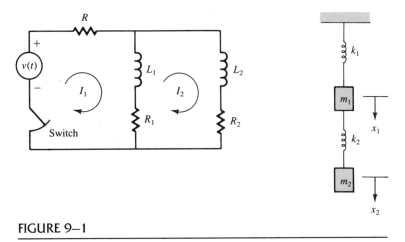

FIGURE 9–1

Our discussion of systems of differential equations is limited to those problems that result in a system of **linear** equations in which the number of unknown functions and the number of equations is the same. Furthermore, except for an easy exercise or two to the contrary, we restrict our coverage to linear differential equations with constant coefficients.

DEFINITION

A **solution** of a system of two linear differential equations in two unknown functions is a pair of functions y_1 and y_2 which simultaneously satisfy both equations.

Example 1

The system

$$3y_1' + 2y_1 - y_2' = 2$$
$$y_1' + y_2 = 0$$

has the solution given by

$$y_1 = c_1 e^{-x} + c_2 e^{-2x} + 1$$
$$y_2 = c_1 e^{-x} + 2c_2 e^{-2x}$$

where c_1 and c_2 are constants. This may be verified by computing

$$y_1' = -c_1 e^{-x} - 2c_2 e^{-2x} \qquad \text{and} \qquad y_2' = -c_1 e^{-x} - 4c_2 e^{-2x}$$

Then substituting into the given system, we get

$$3(-c_1 e^{-x} - 2c_2 e^{-2x}) + 2(c_1 e^{-x} + c_2 e^{-2x} + 1) - (-c_1 e^{-x} - 4c_2 e^{-2x}) = 2$$

and

$$-c_1 e^{-x} - 2c_2 e^{-2x} + c_1 e^{-x} + 2c_2 e^{-2x} = 0 \qquad\qquad ■$$

9–1   THE METHOD OF ELIMINATION

In this section we shall show a technique for solving systems of differential equations that is similar to the method of elimination of a variable used for solving systems of algebraic equations. However, before we can explain this technique, we must introduce some specialized notation.

Polynomial Operators*

To simplify writing and solving systems of nth-order linear differential equations with constant coefficients, we introduce the concept of a differential polynomial operator.

DEFINITION

By the operator D^n is meant the nth derivative operator d^n/dx^n. The positive integer n is called the **power** of the operator.

The differential operator D^n must operate on some n-times differentiable function. Thus when D^2 operates on $x^3 + 3x$, the expression $6x$ is obtained, and we write

$$D^2(x^3 + 3x) = 6x$$

DEFINITION

A linear combination of differential operators of the form

$$a_n D^n + a_{n-1} D^{n-1} + \cdots + a_1 D + a_0$$

where $a_0, a_1, \ldots, a_n$ are constants is called an ***n*th-order polynomial operator** and is denoted $P(D)$.

To indicate that $P(D)$ is being applied to an n-times differentiable function y, we write $P(D)y$, and observe that

$$\begin{aligned} P(D)y &= (a_n D^n + a_{n-1} D^{n-1} + \cdots + a_1 D + a_0)y \\ &= a_n D^n y + a_{n-1} D^{n-1}y + \cdots + a_1 Dy + a_0 y \end{aligned}$$

Recalling that $D^n = d^n/dx^n$, we see that $P(D)y$ means

$$P(D)y = a_n \frac{d^n y}{dx^n} + a_{n-1}\frac{d^{n-1}y}{dx^{n-1}} + \cdots + a_1 \frac{dy}{dx} + a_0 y$$

The material on polynomial operators was covered in Section 5.2 as an alternate approach to the method of undetermined coefficients and may be omitted if covered at that time.

COMMENT: Any linear differential equation with constant coefficients may be expressed in terms of a polynomial operator. For example,

$y'' + 3y' - y = 0$ *may be written as* $(D^2 + 3D - 1)y = 0$

and

$y''' - 4y' = \cos x$ *may be written as* $(D^3 - 4D)y = \cos x.$

In general, the nth-order linear differential equation with constant coefficients may be written

$P(D)y = f(x)$

There are limitations on the kinds of functions for which $P(D)y$ has meaning, such as the existence or continuity of some derivative. Functions for which $P(D)y$ has meaning are said to be "admissible" for that operator. For example, only twice differentiable functions are admissible for the operator $(D^2 + 1)$.

The differential operator $P(D)$ has the two properties

$P(D)(y_1 + y_2) = P(D)y_1 + P(D)y_2$

$P(D)(cy) = cP(D)y$

DEFINITION

Two operators $P_1(D)$ and $P_2(D)$ are said to be equal if and only if $P_1(D)y = P_2(D)y$, for all admissible functions y.

DEFINITION

The **sum** of two operators $P_1(D) + P_2(D)$ is obtained by first expressing P_1 and P_2 as linear combinations of the D operator and adding coefficients of like powers of D.

DEFINITION

By the **product** of two operators, $P_1(D)P_2(D)$, is meant the equivalent operator obtained by using the operator $P_2(D)$ followed by $P_1(D)$. Thus, the product $P_1(D)P_2(D)$ is interpreted to mean

$[P_1(D)P_2(D)]y = P_1(D)[P_2(D)y]$

where y, y_1, and y_2 are admissible functions for $P(D)$ and c is any constant. These two properties are called the **linearity properties.** Hence, $P(D)$ is called a **linear** operator. The algebraic laws of polynomial operators are included in the following definitions.

To **expand** a product of operators, determine the effect of the product operator on any admissible function and then express it as a linear combination of powers of D. The technique is illustrated in the following example.

Example 1

Let $P_1 = 2D + 3$ and $P_2 = D - 5$. Find the expansion of the product operator P_1P_2.

SOLUTION   Applying the product operator P_1P_2 to a function y we have

$$P_1P_2(y) = P_1[P_2(y)]$$

Here

$$P_2(y) = (D - 5)y = \frac{dy}{dx} - 5y$$

and

$$P_1[P_2(y)] = (2D + 3)\left[\frac{dy}{dx} - 5y\right]$$

$$= 2D\left[\frac{dy}{dx} - 5y\right] + 3\left[\frac{dy}{dx} - 5y\right]$$

$$= 2\frac{d^2y}{dx^2} - 7\frac{dy}{dx} - 15y$$

$$= (2D^2 - 7D - 15)y$$

which means that

$$(2D + 3)(D - 5) = 2D^2 - 7D - 15 \qquad \blacksquare$$

COMMENT: It can be shown that polynomial operators satisfy all of the laws of elementary algebra with regard to addition and multiplication, which means that polynomial operators may be manipulated in the same way we manipulate algebraic expressions.

Example 2

(a) If $P_1(D) = 3D^2 + 7D - 5$ and $P_2(D) = D^3 + 6D^2 - 2D - 3$, then

$$P_1(D) + P_2(D) = D^3 + 9D^2 + 5D - 8$$

(b) The product operator $(D^2 - 4)(D + 2)$ may be expanded and written as

$$(D^2 - 4)(D + 2) = D^3 + 2D^2 - 4D - 8$$

(c) The operator $D^3 - D^2 - 6D$ may be factored and written as

$$D^3 - D^2 - 6D = D(D - 3)(D + 2) \qquad \blacksquare$$

A system of two linear differential equations with constant coefficients in the unknown functions y_1 and y_2 may be written in operator form as

$$\begin{aligned} P_{11}(D)y_1 + P_{12}(D)y_2 &= f_1(x) \\ P_{21}(D)y_1 + P_{22}(D)y_2 &= f_2(x) \end{aligned} \qquad (9\text{--}1)$$

The polynomial operators, denoted by $P_{ij}(D)$, are linear combinations of the D operator. For example, in the system

$$\begin{aligned} (D^2 - 1)y_1 + Dy_2 &= x \\ (D^2 + D)y_1 + D(D - 1)y_2 &= \sin x \end{aligned}$$

we note that

$$P_{11} = D^2 - 1, \quad P_{12} = D, \quad P_{21} = D^2 + D, \quad P_{22} = D^2 - D, \quad f_1(x) = x,$$
and $f_2(x) = \sin x$

The technique of solution shown in this section parallels the method of solving simple algebraic systems. Through a series of operator applications we try to eliminate one of the unknown functions, thus arriving at one differential equation (usually of higher order) in one unknown. The basic approach is shown in the next example.

Example 3

Solve the system

$$\begin{aligned} y_1' - y_2 &= x^2 \\ y_2' + 4y_1 &= x \end{aligned}$$

SOLUTION We first write the system, using operator notation, as

$$\begin{aligned} Dy_1 - y_2 &= x^2 \\ 4y_1 + Dy_2 &= x \end{aligned}$$

Applying the operator D to the first of these equations, the system becomes

$$\begin{aligned} D^2y_1 - Dy_2 &= 2x \\ 4y_1 + Dy_2 &= x \end{aligned}$$

Adding these two equations, we obtain

$$D^2y_1 + 4y_1 = 3x$$

The solution of the corresponding homogeneous equation* $(D^2 + 4)y_1 = 0$ is

$$y_{1c} = c_1 \cos 2x + c_2 \sin 2x$$

The form of the y_{1p} is $Ax + B$. Using the method of undetermined coefficients, we get

$$4Ax + 4B = 3x$$

from which $B = 0$ and $A = \frac{3}{4}$. Thus

$$y_1 = c_1 \cos 2x + c_2 \sin 2x + \frac{3}{4}x$$

To obtain y_2, we substitute this expression into the second equation of the given system. This yields

$$\begin{aligned}
y_2' &= x - 4y_1 \\
&= x - 4c_1 \cos 2x - 4c_2 \sin 2x - 3x \\
&= -2x - 4c_1 \cos 2x - 4c_2 \sin 2x
\end{aligned}$$

Integrating to obtain y_2, we get

$$y_2 = -x^2 - 2c_1 \sin 2x + 2c_2 \cos 2x + c_3$$

To determine the relation, if any, that exists between the coefficients c_1, c_2, and c_3, we substitute the expressions for y_1 and y_2 into the first equation of the system. Thus

$$[-2c_1 \sin 2x + 2c_2 \cos 2x + \tfrac{3}{4}] - [-x^2 - 2c_1 \sin 2x + 2c_2 \cos 2x + c_3] = x^2$$

This leads to $c_3 = \frac{3}{4}$. Thus the solution to the given system contains two arbitrary constants and is given by

$$y_1 = c_1 \cos 2x + c_2 \sin 2x + \frac{3}{4}x$$

$$y_2 = -x^2 - 2c_1 \sin 2x + 2c_2 \cos 2x + \frac{3}{4} \qquad \blacksquare$$

COMMENTS: (1) The arbitrary constants of Example 3 can be specified in a variety of ways. For example, the values of y_1 and y_2 at 0 could be given. Or the values of y_1' and y_1 at 0 could be given.

(2) The expression for y_1 may be substituted into either of the two given equations to obtain an expression for y_2. In Example 3 we chose to

---

*When a homogeneous linear differential equation is written in operator notation, the differential operator and the auxiliary equation have the same form. For example, the auxiliary equation for $(D^2 + 3D + 2)y = 0$ is $m^2 + 3m + 2 = 0$. In general if $P(D)y = 0$, then the auxiliary equation is $P(m) = 0$.

substitute y_1 into the second of the two given equations; had we used the first equation, $y_1' - y_2 = x^2$, the expression for y_2 would have been $y_2 = y_1' - x^2$. This approach is easier than that used in the example because it avoids the need to evaluate c_3.

Example 3 is typical of the manipulations that are used to eliminate one of the unknown functions. Since the process often includes differentiation, constants may be introduced that are not arbitrary. <u>The following is an outline of the method of elimination:</u>

1. Apply an operator to one or both equations so that the operator coefficient of one of the unknown functions is the same in each equation.

2. Subtract the two equations, thereby eliminating one of the unknown functions.

3. Solve the remaining linear equation in the one unknown function.

4. Substitute this solution into one of the equations of the given system. Solve this equation for the second function.

5. Substitute both solution functions into the remaining equation to determine any relationships between the arbitrary constants.

Example 4

Solve the system of equations

$$y_1' - 2y_1 + 2y_2' = 2 - 4e^{2x}$$
$$2y_1' - 3y_1 + 3y_2' - y_2 = 0$$

SOLUTION   In polynomial operator form this is written

$$(D - 2)y_1 + 2Dy_2 = 2 - 4e^{2x}$$
$$(2D - 3)y_1 + (3D - 1)y_2 = 0$$

Operate on the first equation with the operator $(3D - 1)$ and on the second by $-2D$ to obtain

$$(3D - 1)(D - 2)y_1 + 2D(3D - 1)y_2 = (3D - 1)(2 - 4e^{2x})$$
$$-2D(2D - 3)y_1 - 2D(3D - 1)y_2 = 0$$

Adding these two equations and carrying out the indicated operation on $2 - 4e^{2x}$, we obtain

$$(3D^2 - 7D + 2 - 4D^2 + 6D)y_1 = -24e^{2x} - 2 + 4e^{2x}$$

which, after simplification, becomes

$$(D^2 + D - 2)y_1 = 20e^{2x} + 2$$

or

$$(D + 2)(D - 1)y_1 = 20e^{2x} + 2$$

We have from $(D + 2)(D - 1)y_1 = 0$ that

$$y_{1c} = c_1 e^{-2x} + c_2 e^x$$

The particular solution is of the form $Ae^{2x} + B$. Using the condition $y_p'' + y_p' - 2y_p = 20e^{2x} + 2$, we obtain

$$4Ae^{2x} + 2Ae^{2x} - 2Ae^{2x} - 2B = 20e^{2x} + 2$$

Therefore $A = 5$ and $B = -1$, and the general solution for the unknown y_1 is

$$y_1 = c_1 e^{-2x} + c_2 e^x + 5e^{2x} - 1$$

We substitute this into the first of the two equations in the given system to obtain

$$-2c_1 e^{-2x} + c_2 e^x + 10e^{2x} - 2c_1 e^{-2x} - 2c_2 e^x - 10e^{2x} + 2 + 2y_2' = 2 - 4e^{2x}$$

Simplifying, and solving for y_2', we have

$$y_2' = 2c_1 e^{-2x} + \frac{1}{2}c_2 e^x - 2e^{2x}$$

Integrating,

$$y_2 = -c_1 e^{-2x} + \frac{1}{2}c_2 e^x - e^{2x} + c_3$$

We now substitute the expressions for y_1 and y_2 into the second equation of the system to determine if each of the constants c_1, c_2, and c_3 is truly arbitrary. Thus

$$2(-2c_1 e^{-2x} + c_2 e^x + 10e^{2x}) - 3(c_1 e^{-2x} + c_2 e^x + 5e^{2x} - 1)$$
$$+ 3\left(2c_1 e^{-2x} + \frac{1}{2}c_2 e^x - 2e^{2x}\right) - \left(-c_1 e^{-2x} + \frac{1}{2}c_2 e^x - e^{2x} + c_3\right) = 0$$

Expanding and simplifying this expression, we obtain

$$3 - c_3 = 0$$

from which c_1 and c_2 are arbitrary and $c_3 = 3$. Therefore the solution of the system is

$$y_1 = c_1 e^{-2x} + c_2 e^x + 5e^{2x} - 1$$
$$y_2 = -c_1 e^{-2x} + \frac{1}{2}c_2 e^x - e^{2x} + 3$$ ∎

The Number of Arbitrary Constants

Recall that for a single nth-order differential equation in one unknown we expect that there will be n arbitrary constants in the general solution. The case for systems of differential equations is not quite so easy. We first consider a few examples.

Example 5

(a) The general solution of

$$y_1' + y_2' = x$$
$$y_1' + y_2' - y_2 = 1$$

is $y_1 = \frac{1}{2}x^2 - x + c$, $y_2 = x - 1$. That is, there is *one* arbitrary constant in the two solution functions.

(b) The system

$$y_1' - y_2' = x$$
$$y_1' + y_2' - y_2 = 1$$

has almost the same form as the system in (a). However, the general solution of this system is $y_1 = \frac{1}{2}x^2 + c_1 e^{x/2} + x + c_2$, $y_2 = c_1 e^{x/2} + x + 1$. Thus there are *two* arbitrary constants in this solution.

(c) The system

$$y_1' + y_1 + y_2' + 3y_2 = 2$$
$$y_1' - y_1 + y_2' + y_2 = x$$

has the solution $y_1 = \frac{1}{4}(1 - 3x)$, $y_2 = \frac{1}{4}(x + 3)$, in which there are *no* arbitrary constants. ∎

The following theorem gives a test to determine how many arbitrary constants to expect. This doesn't help in solving the system but does give us confidence in our manipulations.

THEOREM 9–1

Consider the linear system of differential equations with polynomial operator coefficients

$$P_{11}(D)y_1 + P_{12}(D)y_2 = f_1(x) \qquad\qquad (9\text{--}2)$$
$$P_{21}(D)y_1 + P_{22}(D)y_2 = f_2(x)$$

Then the number of arbitrary constants is equal to the degree of

$$\Delta = P_{11}P_{22} - P_{21}P_{12}$$

If Δ is identically zero, then the system either has infinitely many solutions or no solutions.

Example 6

(a) In Example 5(a), $P_{11} = D$, $P_{12} = D$, $P_{21} = D$, and $P_{22} = D - 1$. The value of Δ is given by $\Delta = D(D - 1) - D^2 = -D$, which is of degree one. Thus we expect the solution to have *one* arbitrary constant.

(b) In Example 5(b), $P_{11} = D$, $P_{12} = -D$, $P_{21} = D$, and $P_{22} = D - 1$. Thus $\Delta = D(D - 1) - (-D)(D) = 2D^2 - D$, which is of degree two, and we expect the solution to have *two* arbitrary constants.

(c) In Example 5(c), $P_{11} = D + 1$, $P_{12} = D + 3$, $P_{21} = D - 1$, and $P_{22} = D + 1$. Thus $\Delta = (D + 1)(D + 1) - (D + 3)(D - 1) = 4$, which is of degree 0, so the solution has *no* arbitrary constants.

(d) The system

$$(D + 1)y_1 + (D + 1)^2 y_2 = 1$$
$$(D - 1)y_1 + (D^2 - 1)y_2 = x$$

has *either infinitely many solutions or no solution* because $\Delta = (D + 1)(D^2 - 1) - (D + 1)^2(D - 1) = 0$. ∎

The elimination procedure can be systematized and made to have the form of Cramer's rule. Operating on the first of Equations 9–2 with $P_{22}(D)$, the second with $P_{12}(D)$, and subtracting, we obtain

$$(P_{11}(D)P_{22}(D) - P_{12}(D)P_{21}(D))y_1 = P_{22}(D)f_1(x) - P_{12}(D)f_2(x) \tag{9–3}$$

This can be put in the pseudo-determinant format

$$\begin{vmatrix} P_{11} & P_{12} \\ P_{21} & P_{22} \end{vmatrix} y_1 = \begin{vmatrix} f_1 & P_{12} \\ f_2 & P_{22} \end{vmatrix} \tag{9–4}$$

The determinant on the right must, of course, receive the proper interpretation corresponding to Equation 9–3. In a similar manner we have

$$\begin{vmatrix} P_{11} & P_{12} \\ P_{21} & P_{22} \end{vmatrix} y_2 = \begin{vmatrix} P_{11} & f_1 \\ P_{21} & f_2 \end{vmatrix} \tag{9–5}$$

> COMMENT: *Using Equations 9–4 and 9–5, we are now able to add a footnote to Theorem 9–1. If $\Delta = P_{11}P_{22} - P_{12}P_{21} = 0$, then the system has infinitely many solutions if the determinants on the right-hand sides of Equations 9–4 and 9–5 vanish; otherwise, there is no solution. This is called the degenerate case.*

Example 7

Show that the following system has infinitely many solutions.

$$y_1'' - y_1 + y_2'' - y_2' = 0$$
$$y_1' + y_1 + y_2' = 0$$

SOLUTION   In operator format this system is written as

$$(D^2 - 1)y_1 + D(D - 1)y_2 = 0$$
$$(D + 1)y_1 + Dy_2 = 0$$

Thus

$$\Delta = (D^2 - 1)D - D(D - 1)(D + 1) = 0$$

Note that the right-hand sides of Equations 9–4 and 9–5 are both zero since f_1 and f_2 are both zero. Hence there are infinitely many solutions.

The infinitely many solutions may be obtained from either of the equations of the system. We let y_2 be $g(x)$ in the second of the equations. Then

$$y_1' + y_1 = -g'(x)$$

Thus for an arbitrary choice for y_2, we obtain the infinitely many solutions of this first-order linear equation.   ∎

EXERCISES FOR SECTION 9–1

In Exercises 1–5 solve the given system.

1. $y_1' = y_2$
$y_2' = y_1$

2. $y_1' = y_1 + y_2$
$y_2' = 3y_1 - y_2$

3. $y_1' = 3y_1 + 2y_2$
$y_2' = y_1 - 5y_2$

4. $y_1' = 4y_1 + 3y_2$
$y_2' = y_1$

5. $y_1' = 5y_1 - 4y_2$
$y_2' = y_1 + 2y_2$

6. Find the general solution of the system

$$y_1' = ay_1 + by_2$$
$$y_2' = by_1 + ay_2$$

where a and b are nonzero constants.

Solve the systems of equations in Exercises 7–20. Using Theorem 9–1, predict the number of arbitrary constants and compare with your solution.

7. $y_1' + y_2' = 3$
$y_1' - y_2' = x$

8. $y_1' = y_2 + 1$
$y_2' = y_1 - 1$

9. $y_1' - y_2 = x^2$
$y_2' + 2y_1 = x$

10. $y_1' + y_2' = 1$
$y_1' + y_1 + y_2' - y_2 = 0$

11. $y_1' + 2y_1 + y_2' + 2y_2 = 0$
$2y_1' - 2y_1 - 3y_2' + 3y_2 = 0$

12. $3y_1' - 2y_2' + 5y_2 = 0$
$y_1' - 4y_1 - 4y_2' + y_2 = 0$

13. $3y_1' - 2y_2' + 5y_2 = e^{3x}$
$y_1' - 4y_1 - 4y_2' + y_2 = 0$

14. $y_1'' + 6y_1 + y_2' = 0$
$y_1' + 2y_1 + y_2' - 2y_2 = 2$

15. $y_1'' - 2y_2' + y_2 = 1$
$2y_1' + y_1 + y_2'' - 4y_2 = 0$

16. $y_1' + 2y_1 + y_2' - y_2 = 0$
$2y_1' + 3y_1 + 3y_2' + y_2 = \sin 2x$

17. $y_1'' - y_1 - y_2'' = -2 \sin x$
$\quad y_1'' + y_1 + y_2'' = 0$

18. $y_1'' - y_1 + y_2'' = -2 \sin x$
$\quad y_1'' + y_1 + y_2'' = 0$

19. $y_1'' - y_1 + y_2'' - 2y_2 = -2 \sin x$
$\quad y_1'' + y_1 + y_2'' = 0$

20. $y_1'' - y_1 + y_2'' - y_2' = -2 \sin x$
$\quad y_1'' + y_1' + y_2'' = 0$

21. Solve the three systems given in Example 5 to verify the indicated solutions.

In Exercises 22–25 determine if the system has no solutions or infinitely many solutions. If the system has infinitely many solutions, indicate how that set may be obtained.

22. $y_1' + y_2' = 3$
$\quad y_1'' - y_1' + y_2'' - y_2' = x$

23. $y_1' - y_1 + y_2' = x$
$\quad y_1'' - y_1' + y_2'' = 1$

24. $y_1'' + y_1 + y_2' - y_2 = x$
$\quad y_1''' + y_1' + y_2'' - y_2' = 2$

25. $y_1' + y_2' = -1$
$\quad y_1'' + y_2'' = 2$

26. A certain double-feedback servomechanism can be described by the system $y_1' = y_2;\ y_2' = y_3;\ y_3' = y_1.$ Solve this system.

9–2 TRIANGULAR SYSTEMS

As we learned in the previous section, when solving linear systems of differential equations, the most difficult task is to write the solutions with the proper number of essential arbitrary constants. In this section we show a technique that will always yield the correct number of arbitrary constants. We illustrate the method for a system of two equations in two independent variables. First notice that certain systems are inherently easy to solve.

Example 1

Solve the system

$$Dy_1 + (D + 1)y_2 = 1$$
$$(D - 1)y_2 = 0$$

SOLUTION Since the second equation does not involve the unknown function y_1, we can solve that equation for y_2 to obtain

$$y_2 = c_1 e^x$$

Now substitute this into the first equation:

$$Dy_1 = 1 - (D + 1)y_2$$
$$= 1 - 2c_1 e^x$$

Integrating, we obtain

$$y_1 = x - 2c_1 e^x + c_2 \qquad \blacksquare$$

The system solved in Example 1 is an example of a triangular system.

DEFINITION

A system of differential equations is in **triangular form** if each succeeding equation in the system has at least one less unknown function
than the previous equation.

In the case of systems of two equations in two unknown functions the general
triangularized system has a form like

$$P_{11}(D)y_1 + P_{12}(D)y_2 = F_1(t)$$
$$P_{22}(D)y_2 = F_2(t) \qquad (9\text{-}6)$$

The number of arbitrary constants in the solution of this system will be equal
to the degree of Δ, where $\Delta = P_{11}(D) P_{22}(D)$. These solutions are obtained as in
Example 1 by first solving the second equation for y_2 (generating deg P_{22} arbitrary
constants) and then substituting for y_2 in the first equation and solving for y_1, where
an additional deg P_{11} constants are introduced.

Consider now the general 2×2 system

$$P_{11}(D)y_1 + P_{12}(D)y_2 = f_1(t)$$
$$P_{21}(D)y_1 + P_{22}(D)y_2 = f_2(t) \qquad (9\text{-}7)$$

DEFINITION

A system is **equivalent** to that of Equations 9-7 if it has exactly the
same solution.

Example 2

The systems

$$Dy_1 + (D + 1)y_2 = 1 \qquad \text{and} \qquad Dy_1 + 2y_2 = 1$$
$$(D - 1)y_2 = 0 \qquad\qquad\qquad (D - 1)y_2 = 0$$

are equivalent because both systems have the solution

$$y_1 = x - 2c_1e^x + c_2, \quad y_2 = c_1e^x \qquad\blacksquare$$

The key to the technique is to operate on one equation at a time in a system
to sequentially obtain equivalent systems until a triangular form is obtained. The

triangular form of the system is then solved as in Example 1. The following theorem contains the main thrust of the method of triangularization.

THEOREM 9–2

Let M(D) be a polynomial operator. Then System 9–7 and the system

$$P_{11}(D)y_1 + P_{12}(D)y_2 = f_1(t)$$

$$[P_{21}(D) + M(D)P_{11}(D)]y_1 + [P_{22}(D) + M(D)P_{12}(D)]y_2 = f_2(t) + M(D)f_1(t)$$

are equivalent.

As a consequence of the theorem we can outline how to triangularize a system.

(1)   Pick a $P_{ij}(D)$ of lowest order and use it to reduce the order of all other operators in its column. (This will be explained in the examples.)

(2)   Operate on either equation and add this result to the other equation.

(3)   The equation that is operated upon must be included in the new equivalent system in unaltered form.

(4)   Repeat this process until the system is triangularized.

Example 3

Triangularize the system

$$(D - 1)y_1 + 3y_2 = t$$
$$Dy_1 + (D - 2)y_2 = 1$$

SOLUTION   The coefficient of y_2 in the first equation is the operator of lowest order. Therefore we operate on the first equation with the operator $-\frac{1}{3}(D - 2)$. The new system will contain (a) the first equation in unaltered form and (b) the result of the sum of the first equation, after being operated upon, plus the second equation. Thus the new equivalent system is

$$(D - 1)y_1 + 3y_2 = t$$

$$\left[-\frac{1}{3}(D - 2)(D - 1) + D \right]y_1 + 0y_2 = -\frac{1}{3}(D - 2)t + 1$$

Simplifying, the system is

$$(D - 1)y_1 + 3y_2 = t$$
$$(D^2 - 6D + 2)y_1 = -2 - 2t$$

Note that this system is now solved by solving the (second-order) second equation and then using it in the first equation to obtain y_2. The concern about arbitrary constants has been eliminated! ∎

Example 4

The system

$$(D - 2)y_1 + 2Dy_2 = 2 - 4e^{2x}$$
$$(2D - 3)y_1 + (3D - 1)y_2 = 0$$

was solved in Example 4 of the preceding section. To put this system into triangular form, we multiply the first equation by $-\frac{3}{2}$ and add it to the second. Thus

$$(D - 2)y_1 + 2Dy_2 = 2 - 4e^{2x}$$
$$Dy_1 - 2y_2 = -6 + 12e^{2x}$$

Operating on the second equation with D and adding it to the first gives the triangular system

$$(D^2 + D - 2)y_1 = 2 + 20e^{2x}$$
$$Dy_1 - 2y_2 = -6 + 12e^{2x}$$

The first equation may now be solved for y_1 as in the preceding section, but the expression for y_2 is much easier to obtain. ∎

COMMENT: An equivalent triangular system is not unique. In Example 4 an equivalent system is obtained by eliminating y_1 from one of the equations of the system instead of y_2. Multiplying the first equation by -2 and adding to the second and then operating on the new second equation by $-(D - 2)$ and adding to the first yields

$$(D^2 + D - 2)y_2 = -6 - 4e^{2x}$$
$$y_1 - (D + 1)y_2 = -4 + 8e^{2x}$$

Using the system of Example 4 we would arrive at the solution

$$y_1 = c_1 e^{-2x} + c_2 e^x + 5e^{2x} - 1$$
$$y_2 = -c_1 e^{-2x} + \frac{1}{2}c_2 e^x - e^{2x} + 3$$

Using the one with y_1 eliminated, we get

$$y_1 = -c_1' e^{-2x} + 2c_2' e^x + 5e^{2x} - 1$$
$$y_2 = c_1' e^{-2x} + c_2' e^x - e^{2x} + 3$$

These two solution pairs are made equivalent by identifying c_1 with $-c_1'$ and c_2 with $2c_2'$.

Example 5

Triangularize and solve the system

$$(D + 1)y_1 + (D - 1)y_2 = 0$$
$$Dy_1 + (D - 3)y_2 = 1$$

SOLUTION   From Theorem 9–1 we notice that the number of essential arbitrary constants in the solution is equal to the degree of Δ; that is,

$$\deg[(D + 1)(D - 3) - D(D - 1)] = \deg(D + 3) = 1$$

We operate on the second of the equations by (-1) and add it to the first to obtain the equivalent system

$$Dy_1 + (D - 3)y_2 = 1$$
$$y_1 + 2y_2 = -1$$

We operate on the second equation of this system by $(-D)$ and add it to the first, thereby getting the triangularized system

$$y_1 + 2y_2 = -1$$
$$(D + 3)y_2 = -1$$

Solving $(D + 3)y_2 = -1$ for y_2,

$$y_2 = ce^{-3x} - \tfrac{1}{3}$$

Then using the first equation of the triangularized system, we can find y_1:

$$y_1 = -1 - 2y_2$$
$$= -\tfrac{1}{3} - 2ce^{-3x}$$ ∎

The next example shows the real power of the method of triangularization. The minor annoyance of finding an equivalent system in triangular form should be compared to the more major problem of determining the relationship between the coefficients.

Example 6

Triangularize and solve the system

$$(D^2 - 4)y_1 - (2D - 1)y_2 = 0$$
$$(2D + 1)y_1 + D^2y_2 = 0$$

SOLUTION   Operate on the first equation with $\tfrac{1}{2}D$ and add the result to the second equation. The new equivalent system is

$$(D^2 - 4)y_1 - (2D - 1)y_2 = 0$$
$$(\tfrac{1}{2}D^3 + 1)y_1 + \tfrac{1}{2}Dy_2 = 0$$

Now operate on the second equation of the equivalent system with 4 and add to the first:

$$(2D^3 + D^2)y_1 + y_2 = 0$$

$$(\tfrac{1}{2}D^3 + 1)y_1 + \tfrac{1}{2}Dy_2 = 0$$

As the last step, operate on the first equation of this equivalent system with $-\tfrac{1}{2}D$ and add to the second equation:

$$(2D^3 + D^2)y_1 + y_2 = 0$$

$$(D^4 - 1)y_1 = 0$$

This system is in triangularized form. The solution for y_1 is

$$y_1 = c_1 \cos x + c_2 \sin x + c_3 e^x + c_4 e^{-x}$$

From the first equation of the triangularized system,

$$y_2 = -(2D^3 + D^2)y_1$$
$$= (-2c_1 + c_2) \sin x + (c_1 + 2c_2) \cos x - 3c_3 e^x + c_4 e^{-x} \qquad \blacksquare$$

EXERCISES FOR SECTION 9–2

In Exercises 1–20 find a triangularized system equivalent to the given system and then solve.

1. $y_1' = y_2$
$\quad y_2' = y_1$

2. $y_1' = y_1 + y_2$
$\quad y_2' = 3y_1 - y_2$

3. $y_1' = 3y_1 + 2y_2$
$\quad y_2' = y_1 - 5y_2$

4. $y_1' = 4y_1 + 3y_2$
$\quad y_2' = y_1$

5. $y_1' = 5y_1 - 4y_2$
$\quad y_2' = y_1 + 2y_2$

6. $y_1' = ay_1 + by_2$
$\quad y_2' = by_1 + ay_2$

where a and b are nonzero constants.

7. $y_1' + y_2' = 3$
$\quad y_1' - y_2' = x$

8. $y_1' = y_2 + 1$
$\quad y_2' = y_1 - 1$

9. $y_1' - y_2 = x^2$
$\quad y_2' + 2y_1 = x$

10. $y_1' + y_2' = 1$
$\quad y_1' + y_1 + y_2' - y_2 = 0$

11. $y_1' + 2y_1 + y_2' + 2y_2 = 0$
$\quad 2y_1' - 2y_1 - 3y_2' + 3y_2 = 0$

12. $3y_1' - 2y_2' + 5y_2 = 0$
$\quad y_1' - 4y_1 - 4y_2' + y_2 = 0$

13. $3y_1' - 2y_2' + 5y_2 = e^{3x}$
$\quad y_1' - 4y_1 - 4y_2' + y_2 = 0$

14. $y_1'' + 6y_1 + y_2' = 0$
$\quad y_1' + 2y_1 + y_2' - 2y_2 = 2$

15. $y_1'' - 2y_2' + y_2 = 1$
$2y_1' + y_1 + y_2'' - 4y_2 = 0$

16. $y_1' + 2y_1 + y_2' - y_2 = 0$
$2y_1' + 3y_1 + 3y_2' + y_2 = \sin 2x$

17. $y_1'' - y_1 - y_2'' = -2 \sin x$
$y_1'' + y_1 + y_2'' = 0$

18. $y_1'' - y_1 + y_2'' = -2 \sin x$
$y_1'' + y_1 + y_2'' = 0$

19. $y_1'' - y_1 + y_2'' - 2y_2 = -2 \sin x$
$y_1'' + y_1 + y_2'' = 0$

20. $y_1'' - y_1 + y_2'' - y_2' = -2 \sin x$
$y_1'' + y_1' + y_2'' = 0$

21. Find a triangularized form for the three systems given in Example 5, Section 9–1.

9–3   LAPLACE TRANSFORM METHOD

The Laplace transform may be used to solve systems of initial-value problems. The advantage of this method is that the transformed system becomes a system of algebraic equations.

Example 1

Solve the system

$$\left.\begin{array}{c} \dot{y}_1 + y_1 + 3\dot{y}_2 = 1 \\ 3y_1 + \dot{y}_2 + 2y_2 = t \end{array}\right\} \qquad y_1(0) = 0, \; y_2(0) = 0$$

SOLUTION   Taking the Laplace transform of both sides of both equations, we obtain

$$sY_1 + Y_1 + 3sY_2 = \frac{1}{s}$$

$$3Y_1 + sY_2 + 2Y_2 = \frac{1}{s^2}$$

The transforms $Y_1(s)$ and $Y_2(s)$ are

$$Y_1(s) = \frac{\begin{vmatrix} 1/s & 3s \\ 1/s^2 & s+2 \end{vmatrix}}{\begin{vmatrix} s+1 & 3s \\ 3 & s+2 \end{vmatrix}} = \frac{s-1}{s(s^2 - 6s + 2)} = -\frac{1}{2s} + \frac{1}{2}\frac{s-4}{s^2 - 6s + 2}$$

$$Y_2(s) = \frac{\begin{vmatrix} s+1 & 1/s \\ 3 & 1/s^2 \end{vmatrix}}{\begin{vmatrix} s+1 & 3s \\ 3 & s+2 \end{vmatrix}} = \frac{1 - 2s}{s^2(s^2 - 6s + 2)} = \frac{1}{2s} + \frac{1}{2s^2} - \frac{1}{2}\frac{s-5}{s^2 - 6s + 2}$$

where a partial fractions decomposition was used to obtain the last two expansions. By completing the square we can express these as

$$Y_1(s) = -\frac{1}{2s} + \frac{1}{2}\frac{s-3}{(s-3)^2-7} - \frac{1}{2}\frac{1}{(s-3)^2-7}$$

$$Y_2(s) = \frac{1}{2s} + \frac{1}{2s^2} - \frac{1}{2}\frac{s-3}{(s-3)^2-7} + \frac{1}{(s-3)^2-7}$$

Therefore

$$y_1(t) = -\frac{1}{2} + \frac{1}{2}e^{3t}\cosh\sqrt{7}t - \frac{1}{2\sqrt{7}}e^{3t}\sinh\sqrt{7}t$$

$$y_2(t) = \frac{1}{2} + \frac{1}{2}t - \frac{1}{2}e^{3t}\cosh\sqrt{7}t + \frac{1}{\sqrt{7}}e^{3t}\sinh\sqrt{7}t$$ ∎

Example 2

We know from Example 5(a), Section 9–1, that the system

$$y_1' + y_2' = t$$
$$y_1' + y_2' - y_2 = 1$$

has only one arbitrary constant in its solution. To verify this, let $y_1(0) = c_1$ and $y_2(0) = c_2$. Then taking the Laplace transform of both sides of both equations, we obtain

$$sY_1 - c_1 + sY_2 - c_2 = \frac{1}{s^2}$$

$$sY_1 - c_1 + sY_2 - c_2 - Y_2 = \frac{1}{s}$$

The transforms Y_1 and Y_2 are obtained by Cramer's rule. Thus

$$Y_1 = \frac{\begin{vmatrix} 1/s^2 + c_1 + c_2 & s \\ 1/s + c_1 + c_2 & s-1 \end{vmatrix}}{\begin{vmatrix} s & s \\ s & s-1 \end{vmatrix}} = \frac{1/s - 1/s^2 - 1 - c_1 - c_2}{-s}$$

$$= -\frac{1}{s^2} + \frac{1}{s^3} + \frac{c_1 + c_2 + 1}{s}$$

$$Y_2 = \frac{\begin{vmatrix} s & 1/s^2 + c_1 + c_2 \\ s & 1/s + c_1 + c_2 \end{vmatrix}}{\begin{vmatrix} s & s \\ s & s-1 \end{vmatrix}} = \frac{1 - 1/s}{-s} = -\frac{1}{s} + \frac{1}{s^2}$$

Taking the inverse transform of $Y_1 = -1/s^2 + 1/s^3 + (c_1 + c_2 + 1)/s$ and $Y_2 = -1/s + 1/s^2$, we get

$$y_1 = -t + \tfrac{1}{2}t^2 + c_1 + c_2 + 1$$
$$y_2 = -1 + t$$

There is only one arbitrary constant since $c_1 + c_2 + 1$ may be replaced by c'.   ∎

Example 3

Use the method of Laplace transforms to solve the system

$$y_1' + y_1 + y_2' + 3y_2 = 2$$
$$y_1' - y_1 + y_2' + y_2 = t$$

SOLUTION   Letting $y_1(0) = c_1$ and $y_2(0) = c_2$ and taking the Laplace transform of both sides of both equations, we have

$$sY_1 - c_1 + Y_1 + sY_2 - c_2 + 3Y_2 = \frac{2}{s}$$

$$sY_1 - c_1 - Y_1 + sY_2 - c_2 + Y_2 = \frac{1}{s^2}$$

Solving for Y_1 and Y_2, we have

$$Y_1 = \frac{\begin{vmatrix} 2/s + c_1 + c_2 & s + 3 \\ 1/s^2 + c_1 + c_2 & s + 1 \end{vmatrix}}{\begin{vmatrix} s + 1 & s + 3 \\ s - 1 & s + 1 \end{vmatrix}} = \frac{2 + 1/s - 3/s^2 - 2c_1 - 2c_2}{4}$$

$$Y_2 = \frac{\begin{vmatrix} s + 1 & 2/s + c_1 + c_2 \\ s - 1 & 1/s^2 + c_1 + c_2 \end{vmatrix}}{4} = \frac{1/s + 1/s^2 - 2 + 2/s + 2c_1 + 2c_2}{4}$$

For either one of these inverse transforms to exist, $c_1 + c_2 - 1 = 0$. Then

$$Y_1 = \frac{1/s - 3/s^2}{4} \qquad \text{and} \qquad Y_2 = \frac{1/s^2 + 3/s}{4}$$

and

$$y_1 = \tfrac{1}{4}(1 - 3t) \qquad \text{and} \qquad y_2 = \tfrac{1}{4}(t + 3)$$

Notice that $c_1 + c_2 = y_1(0) + y_2(0) = 1$ is a requirement of the two solutions and is not an arbitrary choice.   ∎

COMMENT: *Example 3 demonstrates that the number and kind of initial conditions that may be specified are tied to the degree of* Δ *(see Theorem 9–1). The result of Example 3 shows that the system*

$$\left.\begin{array}{l} y_1' + y_1 + y_2' + 3y_2 = 2 \\ y_1' - y_1 + y_2' + y_2 = t \end{array}\right\} \qquad y_1(0) = 1,\ y_2(0) = 2$$

is not solvable since $y_1(0) + y_2(0) \neq 1$, *but the same system is solvable if* $y_2(0)$ *is specified to be* 0.

EXERCISES FOR SECTION 9–3

Solve the systems in Exercises 1–8 by the method of Laplace transforms.

1. $\left.\begin{array}{l} y_1' = y_2 \\ y_2' = -y_1 \end{array}\right\} \qquad y_1(0) = 0,\ y_2(0) = 1$

2. $\left.\begin{array}{l} y_1' - y_1 = y_2 \\ y_2 - y_2' = y_1 \end{array}\right\} \qquad y_1(0) = y_2(0) = 1$

3. $\left.\begin{array}{l} y_1' - y_2' = t \\ y_1 + y_1' + y_2' = 1 \end{array}\right\} \qquad y_1(0) = y_2(0) = 0$

4. $\left.\begin{array}{l} y_1' - y_1 + y_2' + y_2 = 1 \\ y_1' + y_2' = t \end{array}\right\} \qquad y_1(0) = -1,\ y_2(0) = 0$

5. $\left.\begin{array}{l} y_1' - y_2' = -\sin t \\ y_1' + y_2 = t \end{array}\right\} \qquad y_1(0) = y_2(0) = 0$

6. $\left.\begin{array}{l} y_1'' - 3y_1' - y_2' + 2y_2 = 3 \\ y_1' - 3y_1 + y_2' = t \end{array}\right\} \qquad y_1(0) = 0,\ y_1'(0) = 0,\ y_2(0) = 0$

7. $\left.\begin{array}{l} y_1' + y_2' = 0 \\ y_1 - y_2' = 0 \end{array}\right\} \qquad y_1(0) = 0,\ y_2(0) = 1$

8. $\left.\begin{array}{l} y_1' + y_2 = t^2 + 1 \\ y_1 - y_2' = -t - 1 \end{array}\right\} \qquad y_1(0) = y_2(0) = 0$

In Exercises 9–13 solve the systems by the method of Laplace transforms and determine what limitations there are (if any) on the initial conditions $y_1(0)$ *and* $y_2(0)$.

9. $\begin{array}{l} y_1' - y_2 = 0 \\ y_1 + y_2' = 0 \end{array}$

10. $\begin{array}{l} y_1' - y_2' = 0 \\ y_1 + y_2 = 0 \end{array}$

11. $\begin{array}{l} y_1' - y_2' = 0 \\ y_1 - y_2 = 0 \end{array}$

12. $\begin{array}{l} y_1' + 2y_1 + y_2' = 0 \\ y_1' + y_2' + 2y_2 = 0 \end{array}$

13. $\begin{array}{l} y_1' + 2y_1 + y_2' + 4y_2 = 0 \\ y_1' + y_2' + 2y_2 = 0 \end{array}$

9–4   MATRICES AND LINEAR SYSTEMS

Matrix methods are used in this section to solve systems of linear differential equations. You may wish to review some of the elementary matrix topics in Appendix A before beginning this section.

The **normalized** form of a system of n first-order linear differential equations is

$$\frac{dy_1}{dt} = a_{11}y_1 + a_{12}y_2 + \cdots + a_{1n}y_n + f_1(t)$$

$$\frac{dy_2}{dt} = a_{21}y_1 + a_{22}y_2 + \cdots + a_{2n}y_n + f_2(t)$$

$$\vdots \qquad\qquad \vdots \qquad \vdots$$

$$\frac{dy_n}{dt} = a_{n1}y_1 + a_{n2}y_2 + \cdots + a_{nn}y_n + f_n(t) \qquad\qquad (9\text{–}8)$$

where $y_1, y_2, \ldots, y_n$ are unknown functions and t is the independent variable. The coefficients a_{ij} may be continuous functions but in our discussion are limited to constants. If $f_i(t) = 0$ for all i, then the system is called **homogeneous**; otherwise, the sytem is **nonhomogeneous**.

The normalized system, Equation 9–8, can be written in matrix form as

$$\mathbf{Y}' = \mathbf{AY} + \mathbf{F} \qquad\qquad (9\text{–}9)$$

where $\mathbf{Y}$ is the $n \times 1$ vector of solution functions, $\mathbf{A}$ is the $n \times n$ coefficient matrix and $\mathbf{F}$ is the $n \times 1$ vector of driving functions. Specifically,

$$\mathbf{Y} = \begin{pmatrix} y_1 \\ y_2 \\ \vdots \\ y_n \end{pmatrix} \qquad \mathbf{A} = \begin{pmatrix} a_{11} & a_{12} & \cdots & a_{1n} \\ a_{21} & a_{22} & \cdots & a_{2n} \\ \vdots & & & \vdots \\ a_{n1} & a_{n2} & \cdots & a_{nn} \end{pmatrix} \qquad \mathbf{F} = \begin{pmatrix} f_1 \\ f_2 \\ \vdots \\ f_n \end{pmatrix}$$

The vector function $\mathbf{Y}'$ is the derivative of the vector function $\mathbf{Y}$.

Example 1

The matrix form of the homogeneous system

$$y_1' = 2y_1 + y_2$$

$$y_2' = y_1 - y_2$$

is

$$\mathbf{Y}' = \mathbf{AY}$$

where

$$\mathbf{Y}' = \begin{pmatrix} y_1' \\ y_2' \end{pmatrix} \qquad \mathbf{A} = \begin{pmatrix} 2 & 1 \\ 1 & -1 \end{pmatrix} \qquad \mathbf{Y} = \begin{pmatrix} y_1 \\ y_2 \end{pmatrix}$$

∎

Some systems of linear differential equations that are not given in normalized form can be re-expressed in this form by algebraic means. The next example shows a specific case.

Example 2

Express the given homogeneous system in normalized form and then write it in matrix form.

$$2y_1' + y_2' + y_2 = 0$$
$$y_1' - y_1 + y_2' = 0$$

SOLUTION   To obtain the normalized form for this system we solve for y_1' and y_2'. Eliminating y_2' by subtracting the second equation from the first, we obtain

$$y_1' = -y_1 - y_2$$

Substituting this result in the first equation, we get

$$y_2' = -2(-y_1 - y_2) - y_2 = 2y_1 + y_2$$

Hence, the matrix form of the given system is

$$\begin{pmatrix} y_1' \\ y_2' \end{pmatrix} = \begin{pmatrix} -1 & -1 \\ 2 & 1 \end{pmatrix} \begin{pmatrix} y_1 \\ y_2 \end{pmatrix} \qquad \blacksquare$$

Example 3

Express the given system in normalized form then write it in matrix form (Equation 9–9).

$$y_1' + y_2' = t$$
$$y_1' - 2y_2' + y_2 = \sin t$$

SOLUTION   Multiplying the first equation by 2 and adding to the second, we get

$$y_1' = \frac{1}{3}(-y_2 + 2t + \sin t)$$

If we multiply the first equation by -1 and add it to the second, we get

$$y_2' = \frac{1}{3}(y_2 + t - \sin t)$$

Having written the system in normalized form, the system can be expressed by the matrix equation

$$\begin{pmatrix} y_1' \\ y_2' \end{pmatrix} = \frac{1}{3}\begin{pmatrix} 0 & -1 \\ 0 & 1 \end{pmatrix} \begin{pmatrix} y_1 \\ y_2 \end{pmatrix} + \frac{1}{3}\begin{pmatrix} 2t + \sin t \\ t - \sin t \end{pmatrix} \qquad \blacksquare$$

Not all systems of first-order linear differential equations can be put in normalized form (Equation 9–8). For example, the system

$$y_1' + y_2' = 0$$
$$y_1' + y_2' - y_2 = 0$$

can not be expressed in the form of Equation 9-8 since it is impossible to algebraically solve for y_1' and y_2'. See if you can prove this. (HINT: Use Cramer's rule in attempting to solve for y_1' and y_2'.) Systems like this may be solved using the elimination technique of Section 9–1.

> COMMENT: Linear differential equations involving higher-order derivatives can be expressed in the form of Equation 9–8. For example, the linear second-order differential equation
>
> $$u'' - 3u' + 7u = t$$
>
> can be treated as a first-order system by letting $y_1 = u$ and $y_2 = u'$. Then the second-order equation $u'' - 3u' + 7u = t$ can be expressed as the first-order system
>
> $$y_1' = y_2$$
> $$y_2' = 3y_2 - 7y_1 + t$$
>
> This system can then be written in matrix form as
>
> $$\begin{pmatrix} y_1' \\ y_2' \end{pmatrix} = \begin{pmatrix} 0 & 1 \\ -7 & 3 \end{pmatrix} \begin{pmatrix} y_1 \\ y_2 \end{pmatrix} + \begin{pmatrix} 0 \\ t \end{pmatrix}$$

DEFINITION

A **solution** of the linear system

$$\mathbf{Y}' = \mathbf{A}\mathbf{Y} + \mathbf{F}$$

on an interval $[a, b]$ is any $n \times 1$ vector function $\mathbf{Y}$ that is differentiable on $[a, b]$ and that satisfies the system on that interval.

Example 4

Show that

$$\mathbf{Y}_1 = \begin{pmatrix} -2e^{2t} \\ e^{2t} \end{pmatrix} \quad \text{and} \quad \mathbf{Y}_2 = \begin{pmatrix} e^{3t} \\ -e^{3t} \end{pmatrix}$$

are both solutions of

$$\mathbf{Y}' = \begin{pmatrix} 1 & -2 \\ 1 & 4 \end{pmatrix} \mathbf{Y} = \mathbf{A}\mathbf{Y}$$

SOLUTION   Given $\mathbf{Y}_1 = \begin{pmatrix} -2e^{2t} \\ e^{2t} \end{pmatrix}$, by differentiating we obtain $\mathbf{Y}_1' = \begin{pmatrix} -4e^{2t} \\ 2e^{2t} \end{pmatrix}$.

Also,

$$\mathbf{AY}_1 = \begin{pmatrix} 1 & -2 \\ 1 & 4 \end{pmatrix} \begin{pmatrix} -2e^{2t} \\ e^{2t} \end{pmatrix} = \begin{pmatrix} -2e^{2t} - 2e^{2t} \\ -2e^{2t} + 4e^{2t} \end{pmatrix} = \begin{pmatrix} -4e^{2t} \\ 2e^{2t} \end{pmatrix}$$

Since $\mathbf{Y}_1' = \mathbf{AY}_1$, the given vector $\mathbf{Y}_1$ is a solution. In a similar manner, we can show that $\mathbf{Y}_2 = \begin{pmatrix} e^{3t} \\ -e^{3t} \end{pmatrix}$ is also a solution to the given system.   ∎

In most cases, we are interested in linearly independent solutions of the homogeneous system $\mathbf{Y}' = \mathbf{AY}$. The concept of a linearly independent set of functions introduced in Chapter 4 is extended to sets of vector functions.

DEFINITION

> The vector functions $\mathbf{Y}_1, \mathbf{Y}_2, \ldots, \mathbf{Y}_n$ are **linearly dependent** on some interval if constants $c_1, c_2, \ldots, c_n$, not all zero, can be found such that
>
> $$c_1\mathbf{Y}_1 + c_2\mathbf{Y}_2 + \cdots + c_n\mathbf{Y}_n = \mathbf{0}$$
>
> for all t in the interval. A set of vector functions that is not linearly dependent is said to be **linearly independent.**

Example 5

Show that the vectors

$$\mathbf{Y}_1 = \begin{pmatrix} e^t \\ e^t \end{pmatrix} \quad \mathbf{Y}_2 = \begin{pmatrix} e^{-t} \\ -e^{-t} \end{pmatrix} \quad \mathbf{Y}_3 = \begin{pmatrix} \cosh t \\ \sinh t \end{pmatrix}$$

are linearly dependent.

SOLUTION   To show that $\mathbf{Y}_1$, $\mathbf{Y}_2$, and $\mathbf{Y}_3$ are linearly dependent we can show that constants c_1, c_2, and c_3, not all of which are zero, exist such that

$$c_1 \begin{pmatrix} e^t \\ e^t \end{pmatrix} + c_2 \begin{pmatrix} e^{-t} \\ -e^{-t} \end{pmatrix} + c_3 \begin{pmatrix} \cosh t \\ \sinh t \end{pmatrix} = \mathbf{0}$$

The constants $c_1 = 1$, $c_2 = 1$, and $c_3 = -2$ are three such constants. Therefore, the given vectors are linearly dependent.   ∎

COMMENT: If two vector functions are linearly dependent, then one of them is a constant multiple of the other.

DEFINITION

> Any set $\mathbf{Y}_1, \mathbf{Y}_2, \ldots, \mathbf{Y}_n$ of n linearly independent solution vectors of the homogeneous system $\mathbf{Y}' = \mathbf{AY}$ on an interval $[a, b]$ is called a **fundamental set of solutions** on that interval.

In Chapter 4 we use the concept of the Wronskian determinant as a test for the linear independence of solutions of a linear homogeneous differential equation. Theorem 9–3 shows that the Wronskian determinant can also be used as a test for the linear independence of solution vectors

$$\mathbf{Y}_1 = \begin{pmatrix} y_{11} \\ y_{21} \\ \vdots \\ y_{n1} \end{pmatrix}, \mathbf{Y}_2 = \begin{pmatrix} y_{12} \\ y_{22} \\ \vdots \\ y_{n2} \end{pmatrix}, \ldots, \mathbf{Y}_n = \begin{pmatrix} y_{1n} \\ y_{2n} \\ \vdots \\ y_{nn} \end{pmatrix}$$

of the homogeneous system $\mathbf{Y}' = \mathbf{AY}$.

THEOREM 9–3

> Let $\mathbf{Y}_1, \mathbf{Y}_2, \ldots, \mathbf{Y}_n$ be n solution vectors of the homogeneous system $\mathbf{Y}' = \mathbf{AY}$ on $[a, b]$. Then the set of solution vectors are linearly independent and, therefore, a fundamental set if, and only if, the **Wronskian** determinant
>
> $$W(t) = \det (\mathbf{Y}_1, \mathbf{Y}_2, \ldots, \mathbf{Y}_n) = \begin{vmatrix} y_{11} & y_{12} & \cdots & y_{1n} \\ y_{21} & y_{22} & \cdots & y_{2n} \\ \vdots & \vdots & & \vdots \\ y_{n1} & y_{n2} & \cdots & y_{nn} \end{vmatrix} \neq 0$$
>
> for every t in $[a, b]$.

Example 6

Use Theorem 9–3 to show that

$$\mathbf{Y}_1 = \begin{pmatrix} -2e^{2t} \\ e^{2t} \end{pmatrix} \quad \text{and} \quad \mathbf{Y}_2 = \begin{pmatrix} e^{3t} \\ -e^{3t} \end{pmatrix}$$

form a fundamental set of solutions of

$$\mathbf{Y}' = \begin{pmatrix} 1 & -2 \\ 1 & 4 \end{pmatrix} \mathbf{Y}$$

SOLUTION   In Example 4 we showed that Y_1 and Y_2 are both solution vectors of the given homogeneous system. The Wronskian determinant of these two solution vectors is

$$W(t) = \det(Y_1, Y_2) = \begin{vmatrix} -2e^{2t} & e^{3t} \\ e^{2t} & -e^{3t} \end{vmatrix} = e^{5t} \neq 0$$

We conclude from Theorem 9–3 that Y_1 and Y_2 are linearly independent and, therefore, form a fundamental set of solutions of the given system.   ■

The general solution of the homogeneous system $Y' = AY$ is defined to be a linear combination of the functions in its fundamental set.

DEFINITION

Let Y_1, Y_2, . . . , Y_n be a fundamental set of solutions for $Y' = AY$ on the interval $[a, b]$. The **general solution** of the system is

$$Y = c_1Y_1 + c_2Y_2 + \cdots + c_nY_n$$

where $c_1, c_2, \ldots, c_n$ are arbitrary constants.

Example 7

Write the general solution of the homogeneous system

$$Y' = \begin{pmatrix} 1 & -2 \\ 1 & 4 \end{pmatrix} Y$$

SOLUTION   This is the system given in Example 6. In that example we showed that

$$Y_1 = \begin{pmatrix} -2e^{2t} \\ e^{2t} \end{pmatrix} \quad \text{and} \quad Y_2 = \begin{pmatrix} e^{3t} \\ -e^{3t} \end{pmatrix}$$

form a fundamental set of solutions of the given system. Thus, the general solution of this system is

$$Y = c_1 \begin{pmatrix} -2e^{2t} \\ e^{2t} \end{pmatrix} + c_2 \begin{pmatrix} e^{3t} \\ -e^{3t} \end{pmatrix}$$

■

Example 8

Let $Y_1 = e^{-t} \begin{pmatrix} -1 \\ 1 \\ 0 \end{pmatrix}$, $Y_2 = e^{-t} \begin{pmatrix} -1 \\ 0 \\ 1 \end{pmatrix}$, and $Y_3 = e^{-4t} \begin{pmatrix} 1 \\ 1 \\ 1 \end{pmatrix}$. Show that Y_1, Y_2, and Y_3 form a fundamental set of solutions of

$$Y' = \begin{pmatrix} -2 & -1 & -1 \\ -1 & -2 & -1 \\ -1 & -1 & -2 \end{pmatrix} Y$$

and write the general solution.

SOLUTION   To show that Y_1, Y_2 and Y_3 are solutions, we need only show that each satisfies the given system. To show that they form a fundamental set, we show that the Wronskian determinant is nonzero. Thus

$$W(t) = \det(Y_1, Y_2, Y_3) = \begin{vmatrix} -e^{-t} & -e^{-t} & e^{-4t} \\ e^{-t} & 0 & e^{-4t} \\ 0 & e^{-t} & e^{-4t} \end{vmatrix} = 3e^{-6t}$$

Since e^{-6t} is never zero, Y_1, Y_2, and Y_3 form a fundamental set and the general solution of the given system is

$$Y = c_1 e^{-t} \begin{pmatrix} -1 \\ 1 \\ 0 \end{pmatrix} + c_2 e^{-t} \begin{pmatrix} -1 \\ 0 \\ 1 \end{pmatrix} + c_3 e^{-4t} \begin{pmatrix} 1 \\ 1 \\ 1 \end{pmatrix} \qquad \blacksquare$$

An alternate form of the general solution of $Y' = AY$ involves the product of a matrix, called the fundamental matrix of the system, and an $n \times 1$ column vector of arbitrary constants. The fundamental matrix of $Y' = AY$ is defined as follows.

DEFINITION

If the solution vectors

$$Y_1 = \begin{pmatrix} y_{11} \\ y_{21} \\ \vdots \\ y_{n1} \end{pmatrix}, Y_2 = \begin{pmatrix} y_{12} \\ y_{22} \\ \vdots \\ y_{n2} \end{pmatrix}, \ldots, Y_n = \begin{pmatrix} y_{1n} \\ y_{2n} \\ \vdots \\ y_{nn} \end{pmatrix}$$

form a fundamental set of solutions of $Y' = AY$ on $[a, b]$, then the matrix Φ given by

$$\Phi = (Y_1, Y_2, \ldots, Y_n) = \begin{pmatrix} y_{11} & y_{12} & \cdots & y_{1n} \\ y_{21} & y_{22} & \cdots & y_{2n} \\ \vdots & \vdots & & \vdots \\ y_{n1} & y_{n2} & \cdots & y_{nn} \end{pmatrix}$$

is called a **fundamental matrix of the system.**

Example 9

Find a fundamental matrix for the system

$$Y' = \begin{pmatrix} 1 & -2 \\ 1 & 4 \end{pmatrix} Y$$

SOLUTION   A fundamental set of solutions for this system was shown in Example 6 to be

$$Y_1 = \begin{pmatrix} -2e^{2t} \\ e^{2t} \end{pmatrix} \quad \text{and} \quad Y_2 = \begin{pmatrix} e^{3t} \\ -e^{3t} \end{pmatrix}$$

A fundamental matrix for this system is then

$$\Phi = \begin{pmatrix} -2e^{2t} & e^{3t} \\ e^{2t} & -e^{3t} \end{pmatrix}$$

■

The general solution of a homogeneous system $Y' = AY$ can be written as a product of a fundamental matrix Φ and an $n \times 1$ column vector of arbitrary constants C. To illustrate this, we note that if $Y_1, Y_2, \ldots, Y_n$ form a fundamental set of solutions of $Y' = AY$ on $[a, b]$, then its general solution is

$$Y = c_1 Y_1 + c_2 Y_2 + \cdots + c_n Y_n = c_1 \begin{pmatrix} y_{11} \\ y_{21} \\ \vdots \\ y_{n1} \end{pmatrix} + c_2 \begin{pmatrix} y_{12} \\ y_{22} \\ \vdots \\ y_{n2} \end{pmatrix} + \cdots + c_n \begin{pmatrix} y_{1n} \\ y_{2n} \\ \vdots \\ y_{nn} \end{pmatrix}$$

$$= \begin{pmatrix} c_1 y_{11} + c_2 y_{12} + \cdots + c_n y_{1n} \\ c_1 y_{21} + c_2 y_{22} + \cdots + c_n y_{2n} \\ \vdots & \vdots & & \vdots \\ c_1 y_{n1} + c_2 y_{n2} + \cdots + c_n y_{nn} \end{pmatrix} = \begin{pmatrix} y_{11} & y_{12} & \cdots & y_{1n} \\ y_{21} & y_{22} & \cdots & y_{2n} \\ \vdots & \vdots & & \vdots \\ y_{n1} & y_{n2} & \cdots & y_{nn} \end{pmatrix} \begin{pmatrix} c_1 \\ c_2 \\ \vdots \\ c_n \end{pmatrix}$$

$$Y = \Phi C$$

Example 10

Write the general solution of

$$Y' = \begin{pmatrix} 1 & -2 \\ 1 & 4 \end{pmatrix} Y$$

as a product of Φ and a 2×1 column vector of arbitrary constants C.

SOLUTION   A fundamental matrix for this system is given in Example 9. Thus, the general solution of this system can be written

$$Y = \begin{pmatrix} -2e^{2t} & e^{3t} \\ e^{2t} & -e^{3t} \end{pmatrix} \begin{pmatrix} c_1 \\ c_2 \end{pmatrix}$$

■

COMMENT: The determinant of a fundamental matrix Φ is, interestingly enough, the same determinant encountered in connection with linearly independent solution vectors; that is,

$$\det \Phi = W$$

where W is the Wronskian determinant defined in Theorem 9–3. The columns of Φ are linearly independent on $[a, b]$ since its columns consist of the fundamental set of solutions $Y_1, Y_2, \ldots, Y_n$. Recalling that the Wronskian determinant for a fundamental set of solutions is nonzero, we conclude that

$$\det \Phi \neq 0$$

This means that Φ is nonsingular and guarantees that Φ^{-1} exists for every t in $[a, b]$, a fact that is useful in solving nonhomogeneous systems.

If particular values of the solution vector of a given system are specified for some value t_0 in the interval of solution, then the system of differential equations with this pairing of values is called an **initial-value problem**. For example, the homogeneous system

$$Y' = \begin{pmatrix} 2 & 1 & 0 \\ 1 & 0 & 1 \\ -1 & 1 & 2 \end{pmatrix} Y$$

with the condition $Y(1) = \begin{pmatrix} -2 \\ 1 \\ 0 \end{pmatrix}$ is an initial-value problem.

Example 11

Solve the initial-value problem

$$Y' = \begin{pmatrix} 1 & -2 \\ 1 & 4 \end{pmatrix} Y, \; Y(0) = \begin{pmatrix} 1 \\ 2 \end{pmatrix}$$

SOLUTION   The given system is the same one whose solution we gave in Example 7 and, therefore, the particular solution required for this initial-value problem is obtained from the general solution

$$Y = c_1 \begin{pmatrix} -2e^{2t} \\ e^{2t} \end{pmatrix} + c_2 \begin{pmatrix} e^{3t} \\ -e^{3t} \end{pmatrix} = c_1 \begin{pmatrix} -2 \\ 1 \end{pmatrix} e^{2t} + c_2 \begin{pmatrix} 1 \\ -1 \end{pmatrix} e^{3t}$$

by substituting the conditions $Y(0) = \begin{pmatrix} 1 \\ 2 \end{pmatrix}$ and solving for c_1 and c_2. Hence,

$$c_1 \begin{pmatrix} -2 \\ 1 \end{pmatrix} + c_2 \begin{pmatrix} 1 \\ -1 \end{pmatrix} = \begin{pmatrix} 1 \\ 2 \end{pmatrix}$$

This yields the following two conditions on c_1 and c_2.

$$-2c_1 + c_2 = 1$$
$$c_1 - c_2 = 2$$

The solution of this system of equations yields $c_1 = -3$ and $c_2 = -5$. Hence the solution of the given initial-value problem is

$$\mathbf{Y} = -3\begin{pmatrix} -2 \\ 1 \end{pmatrix}e^{2t} - 5\begin{pmatrix} 1 \\ -1 \end{pmatrix}e^{3t}$$

∎

Theorem 9–4, which is an analog of Theorem 4–1, gives conditions for which the initial-value problem $\mathbf{Y}' = \mathbf{AY}$, $\mathbf{Y}(t_0) = \mathbf{C}$ has a unique solution.

THEOREM 9–4

Let $\mathbf{A}$ be an $n \times n$ matrix of constant coefficients, t_0 any point in the interval $[a, b]$, and $\mathbf{C}$ an $n \times 1$ vector of real constants, then there exists a unique solution of $\mathbf{Y}' = \mathbf{AY}$, $\mathbf{Y}(t_0) = \mathbf{C}$ on the interval.

EXERCISES FOR SECTION 9–4

In Exercises 1–10 write the given system in the form $\mathbf{Y}' = \mathbf{AY} + \mathbf{F}$.

1. $y_1' = 2y_1 + y_2$
   $y_2' = y_1 - y_2$

2. $y_1' = 3y_1 + y_2$
   $y_2' = y_1$

3. $y_1' = y_1 + y_2 + t$
   $y_2' = y_1 + 3y_2 + 5$

4. $y_1' = e^{-t}\sin t$
   $y_2' = y_1 - y_2 + e^{-t}\cos t$

5. $y_1' - y_2' = 1$
   $y_1' + 3y_2' = 0$

6. $y_1' + 2y_2' = t$
   $y_1' - y_2' = 1$

7. $y_1' - y_1 + y_2' + 2y_2 = 0$
   $y_1' + y_1 - 2y_2' = 0$

8. $y_1' + 2y_1 + 2y_2' = 0$
   $y_1' - y_2' - y_2 = 0$

9. $y_1' - y_1 + 3y_2 - y_3 = 0$
   $y_1' - 2y_2 - y_2' + 2y_3 = 0$
   $2y_1 - 3y_2 - y_3' = 0$

10. $y_1' - 2y_1 + y_2' - y_3' = 0$
    $y_1' + y_2' - y_2 + y_3' - y_3 = 0$
    $2y_1' - y_2' + y_3' - 2y_3 = 0$

In Exercises 11–14 write the systems without the use of matrices.

11. $\mathbf{Y}' = \begin{pmatrix} 1 & 2 \\ 0 & 1 \end{pmatrix}\mathbf{Y} + \begin{pmatrix} 1 \\ -1 \end{pmatrix}t$

12. $\mathbf{Y}' = \begin{pmatrix} -1 & 2 \\ 5 & 3 \end{pmatrix}\mathbf{Y} + \begin{pmatrix} 1 \\ 0 \end{pmatrix}\sin t$

13. $Y' = \begin{pmatrix} 1 & 5 & -1 \\ 2 & 0 & 1 \\ 1 & 0 & 0 \end{pmatrix} Y + \begin{pmatrix} 1 \\ 1 \\ 0 \end{pmatrix} t$ **14.** $Y' = \begin{pmatrix} 2 & 0 & 1 \\ 1 & 2 & 0 \\ 0 & 1 & 0 \end{pmatrix} Y + \begin{pmatrix} 1 \\ 0 \\ -1 \end{pmatrix} \sin t$

In Exercises 15–20 verify that the given vector function is a solution to the system.

15. $\begin{aligned} y_1' &= y_1 + 4y_2 \\ y_2' &= y_1 + y_2 \end{aligned}$; $Y = \begin{pmatrix} -2 \\ 1 \end{pmatrix} e^{-t}$

16. $Y' = \begin{pmatrix} 2 & 3 \\ 3 & 2 \end{pmatrix} Y$; $Y = \begin{pmatrix} 1 \\ -1 \end{pmatrix} e^{-t}$

17. $\begin{aligned} y_1' &= 6y_1 + 8y_2 \\ y_2' &= -y_1 + 2y_2 \end{aligned}$; $Y = e^{4t} \begin{pmatrix} 4 \sin 2t \\ \cos 2t - \sin 2t \end{pmatrix}$

18. $Y' = \begin{pmatrix} 3 & -18 \\ 2 & -9 \end{pmatrix} Y$; $Y = \begin{pmatrix} 3 \\ 1 \end{pmatrix} e^{-3t}$

19. $Y' = \begin{pmatrix} 1 & 0 & 1 \\ 1 & 1 & 0 \\ -2 & 0 & -1 \end{pmatrix} Y$; $Y = \begin{pmatrix} -\cos t \\ e^t + \frac{1}{2}\cos t - \frac{1}{2}\sin t \\ \cos t + \sin t \end{pmatrix}$

20. $\begin{aligned} y_1' &= 5y_1 + 2y_2 + 2y_3 \\ y_2' &= 2y_1 + 2y_2 - 4y_3 \\ y_3' &= 2y_1 - 4y_2 + 2y_3 \end{aligned}$; $Y = \begin{pmatrix} 4 \\ 1 \\ 1 \end{pmatrix} e^{6t}$

In Exercises 21–26 each of the given vectors is a solution to the system. Show that each forms a fundamental set, and write the general solution of the system.

21. $Y_1 = \begin{pmatrix} 2e^{3t} \\ e^{3t} \end{pmatrix}$, $Y_2 = \begin{pmatrix} -2e^{-t} \\ e^{-t} \end{pmatrix}$; $Y' = \begin{pmatrix} 1 & 4 \\ 1 & 1 \end{pmatrix} Y$

22. $Y_1 = \begin{pmatrix} e^{-5t} \\ e^{-5t} \end{pmatrix}$, $Y_2 = \begin{pmatrix} -e^t \\ e^t \end{pmatrix}$; $Y' = \begin{pmatrix} -2 & -3 \\ -3 & -2 \end{pmatrix} Y$

23. $Y_1 = e^{4t} \begin{pmatrix} -4\cos 2t \\ \cos 2t + \sin 2t \end{pmatrix}$, $Y_2 = e^{4t} \begin{pmatrix} -4\sin 2t \\ \sin 2t - \cos 2t \end{pmatrix}$; $Y' = \begin{pmatrix} 6 & 8 \\ -1 & 2 \end{pmatrix} Y$

24. $Y_1 = e^{2t} \begin{pmatrix} 0 \\ 1 \end{pmatrix}$, $Y_2 = e^{2t} \begin{pmatrix} \frac{1}{3} \\ t \end{pmatrix}$; $Y' = \begin{pmatrix} 2 & 0 \\ 3 & 2 \end{pmatrix} Y$

25. $Y_1 = \begin{pmatrix} 0 \\ 1 \\ 0 \end{pmatrix}$, $Y_2 = \begin{pmatrix} -\cos t \\ \frac{1}{2}(\cos t - \sin t) \\ \cos t + \sin t \end{pmatrix}$, $Y_3 = \begin{pmatrix} -\sin t \\ \frac{1}{2}(\cos t + \sin t) \\ \sin t - \cos t \end{pmatrix}$;

$Y' = \begin{pmatrix} 1 & 0 & 1 \\ 1 & 1 & 0 \\ -2 & 0 & -1 \end{pmatrix} Y$

26. $Y_1 = \begin{pmatrix} 0 \\ e^{2t} \\ -e^{2t} \end{pmatrix}$, $Y_2 = \begin{pmatrix} e^t\cos t \\ e^t\sin t \\ e^t\cos t \end{pmatrix}$, $Y_3 = \begin{pmatrix} e^t\sin t \\ -e^t\cos t \\ e^t\sin t \end{pmatrix}$; $Y' = \begin{pmatrix} 2 & -1 & -1 \\ 2 & 1 & -1 \\ 0 & -1 & 1 \end{pmatrix} Y$

27–32. For each of the fundamental sets of solutions in Exercises 21–26 write the general solution of the system in terms of the fundamental matrix.

33–36. For each of the systems given in Exercises 21–24 find the particular solution corresponding to $Y(0) = \begin{pmatrix} 1 \\ -1 \end{pmatrix}$.

37–38. For the systems of Exercises 25 and 26 find the particular solution corresponding to $Y(0) = \begin{pmatrix} 0 \\ 1 \\ 0 \end{pmatrix}$.

9–5   HOMOGENEOUS SYSTEMS: DISTINCT REAL EIGENVALUES

In this section we show how to solve the homogeneous system

$$Y' = AY$$

where A is an $n \times n$ matrix of constants. The solution of the nonhomogeneous system

$$Y' = AY + F$$

where F is a nonzero $n \times 1$ column vector is covered in Section 9–7.

We begin by assuming that the solution of a homogeneous system will consist of exponential functions of the form

$$Ve^{\lambda t}$$

where V is an $n \times 1$ constant vector and λ is a constant to be determined. This assumption parallels the one made in Chapter 4 that nth-order linear differential equations have solutions that are exponential functions. To see if $Y' = AY$ has solutions of the form $Y = Ve^{\lambda t}$, we substitute the assumed solution vector into the homogeneous system. Thus,

$$\lambda Ve^{\lambda t} = AVe^{\lambda t}$$

Or, since $e^{\lambda t}$ is nonzero,

$$\lambda V = AV$$

This is equivalent to

$$(A - \lambda I)V = 0 \tag{9–10}$$

where I is the $n \times n$ identity matrix. Hence, the homogeneous system $Y' = AY$ will have solutions of the form $Y = Ve^{\lambda t}$ if Equation 9–10 can be solved for λ and V. We will develop the process for solving Equation 9–10 in the context of a specific homogeneous system.

Example 1

Find the general solution of the homogeneous system

$$y_1' = y_1 + 2y_2$$
$$y_2' = 3y_1 + 2y_2$$

SOLUTION From the previous discussion we know that $Y' = AY$ will have a solution of the form $Y = Ve^{\lambda t}$ if values of λ and V can be found to satisfy the vector equation $(A - \lambda I)V = 0$. Since the given system can be written

$$Y' = \begin{pmatrix} 1 & 2 \\ 3 & 2 \end{pmatrix} Y$$

we have $A = \begin{pmatrix} 1 & 2 \\ 3 & 2 \end{pmatrix}$, so

$$(A - \lambda I)V = \left(\begin{pmatrix} 1 & 2 \\ 3 & 2 \end{pmatrix} - \lambda \begin{pmatrix} 1 & 0 \\ 0 & 1 \end{pmatrix} \right) \begin{pmatrix} v_1 \\ v_2 \end{pmatrix} = \begin{pmatrix} 0 \\ 0 \end{pmatrix}$$

or,

$$\begin{pmatrix} 1 - \lambda & 2 \\ 3 & 2 - \lambda \end{pmatrix} \begin{pmatrix} v_1 \\ v_2 \end{pmatrix} = \begin{pmatrix} 0 \\ 0 \end{pmatrix}$$

Expanding the matrix product, we get the following equivalent system of two equations in v_1 and v_2.

$$(1 - \lambda)v_1 + 2v_2 = 0$$
$$3v_1 + (2 - \lambda)v_2 = 0 \tag{9–11}$$

To solve this system of equations, determine values of λ for which there are nontrivial solutions; that is, values for which there are solutions other than $v_1 = 0$ and $v_2 = 0$. We know that the system in Equation 9–11 will have a nontrivial solution if, and only if, the determinant of the coefficient matrix of the system is equal to zero. As a result, we seek values of λ such that

$$\begin{vmatrix} 1 - \lambda & 2 \\ 3 & 2 - \lambda \end{vmatrix} = 0$$

Expanding the determinant, we get

$$(1 - \lambda)(2 - \lambda) - 3(2) = 0$$

Expanding and simplifying yields

$$(\lambda - 4)(\lambda + 1) = 0$$

Therefore, the system of Equation 9–11 will have nontrivial solutions for $\lambda = -1$ and $\lambda = 4$. Each value of λ is substituted into Equation 9–11 and the system solved for v_1 and v_2.

Using $\lambda = 4$, the system of Equation 9–11 becomes

$$-3v_1 + 2v_2 = 0$$
$$3v_1 - 2v_2 = 0$$

This system has infinitely many solutions characterized by the constraint $v_2 = \frac{3}{2} v_1$. Any convenient choice of v_1 may be used. If we let $v_1 = 2$, then $v_2 = 3$, and one solution of the given system of differential equations is

$$\mathbf{Y}_1 = \begin{pmatrix} 2 \\ 3 \end{pmatrix} e^{4t}$$

Substituting $\lambda = -1$, the system of Equation 9–11 becomes

$$2v_1 + 2v_2 = 0$$
$$3v_1 + 3v_2 = 0$$

This system also has infinitely many solutions characterized by $v_2 = -v_1$. Letting $v_1 = 1$, we get $v_2 = -1$, and another solution of the given system is

$$\mathbf{Y}_2 = \begin{pmatrix} 1 \\ -1 \end{pmatrix} e^{-t}$$

Since $\mathbf{Y}_1$ and $\mathbf{Y}_2$ are linearly independent vectors, the general solution of the given system is

$$\mathbf{Y} = c_1 \begin{pmatrix} 2 \\ 3 \end{pmatrix} e^{4t} + c_2 \begin{pmatrix} 1 \\ -1 \end{pmatrix} e^{-t} \qquad \blacksquare$$

As you can see from Example 1, the procedure for solving the given system is reduced to the algebraic problem of solving the equation

$$\begin{vmatrix} 1 - \lambda & 2 \\ 3 & 2 - \lambda \end{vmatrix} = 0$$

This equation is called the characteristic equation of the matrix of the system.

DEFINITION

Let $\mathbf{A}$ be an $n \times n$ constant matrix and $\mathbf{I}$ be an $n \times n$ identity matrix. The equation

$$\det(\mathbf{A} - \lambda\mathbf{I}) = 0$$

is called the **characteristic equation** of $\mathbf{A}$.

Example 2

Write the characteristic equation for

$$A = \begin{pmatrix} 9 & -5 & 0 \\ 0 & -1 & 5 \\ 0 & -1 & 5 \end{pmatrix}$$

SOLUTION   The characteristic equation for A is

$$\left| \begin{pmatrix} 9 & -5 & 0 \\ 0 & -1 & 5 \\ 0 & -1 & 5 \end{pmatrix} - \lambda \begin{pmatrix} 1 & 0 & 0 \\ 0 & 1 & 0 \\ 0 & 0 & 1 \end{pmatrix} \right| = 0$$

Or,

$$\begin{vmatrix} 9 - \lambda & -5 & 0 \\ 0 & -1 - \lambda & 5 \\ 0 & -1 & 5 - \lambda \end{vmatrix} = 0$$

Expanding this determinant by minors about the first column, we get

$$(9 - \lambda)[(-1 - \lambda)(5 - \lambda) - (-1)(5)] = 0$$
$$(9 - \lambda)(-4\lambda + \lambda^2) = 0$$
$$\lambda(\lambda - 4)(9 - \lambda) = 0 \qquad \blacksquare$$

A solution of the characteristic equation is called an eigenvalue, a word taken from the German word *eigenwert*, meaning "particular value." The next definition formalizes this concept.

DEFINITION

Let A be an $n \times n$ constant matrix. The roots λ_i of the characteristic equation $\det(A - \lambda I) = 0$ are called **eigenvalues** of the matrix A. A nonzero vector V_1, which is a solution of $(A - \lambda I)V = 0$ for a particular eigenvalue λ_1, is called an **eigenvector** of the matrix A associated with the eigenvalue λ_1.

COMMENT: *Referring to Example 1, note that the eigenvalues of the matrix* $A = \begin{pmatrix} 1 & 2 \\ 3 & 2 \end{pmatrix}$ *are* $\lambda_1 = -1$ *and* $\lambda_2 = 4$. *The associated eigenvectors are*

$$V_1 = \begin{pmatrix} 2 \\ 3 \end{pmatrix} \text{ and } V_2 = \begin{pmatrix} 1 \\ -1 \end{pmatrix}$$

Example 3

Find the eigenvalues and eigenvectors for

$$\mathbf{A} = \begin{pmatrix} 9 & -5 & 0 \\ 0 & -1 & 5 \\ 0 & -1 & 5 \end{pmatrix}$$

SOLUTION   We found in Example 2 that the characteristic equation of $\mathbf{A}$ was

$$\lambda(\lambda - 4)(9 - \lambda) = 0$$

The eigenvalues of $\mathbf{A}$, then, are $\lambda_1 = 0$, $\lambda_2 = 4$, and $\lambda_3 = 9$. To find an eigenvector associated with each eigenvalue, we substitute each value of λ into $(\mathbf{A} - \lambda\mathbf{I})\mathbf{V} = \mathbf{0}$ and reduce the resulting system to triangular form. Thus, for $\lambda_1 = 0$, we have

$$(\mathbf{A} - 0\mathbf{I})\mathbf{V} = \begin{pmatrix} 9 & -5 & 0 \\ 0 & -1 & 5 \\ 0 & -1 & 5 \end{pmatrix} \begin{pmatrix} v_1 \\ v_2 \\ v_3 \end{pmatrix} = \begin{pmatrix} 0 \\ 0 \\ 0 \end{pmatrix}$$

The 3×3 matrix can be written in triangular form as follows.

$$\begin{pmatrix} 9 & -5 & 0 \\ 0 & -1 & 5 \\ 0 & -1 & 5 \end{pmatrix} \xrightarrow{R3 - R2 \rightarrow R3} \begin{pmatrix} 9 & -5 & 0 \\ 0 & -1 & 5 \\ 0 & 0 & 0 \end{pmatrix}$$

Solving the system $(\mathbf{A} - 0\mathbf{I})\mathbf{V} = \mathbf{0}$ then reduces to solving the system

$$\begin{pmatrix} 9 & -5 & 0 \\ 0 & -1 & 5 \\ 0 & 0 & 0 \end{pmatrix} \begin{pmatrix} v_1 \\ v_2 \\ v_3 \end{pmatrix} = \begin{pmatrix} 0 \\ 0 \\ 0 \end{pmatrix}$$

We conclude that $9v_1 - 5v_2 = 0$, $-v_2 + 5v_3 = 0$, and v_3 is arbitrary. Choosing $v_3 = 1$, we find that $v_1 = \frac{25}{9}$ and $v_2 = 5$. An associated eigenvector is then

$$\mathbf{V}_1 = \begin{pmatrix} \frac{25}{9} \\ 5 \\ 1 \end{pmatrix}$$

Note that because v_3 is arbitrary, any constant multiple of $\mathbf{V}_1$ is also an eigenvector of $\lambda_1 = 0$.

For $\lambda_2 = 4$, we have

$$(\mathbf{A} - 4\mathbf{I})\mathbf{V} = \left(\begin{pmatrix} 9 & -5 & 0 \\ 0 & -1 & 5 \\ 0 & -1 & 5 \end{pmatrix} - \begin{pmatrix} 4 & 0 & 0 \\ 0 & 4 & 0 \\ 0 & 0 & 4 \end{pmatrix} \right) \begin{pmatrix} v_1 \\ v_2 \\ v_3 \end{pmatrix} = \begin{pmatrix} 0 \\ 0 \\ 0 \end{pmatrix}$$

which simplifies to

$$\begin{pmatrix} 5 & -5 & 0 \\ 0 & -5 & 5 \\ 0 & -1 & 1 \end{pmatrix} \begin{pmatrix} v_1 \\ v_2 \\ v_3 \end{pmatrix} = \begin{pmatrix} 0 \\ 0 \\ 0 \end{pmatrix}$$

Using the row operations $\frac{1}{5} R1 \rightarrow R1$, $-\frac{1}{5} R2 \rightarrow R2$, and $R2 + R3 \rightarrow R3$ on the 3×3 matrix, we get

$$
\begin{pmatrix} 1 & -1 & 0 \\ 0 & 1 & -1 \\ 0 & 0 & 0 \end{pmatrix}\begin{pmatrix} v_1 \\ v_2 \\ v_3 \end{pmatrix} = \begin{pmatrix} 0 \\ 0 \\ 0 \end{pmatrix}
$$

The solution of this system is $v_1 - v_2 = 0$, $v_2 - v_3 = 0$, and v_3 is arbitrary. Choosing $v_3 = 1$, we get $v_2 = 1$ and $v_1 = 1$, so an eigenvector associated with $\lambda = 4$ is

$$
\mathbf{V}_2 = \begin{pmatrix} 1 \\ 1 \\ 1 \end{pmatrix}
$$

Finally, substituting $\lambda_3 = 9$ yields

$$
(\mathbf{A} - 9\mathbf{I})\mathbf{V} = \left(\begin{pmatrix} 9 & -5 & 0 \\ 0 & -1 & 5 \\ 0 & -1 & 5 \end{pmatrix} - \begin{pmatrix} 9 & 0 & 0 \\ 0 & 9 & 0 \\ 0 & 0 & 9 \end{pmatrix} \right)\begin{pmatrix} v_1 \\ v_2 \\ v_3 \end{pmatrix} = \begin{pmatrix} 0 \\ 0 \\ 0 \end{pmatrix}
$$

which simplifies to

$$
\begin{pmatrix} 0 & -5 & 0 \\ 0 & -10 & 5 \\ 0 & -1 & -4 \end{pmatrix}\begin{pmatrix} v_1 \\ v_2 \\ v_3 \end{pmatrix} = \begin{pmatrix} 0 \\ 0 \\ 0 \end{pmatrix}
$$

This yields the solutions $v_2 = 0$, $v_3 = 0$, and v_1 is arbitrary. Choosing $v_1 = 1$, we have

$$
\mathbf{V}_3 = \begin{pmatrix} 1 \\ 0 \\ 0 \end{pmatrix}
$$

■

Example 4

Write the general solution of $\mathbf{Y}' = \mathbf{A}\mathbf{Y}$, where

$$
\mathbf{A} = \begin{pmatrix} 9 & -5 & 0 \\ 0 & -1 & 5 \\ 0 & -1 & 5 \end{pmatrix}
$$

SOLUTION   Using the eigenvalues and corresponding eigenvectors found in Example 3, we know that

$$
\mathbf{Y}_1 = \begin{pmatrix} \frac{25}{9} \\ 5 \\ 1 \end{pmatrix}, \quad \mathbf{Y}_2 = \begin{pmatrix} 1 \\ 1 \\ 1 \end{pmatrix}e^{4t}, \quad \mathbf{Y}_3 = \begin{pmatrix} 1 \\ 0 \\ 0 \end{pmatrix}e^{9t}
$$

form a fundamental set of solutions of the given system. Therefore, the general solution is

$$\mathbf{Y} = c_1 \begin{pmatrix} \frac{25}{9} \\ 5 \\ 1 \end{pmatrix} + c_2 \begin{pmatrix} 1 \\ 1 \\ 1 \end{pmatrix} e^{4t} + c_3 \begin{pmatrix} 1 \\ 0 \\ 0 \end{pmatrix} e^{9t}$$

You can see from Examples 3 and 4 that the first step in solving a homogeneous system $\mathbf{Y}' = \mathbf{A}\mathbf{Y}$ is to find the eigenvalues of $\mathbf{A}$ and their associated eigenvectors. The procedure can be summarized in the following way.

FINDING THE GENERAL SOLUTION OF Y' = AY

To solve the nth-order homogeneous system

$$y_1' = a_{11}y_1 + a_{12}y_2 + \cdots + a_{1n}y_n$$
$$y_2' = a_{21}y_1 + a_{22}y_2 + \cdots + a_{2n}y_n$$
$$\vdots$$
$$y_n' = a_{n1}y_1 + a_{n2}y_2 + \cdots + a_{nn}y_n$$

proceed as follows.

1. *Write the system in the matrix form*

   $$\mathbf{Y}' = \mathbf{A}\mathbf{Y}$$

   where

   $$\mathbf{Y}' = \begin{pmatrix} y_1' \\ y_2' \\ \vdots \\ y_n' \end{pmatrix}, \quad \mathbf{A} = \begin{pmatrix} a_{11} & a_{12} & \cdots & a_{1n} \\ a_{21} & a_{22} & \cdots & a_{2n} \\ \vdots & & & \vdots \\ a_{n1} & a_{n2} & \cdots & a_{nn} \end{pmatrix}, \quad \mathbf{Y} = \begin{pmatrix} y_1 \\ y_2 \\ \vdots \\ y_n \end{pmatrix}$$

2. *Find the eigenvalues of $\mathbf{A}$; that is, solve the characteristic equation*

   $$det(\mathbf{A} - \lambda\mathbf{I}) = 0$$

   for λ.

3. *Substitute each eigenvalue λ_i into*

   $$(\mathbf{A} - \lambda\mathbf{I})\mathbf{V} = \mathbf{0}$$

   and solve for the eigenvectors $\mathbf{V}_i$. The eigenvectors resulting from distinct real eigenvalues are linearly independent vectors. Sometimes a given eigenvalue will yield a set of two or more linearly independent eigenvectors.

4.   *Each distinct real eigenvalue λ_i and associated eigenvector $\mathbf{V}_i$ yields a solution of $\mathbf{Y}' = \mathbf{AY}$ of the form*

$$\mathbf{Y}_i = \mathbf{V}_i e^{\lambda_i t}$$

5.   *If the characteristic equation has n distinct real eigenvalues λ_1, λ_2, . . . , λ_n, the general solution of $\mathbf{Y}' = \mathbf{AY}$ is*

$$\mathbf{Y} = c_1 \mathbf{V}_1 e^{\lambda_1 t} + c_2 \mathbf{V}_2 e^{\lambda_2 t} + \cdots + c_n \mathbf{V}_n e^{\lambda_n t}$$

If the characteristic equation has fewer than n distinct real eigenvalues or has complex eigenvalues, the solution of $\mathbf{Y}' = \mathbf{AY}$ does not have this form and will be discussed in the next section.

Example 5

Determine the general solution of the homogeneous system

$$y_1' + y_1 + 3y_2' = 0$$
$$3y_1 + y_2' + 2y_2 = 0$$

SOLUTION   Begin by solving the two equations simultaneously for y_1' and y_2' to obtain the equivalent normalized system

$$y_1' = 8y_1 + 6y_2$$
$$y_2' = -3y_1 - 2y_2$$

Write this system in the form $\mathbf{Y}' = \mathbf{AY}$, where

$$\mathbf{A} = \begin{pmatrix} 8 & 6 \\ -3 & -2 \end{pmatrix}$$

We can obtain the eigenvalues by solving the characteristic equation

$$\left| \begin{pmatrix} 8 & 6 \\ -3 & -2 \end{pmatrix} - \begin{pmatrix} \lambda & 0 \\ 0 & \lambda \end{pmatrix} \right| = 0$$

This yields

$$\lambda^2 - 6\lambda + 2 = 0$$

The eigenvalues are then $\lambda_1 = 3 + \sqrt{7}$ and $\lambda_2 = 3 - \sqrt{7}$.

Now for $\lambda_1 = 3 + \sqrt{7}$, $(\mathbf{A} - \lambda\mathbf{I})\mathbf{V} = \mathbf{0}$ is equivalent to

$$(5 - \sqrt{7})v_1 + 6v_2 = 0$$
$$-3v_1 - (5 + \sqrt{7})v_2 = 0$$

This yields $v_2 = -\dfrac{5 - \sqrt{7}}{6} v_1$. By choosing $v_1 = -6$, we get the eigenvector

$$\mathbf{V}_1 = \begin{pmatrix} -6 \\ 5 - \sqrt{7} \end{pmatrix}$$

For $\lambda_2 = 3 - \sqrt{7}$, the system $(\mathbf{A} - \lambda\mathbf{I})\mathbf{V} = \mathbf{0}$ is equivalent to

$$(5 + \sqrt{7})v_1 + 6v_2 = 0$$
$$-3v_1 - (5 - \sqrt{7})v_2 = 0$$

We can find an associated eigenvector for λ_2 from this system. Choosing $v_1 = -6$, we get $v_2 = 5 + \sqrt{7}$, so an eigenvector is

$$\mathbf{V}_2 = \begin{pmatrix} -6 \\ 5 + \sqrt{7} \end{pmatrix}$$

Hence, the general solution is

$$\mathbf{Y} = c_1 \begin{pmatrix} -6 \\ 5 - \sqrt{7} \end{pmatrix} e^{(3+\sqrt{7})t} + c_2 \begin{pmatrix} -6 \\ 5 + \sqrt{7} \end{pmatrix} e^{(3-\sqrt{7})t}$$

Notice that this solution can also be written in the form $\mathbf{Y} = \mathbf{\Phi C}$, where $\mathbf{\Phi}$ is the fundamental matrix of solutions and $\mathbf{C}$ is an $n \times 1$ vector of arbitrary constants. In this format, the general solution is

$$\mathbf{Y} = \begin{pmatrix} -6e^{(3+\sqrt{7})t} & -6e^{(3-\sqrt{7})t} \\ (5 - \sqrt{7})e^{(3+\sqrt{7})t} & (5 + \sqrt{7})e^{(3-\sqrt{7})t} \end{pmatrix} \begin{pmatrix} c_1 \\ c_2 \end{pmatrix} \qquad ■$$

Example 6

Find the general solution of $\mathbf{Y}' = \mathbf{AY}$ where

$$\mathbf{A} = \begin{pmatrix} 1 & -1 & 1 \\ 1 & 1 & -1 \\ 2 & -1 & 0 \end{pmatrix}$$

SOLUTION   Note that

$$\det(\mathbf{A} - \lambda\mathbf{I}) = \begin{vmatrix} 1-\lambda & -1 & 1 \\ 1 & 1-\lambda & -1 \\ 2 & -1 & -\lambda \end{vmatrix} = -\lambda^3 + 2\lambda^2 + \lambda - 2$$

$$= -(\lambda + 1)(\lambda - 1)(\lambda - 2)$$

so the eigenvalues are $\lambda_1 = 2$, $\lambda_2 = 1$, and $\lambda_3 = -1$. The eigenvector $\mathbf{V}_1$, corresponding to $\lambda_1 = 2$, must satisfy

$$(\mathbf{A} - 2\mathbf{I})\mathbf{V} = \begin{pmatrix} -1 & -1 & 1 \\ 1 & -1 & -1 \\ 2 & -1 & -2 \end{pmatrix} \begin{pmatrix} v_1 \\ v_2 \\ v_3 \end{pmatrix} = \begin{pmatrix} 0 \\ 0 \\ 0 \end{pmatrix}$$

Using row reduction, the matrix

$$\begin{pmatrix} -1 & -1 & 1 \\ 1 & -1 & -1 \\ 2 & -1 & -2 \end{pmatrix}$$

reduces to

$$\begin{pmatrix} -1 & -1 & 1 \\ 0 & -2 & 0 \\ 0 & 0 & 0 \end{pmatrix}$$

so that v_3 is arbitrary, $v_2 = 0$, and $v_1 = -v_3$. An associated eigenvector is

$$\mathbf{V}_1 = \begin{pmatrix} 1 \\ 0 \\ -1 \end{pmatrix}$$

For $\lambda_2 = 1$, $(\mathbf{A} - \mathbf{I})$ is equal to

$$\begin{pmatrix} 0 & -1 & 1 \\ 1 & 0 & -1 \\ 2 & -1 & -1 \end{pmatrix}$$

which reduces to

$$\begin{pmatrix} 1 & 0 & -1 \\ 0 & -1 & 1 \\ 0 & 0 & 0 \end{pmatrix}$$

This indicates that v_3 is arbitrary, $v_2 = v_3$, and $v_1 = v_3$. Hence, an eigenvector for $\lambda_2 = 1$ is

$$\mathbf{V}_2 = \begin{pmatrix} 1 \\ 1 \\ 1 \end{pmatrix}$$

For $\lambda_3 = -1$, $(\mathbf{A} + \mathbf{I})$ is

$$\begin{pmatrix} 2 & -1 & 1 \\ 1 & 2 & -1 \\ 2 & -1 & 1 \end{pmatrix}$$

which reduces to

$$\begin{pmatrix} 2 & -1 & 1 \\ 0 & \frac{5}{2} & -\frac{3}{2} \\ 0 & 0 & 0 \end{pmatrix}$$

This implies that v_3 is arbitrary, $v_2 = \frac{3}{5}v_3$, and $v_1 = \frac{1}{2}v_2 - \frac{1}{2}v_3$. Choosing $v_3 = 5$ yields an eigenvector of

$$\mathbf{V}_3 = \begin{pmatrix} -1 \\ 3 \\ 5 \end{pmatrix}$$

The general solution of the system is then

$$\mathbf{Y} = c_1 \begin{pmatrix} 1 \\ 0 \\ -1 \end{pmatrix} e^{2t} + c_2 \begin{pmatrix} 1 \\ 1 \\ 1 \end{pmatrix} e^{t} + c_3 \begin{pmatrix} -1 \\ 3 \\ 5 \end{pmatrix} e^{-t}$$ ∎

EXERCISES FOR SECTION 9–5

In Exercises 1–10 find the general solution of the given system.

1. $y_1' = 2y_1 + 3y_2$
       $y_2' = 3y_1 + 2y_2$

2. $y_1' = y_1 + 4y_2$
       $y_2' = y_1 + y_2$

3. $y_1' = 3y_1 + y_2$
       $y_2' = y_1$

4. $\mathbf{Y}' = \begin{pmatrix} 2 & 1 \\ 1 & 1 \end{pmatrix} \mathbf{Y}$

5. $\mathbf{Y}' = \begin{pmatrix} 3 & -2 \\ -1 & 0 \end{pmatrix} \mathbf{Y}$

6. $y_1' - y_2' + y_2 = 0$
       $2y_1' + y_2' = 0$

7. $y_1' + 2y_1 + y_2' + y_2 = 0$
       $y_1' - y_1 - y_2' = 0$

8. $\mathbf{Y}' = \begin{pmatrix} 4 & 0 & 1 \\ -2 & 1 & 0 \\ -2 & 0 & 1 \end{pmatrix} \mathbf{Y}$

9. $y_1' = y_1 + y_2 - y_3$
       $y_2' = 2y_2$
       $y_3' = y_2 - y_3$

10. $y_1' = y_1 + y_2 + y_3$
        $y_2' = y_1 + 4y_2 - y_3$
        $y_3' = -5y_1 - 8y_2 - 3y_3$

In Exercises 11–14 solve the initial-value problems.

11. $y_1' = 2y_1 + y_2$
        $y_2' = y_1 + 2y_2$
        $y_1(0) = 1, y_2(0) = 1$

12. $\mathbf{Y}' = \begin{pmatrix} -1 & 6 \\ 3 & 2 \end{pmatrix} \mathbf{Y}; \quad \mathbf{Y}(0) = \begin{pmatrix} -1 \\ 0 \end{pmatrix}$

13. $\mathbf{Y}' = \begin{pmatrix} 1 & -3 & 2 \\ 0 & -1 & 0 \\ 0 & -1 & -2 \end{pmatrix} \mathbf{Y}; \quad \mathbf{Y}(0) = \begin{pmatrix} 1 \\ 1 \\ 0 \end{pmatrix}$

14. $\mathbf{Y}' = \begin{pmatrix} 2 & 1 & 5 \\ -3 & -2 & -8 \\ 3 & 3 & 9 \end{pmatrix} \mathbf{Y}; \quad \mathbf{Y}(0) = \begin{pmatrix} -1 \\ 0 \\ 0 \end{pmatrix}$

15. The diffusion of a solute between two compartments with volumes V_1 and V_2, which are separated by a permeable membrane, is governed by the homogeneous differential system

$$V_1 y_1' = k(y_1 - y_2)$$
$$V_2 y_2' = k(y_2 - y_1)$$

where y_1 and y_2 are concentrations present in the individual compartments, and k is a positive constant of proportionality. Determine the general solution of this system.

16. Assume that two animal populations $P_1(t)$ and $P_2(t)$ interact according to the differential system

$$P_1' = 3P_1 - 5P_2$$
$$P_2' = 2P_1 - 4P_2$$

Discuss the long-term behavior of these populations.

9–6 HOMOGENEOUS SYSTEMS: COMPLEX AND REPEATED REAL EIGENVALUES

The form of the general solution of a homogeneous system depends on the nature of the eigenvalues. In the previous section we showed how to write the general solution of $\mathbf{Y}' = \mathbf{A}\mathbf{Y}$ when $\mathbf{A}$ has n distinct real eigenvalues. Here we will discuss the solution of a homogeneous system in which the eigenvalues are complex numbers or repeated real numbers.

Complex Eigenvalues

We will discuss the solution of a homogeneous system with complex eigenvalues in the context of Example 1.

Example 1

Solve the homogeneous system

$$\mathbf{Y}' = \begin{pmatrix} 3 & -1 \\ 2 & 1 \end{pmatrix} \mathbf{Y}$$

SOLUTION   The characteristic equation $\det(\mathbf{A} - \lambda\mathbf{I}) = 0$ is

$$\begin{vmatrix} 3 - \lambda & -1 \\ 2 & 1 - \lambda \end{vmatrix} = (\lambda - 3)(\lambda - 1) + 2 = \lambda^2 - 4\lambda + 5 = 0$$

The solution of this quadratic equation gives the complex eigenvalues $\lambda_1 = 2 + i$

and $\lambda_2 = 2 - i$. To find an eigenvector associated with $\lambda_1 = 2 + i$, we must solve the system $(\mathbf{A} - (2 + i)\mathbf{I})\mathbf{V} = 0$ or the equivalent system

$$(1 - i)v_1 - v_2 = 0$$

$$2v_1 - (1 + i)v_2 = 0$$

We note that this is a dependent system since the first equation is $(1 + i)$ times the second. Thus, $v_2 = (1 - i)v_1$ and v_1 is arbitrary. If we let $v_1 = 1$, we get $v_2 = 1 - i$, so an eigenvector corresponding to $2 + i$ is

$$\mathbf{V}_1 = \begin{pmatrix} 1 \\ 1 - i \end{pmatrix}$$

Similarly, by solving the system $(\mathbf{A} - (2 - i)\mathbf{I})\mathbf{V} = \mathbf{0}$, we get $v_1 = 1$ and $v_2 = 1 + i$. An eigenvector for $1 - i$ is

$$\mathbf{V}_2 = \begin{pmatrix} 1 \\ 1 + i \end{pmatrix}$$

It can be verified that $\mathbf{V}_1$ and $\mathbf{V}_2$ are linearly independent eigenvectors. Thus, the general solution may be written

$$\mathbf{Y} = c_1 \begin{pmatrix} 1 \\ 1 - i \end{pmatrix} e^{(2 + i)t} + c_2 \begin{pmatrix} 1 \\ 1 + i \end{pmatrix} e^{(2 - i)t} \qquad \blacksquare$$

The eigenvalues $\lambda_1 = 2 + i$ and $\lambda_2 = 2 - i$ obtained in Example 1 are complex conjugates, a fact that is denoted $\lambda_2 = \bar{\lambda}_1$. Further, we note that

$$\mathbf{V}_1 = \begin{pmatrix} 1 \\ 1 - i \end{pmatrix} \text{ and } \mathbf{V}_2 = \begin{pmatrix} 1 \\ 1 + i \end{pmatrix} \text{ can be written as}$$

$$\mathbf{V}_1 = \begin{pmatrix} 1 \\ 1 \end{pmatrix} + i \begin{pmatrix} 0 \\ -1 \end{pmatrix} \quad \text{and} \quad \mathbf{V}_2 = \begin{pmatrix} 1 \\ 1 \end{pmatrix} + i \begin{pmatrix} 0 \\ 1 \end{pmatrix}$$

In this form it is clear that $\mathbf{V}_1$ and $\mathbf{V}_2$ are complex conjugates. Thus, when $\lambda_2 = \bar{\lambda}_1$, we have $\mathbf{V}_2 = \bar{\mathbf{V}}_1$. This observation is true in general and we have verified the following theorem.

THEOREM 9–5

Let $\mathbf{Y}' = \mathbf{A}\mathbf{Y}$ be a homogeneous system with a real $n \times n$ coefficient matrix $\mathbf{A}$, and let $\mathbf{V}_1$ be an eigenvector of $\mathbf{A}$ associated with the complex eigenvalue $\lambda_1 = a + ib$. Then,

$$\mathbf{Y}_1 = \mathbf{V}_1 e^{\lambda_1 t} \quad \text{and} \quad \mathbf{Y}_2 = \bar{\mathbf{V}}_1 e^{\bar{\lambda}_1 t} \qquad (9\text{-}12)$$

are linearly independent solutions of $\mathbf{Y}' = \mathbf{A}\mathbf{Y}$.

COMMENT: Solutions such as Equation 9–12 that are written with complex number coefficients are difficult to interpret when the system *has real-valued initial conditions. Consequently, we will rewrite Equation 9–12 in terms of real coefficients and real-valued functions. Example 2 shows how this is done with the solution from Example 1.*

Example 2

Express the solution

$$Y = c_1 \begin{pmatrix} 1 \\ 1 - i \end{pmatrix} e^{(2 + i)t} + c_2 \begin{pmatrix} 1 \\ 1 + i \end{pmatrix} e^{(2 - i)t}$$

as a linear combination of two real-valued vector functions. (This is the solution obtained in Example 1.)

SOLUTION   Using Euler's formula, we have

$$e^{(2 \pm i)t} = e^{2t}(\cos t \pm i \sin t)$$

The given solution can then be written

$$Y = c_1 \begin{pmatrix} 1 \\ 1 - i \end{pmatrix} e^{2t}(\cos t + i \sin t) + c_2 \begin{pmatrix} 1 \\ 1 + i \end{pmatrix} e^{2t}(\cos t - i \sin t)$$

$$= e^{2t} \begin{pmatrix} c_1 \cos t + ic_1 \sin t + c_2 \cos t - ic_2 \sin t \\ (1 - i)c_1 \cos t + i(1 - i)c_1 \sin t + (1 + i)c_2 \cos t - i(1 + i)c_2 \sin t \end{pmatrix}$$

$$= e^{2t} \begin{pmatrix} (c_1 + c_2) \cos t + i(c_1 - c_2) \sin t \\ [(c_1 + c_2) - i(c_1 - c_2)] \cos t + [(c_1 + c_2) + i(c_1 - c_2)] \sin t \end{pmatrix}$$

By letting $C_1 = c_1 + c_2$ and $C_2 = i(c_1 - c_2)$, we get

$$Y = e^{2t} \begin{pmatrix} C_1 \cos t + C_2 \sin t \\ (C_1 - C_2) \cos t + (C_1 + C_2) \sin t \end{pmatrix}$$

$$= e^{2t} \begin{pmatrix} C_1 \cos t + C_2 \sin t \\ C_1(\cos t + \sin t) + C_2(\sin t - \cos t) \end{pmatrix}$$

$$= C_1 e^{2t} \begin{pmatrix} \cos t \\ \cos t + \sin t \end{pmatrix} + C_2 e^{2t} \begin{pmatrix} \sin t \\ \sin t - \cos t \end{pmatrix}$$

$$= C_1 e^{2t} \left[\begin{pmatrix} 1 \\ 1 \end{pmatrix} \cos t - \begin{pmatrix} 0 \\ -1 \end{pmatrix} \sin t \right] + C_2 e^{2t} \left[\begin{pmatrix} 0 \\ -1 \end{pmatrix} \cos t + \begin{pmatrix} 1 \\ 1 \end{pmatrix} \sin t \right]$$

Finally, we recall that the eigenvector V_1 in Example 1 can be written

$$V_1 = \begin{pmatrix} 1 \\ 1 - i \end{pmatrix} = \begin{pmatrix} 1 \\ 1 \end{pmatrix} + \begin{pmatrix} 0 \\ -1 \end{pmatrix} i$$

Thus, the solution can be written in terms of the real component $V_R = \begin{pmatrix} 1 \\ 1 \end{pmatrix}$ and the imaginary component $\mathbf{V}_I = \begin{pmatrix} 0 \\ -1 \end{pmatrix}$ of $\mathbf{V}_1$; that is,

$$\mathbf{Y} = C_1 e^{2t}(\mathbf{V}_R \cos t - \mathbf{V}_I \sin t) + C_2 e^{2t}(\mathbf{V}_I \cos t + \mathbf{V}_R \sin t)$$ ■

The final form of the real-valued vector solution obtained in Example 2 for the complex eigenvalues $2 \pm i$ illustrates the following general result.

GENERAL SOLUTION OF Y' = AY: COMPLEX EIGENVALUES

Let $\mathbf{Y}' = \mathbf{AY}$ be a homogeneous system with a real $n \times n$ coefficient matrix $\mathbf{A}$, and let $\lambda = a + bi$ be a complex eigenvalue of $\mathbf{A}$ with associated eigenvector $\mathbf{V} = \mathbf{V}_R + \mathbf{V}_I i$. Then,

$$\mathbf{Y}_1 = e^{at}[\mathbf{V}_R \cos bt - \mathbf{V}_I \sin bt] \tag{9–13}$$

and

$$\mathbf{Y}_2 = e^{at}[\mathbf{V}_I \cos bt + \mathbf{V}_R \sin bt] \tag{9–14}$$

are linearly independent solutions of $\mathbf{Y}' = \mathbf{AY}$. The general solution of the system contains a linear combination of these two solutions.

COMMENT: Notice that in writing the solution vectors $\mathbf{Y}_1$ and $\mathbf{Y}_2$ it is not necessary to calculate the eigenvector corresponding to $\lambda = a - bi$.

Example 3

Solve the system $\mathbf{Y}' = \begin{pmatrix} 1 & 10 \\ -1 & -1 \end{pmatrix} \mathbf{Y}$.

SOLUTION   The characteristic equation is

$$-(1 - \lambda)(1 + \lambda) + 10 = 0$$
$$\lambda^2 + 9 = 0$$
$$\lambda = \pm 3i$$

To find the eigenvector corresponding to $3i$, we solve the system

$$\begin{pmatrix} 1 - 3i & 10 \\ -1 & -1 - 3i \end{pmatrix} \begin{pmatrix} v_1 \\ v_2 \end{pmatrix} = \begin{pmatrix} 0 \\ 0 \end{pmatrix}$$

which is equivalent to

$$(1 - 3i) v_1 + 10 v_2 = 0$$
$$-v_1 - (1 + 3i) v_2 = 0$$

If we let $v_1 = 10$, then $v_2 = -1 + 3i$ so that an eigenvector corresponding to $3i$ is

$$\mathbf{V} = \begin{pmatrix} 10 \\ -1 + 3i \end{pmatrix} = \begin{pmatrix} 10 \\ -1 \end{pmatrix} + i \begin{pmatrix} 0 \\ 3 \end{pmatrix}$$

Using $\mathbf{V}_R = \begin{pmatrix} 10 \\ -1 \end{pmatrix}$ and $\mathbf{V}_I = \begin{pmatrix} 0 \\ 3 \end{pmatrix}$ in Equations 9–13 and 9–14, we get

$$\mathbf{Y}_1 = \begin{pmatrix} 10 \\ -1 \end{pmatrix} \cos 3t - \begin{pmatrix} 0 \\ 3 \end{pmatrix} \sin 3t \quad \text{and} \quad \mathbf{Y}_2 = \begin{pmatrix} 0 \\ 3 \end{pmatrix} \cos 3t + \begin{pmatrix} 10 \\ -1 \end{pmatrix} \sin 3t$$

The general solution is then

$$\mathbf{Y} = c_1 \left[\begin{pmatrix} 10 \\ -1 \end{pmatrix} \cos 3t - \begin{pmatrix} 0 \\ 3 \end{pmatrix} \sin 3t \right] + c_2 \left[\begin{pmatrix} 0 \\ 3 \end{pmatrix} \cos 3t + \begin{pmatrix} 10 \\ -1 \end{pmatrix} \sin 3t \right]$$

$$= c_1 \begin{pmatrix} 10 \cos 3t \\ -\cos 3t - 3 \sin 3t \end{pmatrix} + c_2 \begin{pmatrix} 10 \sin 3t \\ 3 \cos 3t - \sin 3t \end{pmatrix} \qquad \blacksquare$$

Example 4

Solve the system $\mathbf{Y}' = \begin{pmatrix} 2 & -1 & -1 \\ 2 & 1 & -1 \\ 0 & -1 & 1 \end{pmatrix} \mathbf{Y}$.

SOLUTION $\quad \det(\mathbf{A} - \lambda\mathbf{I}) = \begin{vmatrix} 2 - \lambda & -1 & -1 \\ 2 & 1-\lambda & -1 \\ 0 & -1 & 1-\lambda \end{vmatrix} = 0$

Expanding the determinant we obtain the characteristic equation

$$-\lambda^3 + 4\lambda^2 - 6\lambda + 4 = 0$$

The roots are 2, $1 + i$, and $1 - i$. Using $\lambda = 2$,

$$(\mathbf{A} - 2\mathbf{I})\mathbf{V} = \begin{pmatrix} 0 & -1 & -1 \\ 2 & -1 & -1 \\ 0 & -1 & -1 \end{pmatrix} \begin{pmatrix} v_1 \\ v_2 \\ v_3 \end{pmatrix} = \begin{pmatrix} 0 \\ 0 \\ 0 \end{pmatrix}$$

which yields an eigenvector $\mathbf{V}_1 = \begin{pmatrix} 0 \\ 1 \\ -1 \end{pmatrix}$. The vector $\mathbf{Y}_1 = \begin{pmatrix} 0 \\ 1 \\ -1 \end{pmatrix} e^{2t}$ is therefore

a solution of the given system. Using $\lambda = 1 + i$,

$$(\mathbf{A} - (1 + i)\mathbf{I})\mathbf{V} = \begin{pmatrix} 1 - i & -1 & -1 \\ 2 & -i & -1 \\ 0 & -1 & -i \end{pmatrix} \begin{pmatrix} v_1 \\ v_2 \\ v_3 \end{pmatrix} = \begin{pmatrix} 0 \\ 0 \\ 0 \end{pmatrix}$$

which yields an eigenvector $V_2 = \begin{pmatrix} 1 \\ -i \\ 1 \end{pmatrix}$. Thus, $V_R = \begin{pmatrix} 1 \\ 0 \\ 1 \end{pmatrix}$ and $V_I = \begin{pmatrix} 0 \\ -1 \\ 0 \end{pmatrix}$.

Using these values for V_R and V_I with $a = b = 1$ in Equations 9–13 and 9–14, we get

$$Y_2 = e^t \left[\begin{pmatrix} 1 \\ 0 \\ 1 \end{pmatrix} \cos t - \begin{pmatrix} 0 \\ -1 \\ 0 \end{pmatrix} \sin t \right] \text{ and } Y_3 = e^t \left[\begin{pmatrix} 0 \\ -1 \\ 0 \end{pmatrix} \cos t + \begin{pmatrix} 1 \\ 0 \\ 1 \end{pmatrix} \sin t \right]$$

The general solution is then $Y = c_1 Y_1 + c_2 Y_2 + c_3 Y_3$; that is,

$$Y = c_1 \begin{pmatrix} 0 \\ 1 \\ -1 \end{pmatrix} e^{2t} + c_2 e^t \left[\begin{pmatrix} 1 \\ 0 \\ 1 \end{pmatrix} \cos t - \begin{pmatrix} 0 \\ -1 \\ 0 \end{pmatrix} \sin t \right]$$

$$+ c_3 e^t \left[\begin{pmatrix} 0 \\ -1 \\ 0 \end{pmatrix} \cos t + \begin{pmatrix} 1 \\ 0 \\ 1 \end{pmatrix} \sin t \right] \qquad \blacksquare$$

Repeated Real Eigenvalues

The homogeneous systems considered so far have all had n distinct eigenvalues, but systems do arise for which there are fewer than n eigenvalues. For example, the matrix

$$A = \begin{pmatrix} 0 & 1 \\ -1 & -2 \end{pmatrix}$$

has a characteristic equation

$$\left| \begin{pmatrix} 0 & 1 \\ -1 & -2 \end{pmatrix} - \begin{pmatrix} \lambda & 0 \\ 0 & \lambda \end{pmatrix} \right| = \begin{vmatrix} -\lambda & 1 \\ -1 & -2-\lambda \end{vmatrix}$$
$$= \lambda^2 + 2\lambda + 1 = (\lambda + 1)^2 = 0$$

and therefore $\lambda_1 = \lambda_2 = -1$. In this case, we say that -1 is a repeated eigenvalue or an eigenvalue of multiplicity 2.

If the matrix A of an nth-order homogeneous system $Y' = AY$ has fewer than n distinct eigenvalues, then the question must be asked: How does one generate the n linearly independent eigenvectors required to write the general solution of the system? Example 5 shows a case in which an eigenvalue of multiplicity 2 generates two linearly independent eigenvectors.

Example 5

Solve the homogeneous system

$$\mathbf{Y}' = \begin{pmatrix} 3 & -2 & 0 \\ -2 & 3 & 0 \\ 0 & 0 & 5 \end{pmatrix} \mathbf{Y}$$

SOLUTION   The characteristic equation $|\mathbf{A} - \lambda\mathbf{I}| = 0$ can be expanded to give

$$\begin{vmatrix} 3 - \lambda & -2 & 0 \\ -2 & 3 - \lambda & 0 \\ 0 & 0 & 5 - \lambda \end{vmatrix} = (5 - \lambda)^2(1 - \lambda) = 0$$

This yields the eigenvalues $\lambda_1 = 1$ and $\lambda_2 = \lambda_3 = 5$. Substituting $\lambda_1 = 1$ into $(\mathbf{A} - \lambda\mathbf{I})\mathbf{V} = \mathbf{0}$, we have

$$\begin{pmatrix} 2 & -2 & 0 \\ -2 & 2 & 0 \\ 0 & 0 & 4 \end{pmatrix}\begin{pmatrix} v_1 \\ v_2 \\ v_3 \end{pmatrix} = \begin{pmatrix} 0 \\ 0 \\ 0 \end{pmatrix}$$

Solving this system yields $v_3 = 0$ and $v_1 = v_2$. Choosing $v_2 = 1$ gives $v_1 = 1$, so an associated eigenvector is

$$\mathbf{V}_1 = \begin{pmatrix} 1 \\ 1 \\ 0 \end{pmatrix}$$

and a solution of the system is

$$\mathbf{Y}_1 = \begin{pmatrix} 1 \\ 1 \\ 0 \end{pmatrix} e^t$$

For $\lambda = 5$, the equation $(\mathbf{A} - 5\mathbf{I})\mathbf{V} = \mathbf{0}$ becomes

$$\begin{pmatrix} -2 & -2 & 0 \\ -2 & -2 & 0 \\ 0 & 0 & 0 \end{pmatrix}\begin{pmatrix} v_1 \\ v_2 \\ v_3 \end{pmatrix} = \begin{pmatrix} 0 \\ 0 \\ 0 \end{pmatrix}$$

In this system, v_3 is arbitrary and $v_2 = -v_1$. Choosing $v_3 = 1$ and $v_1 = 0$, we get $v_2 = 0$. It follows that

$$\mathbf{V}_2 = \begin{pmatrix} 0 \\ 0 \\ 1 \end{pmatrix}$$

is an eigenvector. Another linearly independent eigenvector associated with $\lambda = 5$ is obtained by choosing $v_3 = 0$ and $v_1 = 1$. In this case, $v_2 = -1$ and a second eigenvector is

$$\mathbf{V}_3 = \begin{pmatrix} 1 \\ -1 \\ 0 \end{pmatrix}$$

$\mathbf{V}_2$ and $\mathbf{V}_3$ are linearly independent since $\mathbf{V}_2$ is clearly not a constant multiple of $\mathbf{V}_3$. Thus,

$$\mathbf{Y}_2 = \begin{pmatrix} 0 \\ 0 \\ 1 \end{pmatrix} e^{5t} \quad \text{and} \quad \mathbf{Y}_3 = \begin{pmatrix} 1 \\ -1 \\ 0 \end{pmatrix} e^{5t}$$

are linearly independent solutions. Finally, the general solution of the given system is

$$\mathbf{Y} = c_1 \begin{pmatrix} 1 \\ 1 \\ 0 \end{pmatrix} e^t + c_2 \begin{pmatrix} 0 \\ 0 \\ 1 \end{pmatrix} e^{5t} + c_3 \begin{pmatrix} 1 \\ -1 \\ 0 \end{pmatrix} e^{5t} \qquad \blacksquare$$

Example 6 illustrates a homogeneous system in which an eigenvalue of multiplicity 2 generates only one associated eigenvector.

Example 6

Show that the homogeneous system

$$\mathbf{Y}' = \begin{pmatrix} 2 & 0 \\ 3 & 2 \end{pmatrix} \mathbf{Y}$$

has an eigenvalue $\lambda = 2$ of multiplicity 2 and that only one eigenvector is associated with $\lambda = 2$.

SOLUTION   The characteristic equation $|\mathbf{A} - \lambda \mathbf{I}| = 0$ is equivalently

$$\begin{vmatrix} 2 - \lambda & 0 \\ 3 & 2 - \lambda \end{vmatrix} = (2 - \lambda)^2 = 0$$

Therefore $\lambda_1 = \lambda_2 = 2$ is an eigenvalue of multiplicity 2. The eigenequation $(\mathbf{A} - 2\mathbf{I})\mathbf{V} = \mathbf{0}$ simplifies to

$$\begin{pmatrix} 0 & 0 \\ 3 & 0 \end{pmatrix} \begin{pmatrix} v_1 \\ v_2 \end{pmatrix} = \begin{pmatrix} 0 \\ 0 \end{pmatrix}$$

We infer from this system that $v_1 = 0$ and v_2 is arbitrary. Choosing $v_2 = 1$, an associated eigenvector is

$$\mathbf{V}_1 = \begin{pmatrix} 0 \\ 1 \end{pmatrix}$$

Since v_2 is arbitrary, we could choose other values for v_2 but the resulting eigenvectors would all be constant multiples of $\mathbf{V}_1$. Hence, $\mathbf{V}_1 = \begin{pmatrix} 0 \\ 1 \end{pmatrix}$ is the only linearly independent eigenvector associated with $\lambda = 2$. ∎

To find a second linearly independent solution of the system in Example 6, we might be tempted to let $\mathbf{Y}_2 = \mathbf{C}te^{2t}$, since that is what we did in Chapter 4 to generate a second linearly independent solution of a homogeneous linear differential equation having repeated real roots to the auxiliary equation. To show that this approach does not lead to a second linearly independent solution, substitute $\mathbf{Y}_2 = \mathbf{C}te^{2t}$ into the system given in Example 6. Thus,

$$(\mathbf{C}te^{2t})' = \mathbf{A}\mathbf{C}te^{2t}$$

Carrying out the differentiation on the left side, we have

$$\mathbf{C}e^{2t} + 2\mathbf{C}te^{2t} = \mathbf{A}\mathbf{C}te^{2t}$$

Equating coefficients of e^{2t} and te^{2t} yields the system

$$\mathbf{C} = \mathbf{0}, \, 2\mathbf{C} = \mathbf{A}\mathbf{C}$$

The only possible solution of this system is $\mathbf{C} = \mathbf{0}$, which means that $\mathbf{Y}_2 = \mathbf{C}te^{2t}$ is not a second linearly independent solution of the system in Example 6.

Suppose instead of $\mathbf{Y}_2 = \mathbf{C}te^{2t}$ we try $\mathbf{Y}_2 = (\mathbf{C} + \mathbf{D}t)e^{2t}$, where $\mathbf{C}$ and $\mathbf{D}$ are constant vectors to be determined. Since $\mathbf{Y}_2' = \mathbf{D}e^{2t} + 2(\mathbf{C} + \mathbf{D}t)e^{2t}$, we have

$$\mathbf{D}e^{2t} + 2(\mathbf{C} + \mathbf{D}t)e^{2t} = \mathbf{A}(\mathbf{C} + \mathbf{D}t)e^{2t}$$

Equating the coefficients of e^{2t} and te^{2t} yields the system

$$2\mathbf{C} + \mathbf{D} = \mathbf{A}\mathbf{C}, \, 2\mathbf{D} = \mathbf{A}\mathbf{D}$$

The second equation can be written $\mathbf{A}\mathbf{D} - 2\mathbf{D} = (\mathbf{A} - 2\mathbf{I})\mathbf{D} = \mathbf{0}$, which implies that $\mathbf{D}$ is an eigenvector of $\mathbf{A}$ corresponding to the eigenvalue $\lambda = 2$. Thus, from Example 6, we let $\mathbf{D} = \begin{pmatrix} 0 \\ 1 \end{pmatrix}$. Since the first equation can be written in the form $\mathbf{A}\mathbf{C} - 2\mathbf{C} = (\mathbf{A} - 2\mathbf{I})\mathbf{C} = \mathbf{D}$, we have

$$(\mathbf{A} - 2\mathbf{I})\mathbf{C} = \begin{pmatrix} 0 \\ 1 \end{pmatrix}$$

Substituting $\mathbf{A}$ from Example 6 and $\mathbf{I}$ gives

$$\left[\begin{pmatrix} 2 & 0 \\ 3 & 2 \end{pmatrix} - \begin{pmatrix} 2 & 0 \\ 0 & 2 \end{pmatrix} \right] \begin{pmatrix} c_1 \\ c_2 \end{pmatrix} = \begin{pmatrix} 0 \\ 1 \end{pmatrix}$$

$$\begin{pmatrix} 0 & 0 \\ 3 & 0 \end{pmatrix} \begin{pmatrix} c_1 \\ c_2 \end{pmatrix} = \begin{pmatrix} 0 \\ 1 \end{pmatrix}$$

A solution to this equation is $c_1 = \frac{1}{3}$ and $c_2 = 0$. Notice that other solutions are possible; we chose an obvious one. A second linearly independent solution may now be written as

$$\mathbf{Y}_2 = \left[\begin{pmatrix} \frac{1}{3} \\ 0 \end{pmatrix} + \begin{pmatrix} 0 \\ 1 \end{pmatrix} t \right] e^{2t}$$

From this we can see a general procedure for finding a second linearly independent eigenvector for eigenvalues of multiplicity 2.

Eigenvalues of Multiplicity 2 with Only One Eigenvector

Let $\mathbf{Y}' = \mathbf{AY}$ be a homogeneous system with an eigenvalue λ of multiplicity 2 that generates only one eigenvector $\mathbf{V}$. To find a second linearly independent solution of the system we assume the existence of a solution of the form

$$\mathbf{Y}_2 = (\mathbf{C} + \mathbf{D}t)e^{\lambda t}$$

where $\mathbf{C}$ and $\mathbf{D}$ are $n \times 1$ vectors to be determined. (Don't forget that $\mathbf{Y}_1 = \mathbf{V}e^{\lambda t}$ is a solution of the system.) To find vectors $\mathbf{C}$ and $\mathbf{D}$ such that $\mathbf{Y}_2$ will be a linearly independent solution of $\mathbf{Y}' = \mathbf{AY}$, we substitute $\mathbf{Y}_2 = (\mathbf{C} + \mathbf{D}t)e^{\lambda t}$ into the given system. Observing that $\mathbf{Y}_2'$ is given by

$$\mathbf{Y}_2' = \lambda(\mathbf{C} + \mathbf{D}t)e^{\lambda t} + \mathbf{D}e^{\lambda t}$$

the substitution of $\mathbf{Y}_2$ into $\mathbf{Y}' = \mathbf{AY}$ yields

$$\lambda(\mathbf{C} + \mathbf{D}t)e^{\lambda t} + \mathbf{D}e^{\lambda t} = \mathbf{A}(\mathbf{C} + \mathbf{D}t)e^{\lambda t}$$

Since $e^{\lambda t} \neq 0$, we can write this equation as

$$\lambda\mathbf{C} + \lambda t\mathbf{D} + \mathbf{D} = \mathbf{AC} + t\mathbf{AD} \tag{9–15}$$

Equating the coefficients of t, we get the equation

$$\lambda\mathbf{D} = \mathbf{AD}$$

which can be written as

$$(\mathbf{A} - \lambda\mathbf{I})\mathbf{D} = \mathbf{0}$$

This equation states that $\mathbf{D}$ is an eigenvector of $\mathbf{A}$ associated with the eigenvalue λ. Since we know one eigenvector for λ is $\mathbf{V}$, we choose $\mathbf{D} = \mathbf{V}$.

Similarly, equating the constant matrices in Equation 9–15 yields

$$\lambda\mathbf{C} + \mathbf{D} = \mathbf{AC}$$

which can be written as

$$(\mathbf{A} - \lambda\mathbf{I})\mathbf{C} = \mathbf{D}$$

Using the fact that $\mathbf{D} = \mathbf{V}$ in this equation we can solve for $\mathbf{C}$ since λ and $\mathbf{A}$ are known.

The preceding discussion is summarized by the following rule for writing linearly independent solutions of $\mathbf{Y}' = \mathbf{AY}$ when a real eigenvalue of multiplicity 2 exists that generates only one eigenvector.

SOLUTIONS OF Y' = AY: REAL EIGENVALUES OF MULTIPLICITY 2

Let $\mathbf{Y}' = \mathbf{AY}$ be a homogeneous system with a real $n \times n$ coefficient matrix $\mathbf{A}$ and let $\lambda_1 = \lambda_2 = \lambda$ be a real eigenvalue of multiplicity 2 with a single eigenvector $\mathbf{V}$. Then,

$$\mathbf{Y}_1 = \mathbf{V}e^{\lambda t} \qquad (9\text{–}16)$$

and

$$\mathbf{Y}_2 = (\mathbf{C} + t\mathbf{V})e^{\lambda t}, \qquad (9\text{–}17)$$

where $\mathbf{C}$ is an $n \times 1$ solution vector of $(\mathbf{A} - \lambda\mathbf{I})\mathbf{C} = \mathbf{V}$, are linearly independent solutions of $\mathbf{Y}' = \mathbf{AY}$.

COMMENT: The procedure described for eigenvalues of multiplicity 2 can be extended to eigenvalues of a higher multiplicity. For example, if λ is of multiplicity 3 and generates only one eigenvector $\mathbf{V}$, then three linearly independent solutions of the system $\mathbf{Y}' = \mathbf{AY}$ are

$$\mathbf{Y}_1 = \mathbf{V}e^{\lambda t}$$
$$\mathbf{Y}_2 = (\mathbf{C} + \mathbf{D}t)e^{\lambda t}$$
$$\mathbf{Y}_3 = (\mathbf{C} + \mathbf{D}t + \mathbf{E}t^2)e^{\lambda t}$$

where $\mathbf{C}$, $\mathbf{D}$, and $\mathbf{E}$ are constant vectors to be determined by substitution into the given system.

EXERCISES FOR SECTION 9–6

Find the general solution for each of the first-order homogeneous linear differential systems in Exercises 1–10.

1. $y_1' = 6y_1 + 8y_2$
$\quad y_2' = -y_1 + 2y_2$

2. $\mathbf{Y}' = \mathbf{AY}$ for $\mathbf{A} = \begin{pmatrix} 2 & -1 \\ 9 & 2 \end{pmatrix}$

3. $\mathbf{Y}' = \mathbf{AY}$ for $\mathbf{A} = \begin{pmatrix} 3 & -18 \\ 2 & -9 \end{pmatrix}$

4. $y_1' = 2y_1 - y_2$
$\quad y_2' = 4y_1 + 6y_2$

5. $y_1' = y_1 + 2y_2$
$\quad y_2' = -\frac{1}{2}y_1 + y_2$

6. $\mathbf{Y}' = \mathbf{AY}$ for $\mathbf{A} = \begin{pmatrix} 1 & 0 & 1 \\ 1 & 1 & 0 \\ -2 & 0 & -1 \end{pmatrix}$

7. $\mathbf{Y}' = \mathbf{AY}$ for $\mathbf{A} = \begin{pmatrix} 5 & -5 & -5 \\ -1 & 4 & 2 \\ 3 & -5 & -3 \end{pmatrix}$

8. $y_1' = 5y_1 + 2y_2 + 2y_3$
$y_2' = 2y_1 + 2y_2 - 4y_3$
$y_3' = 2y_1 - 4y_2 + 2y_3$

9. $y_1' = -3y_1 - 3y_3$
$y_2' = 3y_1 - 2y_2 + y_3$
$y_3' = y_1 + y_3$

10. $y_1' = y_2$
$y_2' = y_3$
$y_3' = y_1$

In each of Exercises 11–14 re-express the given system of equations in the form $\mathbf{Y}' = \mathbf{AY}$ and then determine the general solution.

11. $4y_1' + y_1 + 2y_2' + 7y_2 = 0$
$y_1' - y_1 + y_2' + y_2 = 0$

12. $y_1' + y_1 + 2y_2' + 3y_2 = 0$
$y_1' - 2y_1 + 5y_2' = 0$

13. $y_1' + 2y_1 + y_2' + 3y_2 = 0$
$y_1' + 3y_2' + y_2 = 0$

14. $y_1' - y_1 + 3y_2 - y_3 = 0$
$y_2' - 2y_2 - y_3' + 2y_3 = 0$
$2y_1 - 3y_2 - y_3' = 0$

15. Solve the inital-value problem.

$y_1' = 3y_1 - y_2$ $y_1(0) = 1$
$y_2' = -4y_1 + y_2 + 2y_3$ $y_2(0) = 3$
$y_3' = 8y_1 - 3y_2$ $y_3(0) = 1$

16. Solve the initial-value problem.

$$\mathbf{Y}' = \mathbf{AY} \quad \text{for} \quad \mathbf{A} = \begin{pmatrix} 0 & 4 & 0 \\ -1 & 0 & 0 \\ 1 & 4 & -1 \end{pmatrix} \quad \text{and} \quad \mathbf{Y}(0) = \begin{pmatrix} -2 \\ 0 \\ 0 \end{pmatrix}$$

9–7 NONHOMOGENEOUS SYSTEMS

Recall that the first-order nonhomogeneous system of differential equations,

$y_1' = a_{11}y_1 + a_{12}y_2 + \cdots + a_{1n}y_n + f_1(t)$
$y_2' = a_{21}y_1 + a_{22}y_2 + \cdots + a_{2n}y_n + f_2(t)$
$\vdots \qquad\qquad\qquad\qquad \vdots$
$y_n' = a_{n1}y_1 + a_{n2}y_2 + \cdots + a_{nn}y_n + f_n(t)$

is represented by the matrix equation

$$\mathbf{Y}' = \mathbf{AY} + \mathbf{F}$$

where $\mathbf{F} \neq \mathbf{0}$. It is easy to show that the general solution of this nonhomogeneous system takes the form

$$\mathbf{Y} = \mathbf{Y}_c + \mathbf{Y}_p$$

where $\mathbf{Y}_c$ is the general solution of the corresponding homogeneous system $\mathbf{Y}' = \mathbf{AY}$ and $\mathbf{Y}_p$ is any particular solution vector of $\mathbf{Y}' = \mathbf{AY} + \mathbf{F}$. Notice that the form of the general solution of a nonhomogeneous system parallels that of nonhomogeneous differential equations in one variable (see Section 5–1).

We find $\mathbf{Y}_c$ using the method just discussed for solving $\mathbf{Y}' = \mathbf{AY}$. To find a particular solution $\mathbf{Y}_p$, we use either the method of undetermined coefficients or the method of variation of parameters.

Method of Undetermined Coefficients

As in Section 5–2, we assume that the solution vector $\mathbf{Y}_p$ has the same general form as $\mathbf{F}$. The assumed form of $\mathbf{Y}_p$ is then substituted into $\mathbf{Y}' = \mathbf{AY} + \mathbf{F}$ in order to solve for the undetermined coefficients. Because of the complexities involved when a function of $\mathbf{Y}_p$ occurs in $\mathbf{Y}_c$, we will not discuss such systems in general. Example 1(d) gives an approach that may be used in a specific case, however.

Example 1

Given the nonhomogeneous system $\mathbf{Y}' = \mathbf{AY} + \mathbf{F}$, where

$$\mathbf{A} = \begin{pmatrix} 1 & 0 \\ 6 & -1 \end{pmatrix}$$

solve this system for each of the following driving vectors.

(a) $\mathbf{F}(t) = \begin{pmatrix} 2 \\ 1 \end{pmatrix}$ (b) $\mathbf{F}(t) = \begin{pmatrix} 2 \\ 1 \end{pmatrix} e^{2t}$

(c) $\mathbf{F}(t) = \begin{pmatrix} 2 \\ 1 \end{pmatrix} \sin t$ (d) $\mathbf{F}(t) = \begin{pmatrix} 2 \\ 1 \end{pmatrix} e^{t}$

SOLUTION First, solve the corresponding homogeneous system $\mathbf{Y}' = \mathbf{AY}$. The characteristic equation is

$$\begin{vmatrix} 1 - \lambda & 0 \\ 6 & -1 - \lambda \end{vmatrix} = 1 - \lambda^2 = 0$$

The eigenvalues are $\lambda = \pm 1$. It is easy to show that $\begin{pmatrix} 1 \\ 3 \end{pmatrix}$ is an eigenvector for $\lambda = 1$ and $\begin{pmatrix} 0 \\ 1 \end{pmatrix}$ is an eigenvector for $\lambda = -1$. Thus, $\mathbf{Y}_c$ is given by

$$\mathbf{Y}_c = c_1 \begin{pmatrix} 1 \\ 3 \end{pmatrix} e^{t} + c_2 \begin{pmatrix} 0 \\ 1 \end{pmatrix} e^{-t}$$

We now proceed to find the $\mathbf{Y}_p$ for each of the given cases.

(a) $\mathbf{F}(t) = \begin{pmatrix} 2 \\ 1 \end{pmatrix}$. In this case, we choose $\mathbf{Y}_p$ to have the form of a constant vector

$$\mathbf{Y}_p = \mathbf{P} = \begin{pmatrix} p_1 \\ p_2 \end{pmatrix}$$

Substituting $\mathbf{P}$ into $\mathbf{Y}' = \mathbf{AY} + \mathbf{F}$, we obtain

$$\begin{pmatrix} 0 \\ 0 \end{pmatrix} = \begin{pmatrix} 1 & 0 \\ 6 & -1 \end{pmatrix} \begin{pmatrix} p_1 \\ p_2 \end{pmatrix} + \begin{pmatrix} 2 \\ 1 \end{pmatrix}$$

from which we get the system

$$0 = p_1 + 2$$
$$0 = 6p_1 - p_2 + 1$$

This gives $p_1 = -2$, $p_2 = -11$, and $\mathbf{Y}_p = -\begin{pmatrix} 2 \\ 11 \end{pmatrix}$.

Therefore, the general solution in this case is

$$\mathbf{Y} = c_1 \begin{pmatrix} 1 \\ 3 \end{pmatrix} e^t + c_2 \begin{pmatrix} 0 \\ 1 \end{pmatrix} e^{-t} - \begin{pmatrix} 2 \\ 11 \end{pmatrix}$$

(b) $\mathbf{F}(t) = \begin{pmatrix} 2 \\ 1 \end{pmatrix} e^{2t}$. We choose

$$\mathbf{Y}_p = \mathbf{P}e^{2t} = \begin{pmatrix} p_1 \\ p_2 \end{pmatrix} e^{2t}$$

Substituting into the given system, we get

$$2e^{2t} \begin{pmatrix} p_1 \\ p_2 \end{pmatrix} = \begin{pmatrix} p_1 \\ 6p_1 - p_2 \end{pmatrix} e^{2t} + \begin{pmatrix} 2 \\ 1 \end{pmatrix} e^{2t}$$

This gives the system

$$2p_1 = p_1 + 2$$
$$2p_2 = 6p_1 - p_2 + 1$$

from which

$$p_1 = 2, \quad p_2 = \tfrac{13}{3}$$

A particular solution is then

$$\mathbf{Y}_p = \begin{pmatrix} 2 \\ \tfrac{13}{3} \end{pmatrix} e^{2t}$$

and the general solution is

$$\mathbf{Y} = c_1 \begin{pmatrix} 1 \\ 3 \end{pmatrix} e^t + c_2 \begin{pmatrix} 0 \\ 1 \end{pmatrix} e^{-t} + \begin{pmatrix} 2 \\ \tfrac{13}{3} \end{pmatrix} e^{2t}$$

(c) $\mathbf{F}(t) = \begin{pmatrix} 2 \\ 1 \end{pmatrix} \sin t$. Choose $\mathbf{Y}_p$ to be the vector generalization of $\sin t$ which is $\mathbf{P} \sin t + \mathbf{Q} \cos t$. Substituting into the differential equation,

$$\mathbf{P} \cos t - \mathbf{Q} \sin t = \begin{pmatrix} 1 & 0 \\ 6 & -1 \end{pmatrix} \mathbf{P} \sin t + \begin{pmatrix} 1 & 0 \\ 6 & -1 \end{pmatrix} \mathbf{Q} \cos t + \begin{pmatrix} 2 \\ 1 \end{pmatrix} \sin t$$

Equating first components, we get

$$p_1 \cos t - q_1 \sin t = p_1 \sin t + q_1 \cos t + 2 \sin t$$

Equating second components, we get

$$p_2 \cos t - q_2 \sin t = 6p_1 \sin t - p_2 \sin t + 6q_1 \cos t - q_2 \cos t + \sin t$$

In each of these equations we equate the coefficients of $\cos t$ to obtain

$$p_1 = q_1; \quad p_2 = 6q_1 - q_2$$

Similarly, equating the coefficients of $\sin t$ yields

$$-q_1 = p_1 + 2; \quad -q_2 = 6p_1 - p_2 + 1$$

These four equations can be solved to yield the components of $\mathbf{P}$ and $\mathbf{Q}$. Thus, $p_1 = q_1 = -1, p_2 = -\frac{11}{2}$, and $q_2 = -\frac{1}{2}$, and the desired particular solution is

$$\mathbf{Y}_p = -\begin{pmatrix} 1 \\ \frac{11}{2} \end{pmatrix} \sin t - \begin{pmatrix} 1 \\ \frac{1}{2} \end{pmatrix} \cos t$$

The general solution is then

$$\mathbf{Y} = c_1 \begin{pmatrix} 1 \\ 3 \end{pmatrix} e^t + c_2 \begin{pmatrix} 0 \\ 1 \end{pmatrix} e^{-t} - \begin{pmatrix} 1 \\ \frac{11}{2} \end{pmatrix} \sin t - \begin{pmatrix} 1 \\ \frac{1}{2} \end{pmatrix} \cos t$$

(d) $\mathbf{F}(t) = \begin{pmatrix} 2 \\ 1 \end{pmatrix} e^t$. If we substitute $\mathbf{Y}_p = \mathbf{P}e^t$ into the given system, we get

$$\begin{pmatrix} p_1 \\ p_2 \end{pmatrix} e^t = \begin{pmatrix} 1 & 0 \\ 6 & -1 \end{pmatrix} \begin{pmatrix} p_1 \\ p_2 \end{pmatrix} e^t + \begin{pmatrix} 2 \\ 1 \end{pmatrix} e^t$$

The equivalent system is

$$p_1 = p_1 + 2$$
$$p_2 = 6p_1 - p_2 + 1$$

This system is inconsistent, which means there is no solution, so our choice for $\mathbf{Y}_p$ has failed to produce a particular solution of $\mathbf{Y}' = \mathbf{AY} + \mathbf{F}$. The problem is that a term of the form $\mathbf{P}e^t$ is contained in $\mathbf{Y}_c$ and therefore the assumption that $\mathbf{Y}_p = \mathbf{P}e^t$ fails to produce a linearly independent solution. To generate a linearly independent $\mathbf{Y}_p$, multiply e^t by a first-order vector polynomial of the form $\mathbf{P} + \mathbf{Q}t$, where $\mathbf{P}$ and $\mathbf{Q}$ are 2×1 constant vectors to be determined. Substituting $\mathbf{Y}_p = (\mathbf{P} + \mathbf{Q}t) e^t$ into the given system yields

$$(\mathbf{P} + \mathbf{Q}t)e^t + \mathbf{Q}e^t = \begin{pmatrix} 1 & 0 \\ 6 & -1 \end{pmatrix}(\mathbf{P} + \mathbf{Q}t)e^t + \begin{pmatrix} 2 \\ 1 \end{pmatrix}e^t$$

Or, since $e^t \neq 0$,

$$\begin{pmatrix} p_1 \\ p_2 \end{pmatrix} + \begin{pmatrix} q_1 \\ q_2 \end{pmatrix}t + \begin{pmatrix} q_1 \\ q_2 \end{pmatrix} = \begin{pmatrix} 1 & 0 \\ 6 & -1 \end{pmatrix}\begin{pmatrix} p_1 + q_1 t \\ p_2 + q_2 t \end{pmatrix} + \begin{pmatrix} 2 \\ 1 \end{pmatrix}$$

$$\begin{pmatrix} p_1 \\ p_2 \end{pmatrix} + \begin{pmatrix} q_i \\ q_2 \end{pmatrix}t + \begin{pmatrix} q_1 \\ q_2 \end{pmatrix} = \begin{pmatrix} p_1 \\ 6p_1 - p_2 \end{pmatrix} + \begin{pmatrix} q_1 \\ 6q_1 - q_2 \end{pmatrix}t + \begin{pmatrix} 2 \\ 1 \end{pmatrix} \qquad (9\text{--}18)$$

Equating coefficients of t in Equation 9–18 yields the system

$$q_1 = q_1$$
$$q_2 = 6q_1 - q_2$$

And, equating the constant vectors in Equation 9–18 yields the system

$$p_1 + q_1 = p_1 + 2$$
$$p_2 + q_2 = 6p_1 - p_2 + 1$$

Choosing $p_1 = 1$ gives $q_1 = 2$, $q_2 = 6$, and $p_2 = \frac{1}{2}$. Substituting these values into $\mathbf{Y}_p = (\mathbf{P} + \mathbf{Q}t)e^t$, a particular solution is

$$\mathbf{Y}_p = \left[\begin{pmatrix} 1 \\ \frac{1}{2} \end{pmatrix} + \begin{pmatrix} 2 \\ 6 \end{pmatrix}t \right]e^t$$

The general solution for this system is then

$$\mathbf{Y} = c_1\begin{pmatrix} 1 \\ 3 \end{pmatrix}e^t + c_2\begin{pmatrix} 0 \\ 1 \end{pmatrix}e^{-t} + \begin{pmatrix} 1 \\ \frac{1}{2} \end{pmatrix}e^t + \begin{pmatrix} 2 \\ 6 \end{pmatrix}te^t$$

Notice that different choices of values for p_1 lead to differrent forms of $\mathbf{Y}_p$, but in any case, the general solution will be $\mathbf{Y}_c + \mathbf{Y}_p$.   ■

COMMENT: The superposition principle *assures us that if* $\mathbf{Y}_{p1}$ *and* $\mathbf{Y}_{p2}$ *are particular solutions of* $\mathbf{Y}' = \mathbf{A}\mathbf{Y} + \mathbf{F}_1$ *and* $\mathbf{Y}' = \mathbf{A}\mathbf{Y} + \mathbf{F}_2$ *respectively, then*

$$\mathbf{Y}_p = \mathbf{Y}_{p1} + \mathbf{Y}_{p2}$$

is a particular solution of

$$\mathbf{Y}' = \mathbf{A}\mathbf{Y} + \mathbf{F}_1 + \mathbf{F}_2$$

Example 2

Use the results of Example 1 to find a particular solution of the nonhomogeneous system

$$\mathbf{Y}' = \begin{pmatrix} 1 & 0 \\ 6 & -1 \end{pmatrix}\mathbf{Y} + \begin{pmatrix} 2 \\ 1 \end{pmatrix} + \begin{pmatrix} 2 \\ 1 \end{pmatrix}e^t$$

SOLUTION Recall from Example 1(a) that

$$\mathbf{Y}_p = -\begin{pmatrix} 2 \\ 11 \end{pmatrix}$$

is a particular solution of

$$\mathbf{Y}' = \begin{pmatrix} 1 & 0 \\ 6 & -1 \end{pmatrix}\mathbf{Y} + \begin{pmatrix} 2 \\ 1 \end{pmatrix}$$

Also, in Example 1(b) we found that

$$\mathbf{Y}_p = \begin{pmatrix} 2 \\ \frac{13}{3} \end{pmatrix}e^{2t}$$

is a particular solution of

$$\mathbf{Y}' = \begin{pmatrix} 1 & 0 \\ 6 & -1 \end{pmatrix}\mathbf{Y} + \begin{pmatrix} 2 \\ 1 \end{pmatrix}e^{2t}$$

Then, by the superposition principle, we know that

$$\mathbf{Y}_p = -\begin{pmatrix} 2 \\ 11 \end{pmatrix} + \begin{pmatrix} 2 \\ \frac{13}{3} \end{pmatrix}e^{2t}$$

is a particular solution of

$$\mathbf{Y}' = \begin{pmatrix} 1 & 0 \\ 6 & -1 \end{pmatrix}\mathbf{Y} + \begin{pmatrix} 2 \\ 1 \end{pmatrix} + \begin{pmatrix} 2 \\ 1 \end{pmatrix}e^{2t}$$ ∎

Example 3

Solve the initial-value problem

$$y_1' + y_1 + 3y_2' = 1, \quad y_1(0) = 0$$
$$3y_1 + y_2' + 2y_2 = t, \quad y_2(0) = 0$$

SOLUTION The given system is equivalent to

$$\mathbf{Y}' = \mathbf{AY} + \mathbf{F}$$

with $\mathbf{A} = \begin{pmatrix} 8 & 6 \\ -3 & -2 \end{pmatrix}$ and $\mathbf{F} = \begin{pmatrix} 1 - 3t \\ t \end{pmatrix}$. From Example 5 of Section 9–5, the

solution of the homogeneous system is

$$\mathbf{Y}_c(t) = c_1 \begin{pmatrix} -6 \\ 5 - \sqrt{7} \end{pmatrix}e^{(3+\sqrt{7})t} + c_2 \begin{pmatrix} -6 \\ 5 + \sqrt{7} \end{pmatrix}e^{(3-\sqrt{7})t}$$

As a particular solution of the nonhomogeneous equation, we try

$$\mathbf{Y}_p(t) = (\mathbf{P}t + \mathbf{Q})$$

Substitution and matching of like terms then gives us

$$Q = \begin{pmatrix} -\frac{1}{2} \\ \frac{1}{2} \end{pmatrix}, \; P = \begin{pmatrix} 0 \\ \frac{1}{2} \end{pmatrix}, \text{ and } Y_p(t) = \begin{pmatrix} -\frac{1}{2} \\ \frac{1}{2}(t + 1) \end{pmatrix}$$

When we add together $Y_c(t)$ and $Y_p(t)$ and apply the initial conditions, we find $c_1 = (\sqrt{7} - 7)/168$ and $c_2 = -(7 + \sqrt{7})/168$. The desired solution is

$$Y(t) = \frac{1}{28}\begin{pmatrix} 7 - \sqrt{7} \\ -7 + 2\sqrt{7} \end{pmatrix}e^{(3+\sqrt{7})t} + \frac{1}{28}\begin{pmatrix} 7 + \sqrt{7} \\ -7 - 2\sqrt{7} \end{pmatrix}e^{(3-\sqrt{7})t} + \frac{1}{2}\begin{pmatrix} -1 \\ t + 1 \end{pmatrix}$$

∎

Method of Variation of Parameters

Recall from Section 9–4 that the solution vector of the homogeneous system $Y' = AY$ can be written in the form $Y = \Phi C$, where C is an $n \times 1$ column vector of arbitrary constants and Φ is a fundamental matrix. Understanding this, we assume that Y_p can be expressed as

$$Y_p = \Phi U$$

where U is an $n \times 1$ vector function to be determined through substitution of Y_p into the nonhomogeneous system $Y' = AY + F$. Carrying out this substitution, we get

$$(\Phi U)' = A\Phi U + F$$

or, by expanding the left side,

$$\Phi' U + \Phi U' = A\Phi U + F$$

Since Φ is a fundamental matrix of solutions it follows that $\Phi' = A\Phi$, so that this equation reduces to

$$\Phi U' = F$$

Multiplying both sides by Φ^{-1} yields

$$U' = \Phi^{-1}F$$

Integration of this equation gives the vector function U; that is,

$$U = \int \Phi^{-1}F \, dt \qquad (9–19)$$

where the integration of $\Phi^{-1}F$ takes place on each entry of the matrix. The general solution of $Y' = AY + F$ is then $Y = Y_c + Y_p$, or

$$Y = \Phi C + \Phi \int \Phi^{-1}F \, dt \qquad (9–20)$$

Example 4

Solve the nonhomogeneous system

$$\mathbf{Y}' = \begin{pmatrix} 1 & 0 \\ 6 & -1 \end{pmatrix} \mathbf{Y} + \begin{pmatrix} e^t \\ t \end{pmatrix}$$

SOLUTION   The general solution of the corresponding homogeneous system is

$$\mathbf{Y}_c = c_1 \begin{pmatrix} 1 \\ 3 \end{pmatrix} e^t + c_2 \begin{pmatrix} 0 \\ 1 \end{pmatrix} e^{-t}$$

This can be verified by referring to Example 1. The fundamental matrix Φ is then

$$\Phi = \begin{pmatrix} e^t & 0 \\ 3e^t & e^{-t} \end{pmatrix}$$

The inverse of Φ is given by

$$\Phi^{-1} = \begin{pmatrix} e^{-t} & 0 \\ -3e^t & e^t \end{pmatrix} \quad \begin{array}{l} \textit{(See the Appendix for a review of how to com-} \\ \textit{pute the inverse of a matrix.)} \end{array}$$

and

$$\Phi^{-1}\mathbf{F} = \begin{pmatrix} e^{-t} & 0 \\ -3e^t & e^t \end{pmatrix}\begin{pmatrix} e^t \\ t \end{pmatrix} = \begin{pmatrix} 1 \\ -3e^{2t} + te^t \end{pmatrix}$$

We then compute $\mathbf{U}$ by Formula 9–19; that is,

$$\mathbf{U} = \int \Phi^{-1}\mathbf{F}\,dt = \begin{pmatrix} t \\ -\dfrac{3}{2}e^{2t} + te^t - e^t \end{pmatrix}$$

Substituting into $\mathbf{Y}_p = \Phi\mathbf{U}$ yields

$$\mathbf{Y}_p = \Phi\mathbf{U} = \begin{pmatrix} e^t & 0 \\ 3e^t & e^{-t} \end{pmatrix}\begin{pmatrix} t \\ -\frac{3}{2}e^{2t} + te^t - e^t \end{pmatrix} = \begin{pmatrix} te^t \\ 3te^t - \frac{3}{2}te^t + t - 1 \end{pmatrix}$$

$$= \begin{pmatrix} 1 \\ 1 \end{pmatrix} te^t + \begin{pmatrix} 0 \\ -\frac{3}{2} \end{pmatrix} e^t + \begin{pmatrix} 0 \\ 1 \end{pmatrix} t + \begin{pmatrix} 0 \\ -1 \end{pmatrix}$$

Finally, the general solution is

$$\mathbf{Y} = \mathbf{Y}_c + \mathbf{Y}_p = c_1 \begin{pmatrix} 1 \\ 3 \end{pmatrix} e^t + c_2 \begin{pmatrix} 0 \\ 1 \end{pmatrix} e^{-t} + \begin{pmatrix} 1 \\ 1 \end{pmatrix} te^t + \begin{pmatrix} 0 \\ -\frac{3}{2} \end{pmatrix} e^t + \begin{pmatrix} 0 \\ 1 \end{pmatrix} t + \begin{pmatrix} 0 \\ -1 \end{pmatrix}$$

∎

Example 5

Find the solution of $\mathbf{Y}' = \begin{pmatrix} 3 & 2 \\ 1 & 2 \end{pmatrix}\mathbf{Y} + \begin{pmatrix} 4e^{5t} \\ 0 \end{pmatrix}$, where $\mathbf{Y}(0) = \begin{pmatrix} 1 \\ -1 \end{pmatrix}$

SOLUTION $\det(\mathbf{A} - \lambda\mathbf{I}) = \lambda^2 - 5\lambda + 4 = (\lambda - 4)(\lambda - 1)$, so the eigenvalues are $\lambda = 1$ and 4. The solution $\mathbf{Y}_1 = \begin{pmatrix} 2 \\ 1 \end{pmatrix} e^{4t}$ is obtained from $\lambda = 4$ and the

solution $\mathbf{Y}_2 = \begin{pmatrix} 1 \\ -1 \end{pmatrix} e^t$ is obtained from $\lambda = 1$. Hence, a fundamental matrix for the given system is

$$\Phi = \begin{pmatrix} 2e^{4t} & e^t \\ e^{4t} & -e^t \end{pmatrix}$$

The inverse of Φ is then

$$\Phi^{-1} = \frac{1}{-3e^{5t}} \begin{pmatrix} -e^t & -e^t \\ -e^{4t} & 2e^{4t} \end{pmatrix} = \frac{1}{3} \begin{pmatrix} e^{-4t} & e^{-4t} \\ e^{-t} & -2e^{-t} \end{pmatrix}$$

And

$$\Phi^{-1}\mathbf{F} = \frac{4}{3} \begin{pmatrix} e^t \\ e^{4t} \end{pmatrix}$$

Integrating

$$\int \Phi^{-1}\mathbf{F}dt = \begin{pmatrix} \frac{4}{3}e^t \\ \frac{1}{3}e^{4t} \end{pmatrix}$$

Therefore,

$$\mathbf{Y}_p = \Phi \int \Phi^{-1}\mathbf{F}\, dt$$

$$= \begin{pmatrix} 2e^{4t} & e^t \\ e^{4t} & -e^t \end{pmatrix} \begin{pmatrix} \frac{4}{3}e^t \\ \frac{1}{3}e^{4t} \end{pmatrix}$$

$$= \begin{pmatrix} 3e^{5t} \\ e^{5t} \end{pmatrix} = \begin{pmatrix} 3 \\ 1 \end{pmatrix} e^{5t}$$

The general solution is

$$\mathbf{Y} = c_1 \begin{pmatrix} 2 \\ 1 \end{pmatrix} e^{4t} + c_2 \begin{pmatrix} 1 \\ -1 \end{pmatrix} e^t + \begin{pmatrix} 3 \\ 1 \end{pmatrix} e^{5t}$$

Imposing the initial condition, $\mathbf{Y}(0) = \begin{pmatrix} 1 \\ -1 \end{pmatrix}$,

$$\begin{pmatrix} 1 \\ -1 \end{pmatrix} = c_1 \begin{pmatrix} 2 \\ 1 \end{pmatrix} + c_2 \begin{pmatrix} 1 \\ -1 \end{pmatrix} + \begin{pmatrix} 3 \\ 1 \end{pmatrix}$$

from which it is easy to obtain $c_1 = -\frac{4}{3}$ and $c_2 = \frac{2}{3}$. Hence, the solution is

$$\mathbf{Y} = -\frac{4}{3} \begin{pmatrix} 2 \\ 1 \end{pmatrix} e^{4t} + \frac{2}{3} \begin{pmatrix} 1 \\ -1 \end{pmatrix} e^t + \begin{pmatrix} 3 \\ 1 \end{pmatrix} e^{5t}$$

$$= \begin{pmatrix} -\frac{8}{3}e^{4t} + \frac{2}{3}e^t + 3e^{5t} \\ -\frac{4}{3}e^{4t} - \frac{2}{3}e^t + e^{5t} \end{pmatrix}$$

■

COMMENT: In the scalar case, the method of undetermined coefficients is almost always the method of choice, if it is applicable. The
same is not true for systems because the method of undetermined coefficients often may lead to large algebraic systems quickly making it a more
unwieldy technique than the variation of parameters method.

EXERCISES FOR SECTION 9–7

In Exercises 1–6 find the general solution of the given nonhomogeneous system. Use
the method of undetermined coefficients to find a particular solution.

1. $\mathbf{Y}' = \begin{pmatrix} 1 & 4 \\ 1 & 1 \end{pmatrix} \mathbf{Y} + \begin{pmatrix} 1 \\ 4 \end{pmatrix}$

2. $\mathbf{Y}' = \begin{pmatrix} 1 & 4 \\ 1 & 1 \end{pmatrix} \mathbf{Y} + \begin{pmatrix} 4e^t \\ 0 \end{pmatrix}$

3. $y_1' = y_1 + 4y_2$
$\quad y_2' = y_1 + y_2 + e^{-t}$

4. $\mathbf{Y}' = \begin{pmatrix} 1 & 0 & 1 \\ 1 & 1 & 0 \\ -2 & 0 & -1 \end{pmatrix} \mathbf{Y} + \begin{pmatrix} 2 \\ 1 \\ 0 \end{pmatrix}$

5. $\mathbf{Y}' = \begin{pmatrix} 1 & 0 & 1 \\ 1 & 1 & 0 \\ -2 & 0 & -1 \end{pmatrix} \mathbf{Y} + \begin{pmatrix} e^{-t} \\ 0 \\ 0 \end{pmatrix}$

6. $y_1' = y_1 + y_3$
$\quad y_2' = y_1 + y_2 + 2e^t$
$\quad y_3' = -2y_1 - y_3 - 2e^t$

In Exercises 7–11 find the general solution of the given nonhomogeneous system. Use
the method of variation of parameters to find a particular solution.

7. $y_1' = 2y_1 + y_2 + e^{5t}$
$\quad y_2' = -3y_1 + 6y_2 + e^{5t}$

8. $y_1' = 2y_1 + y_2 - e^t$
$\quad y_2' = 3y_1 + 4y_2 - 7e^t$

9. $\mathbf{Y}' = \mathbf{AY} + \mathbf{F}$ for $\mathbf{A} = \begin{pmatrix} 0 & 4 \\ -1 & 0 \end{pmatrix}$ and $\mathbf{F} = \begin{pmatrix} -3\cos t \\ 0 \end{pmatrix}$

10. $\mathbf{Y}' = \mathbf{AY} + \mathbf{F}$ for $\mathbf{A} = \begin{pmatrix} 2 & 2 \\ -3 & -3 \end{pmatrix}$ and $\mathbf{F} = \begin{pmatrix} 1 \\ 2t \end{pmatrix}$

11. $y_1' = \quad y_1 + y_2 + y_3 + 1$
$\quad y_2' = -y_2$
$\quad y_3' = -2y_1 - y_2 - 2y_3 + 2e^{-t}$

In Exercises 12–15 solve the given initial-value problem.

12. $y_1' = y_2 - 1 \qquad y_1(0) = 0$
$\quad\; y_2' = -y_1 + 1 \quad y_2(0) = 0$

13. $y_1' = y_2 + e^{3t} \qquad y_1(0) = 0$
$\quad\; y_2' = -2y_1 + 3y_2 \quad y_2(0) = 0$

14. $y_1' + 2y_1 + \;\; y_2' + \;\; y_2 = 0 \quad y_1(0) = 0$
$\quad\; y_1' + 4y_1 + 2y_2' + 3y_2 = 1 \quad y_2(0) = 0$

15. $\mathbf{Y}' = \mathbf{AY} + \mathbf{F} \qquad$ for $\mathbf{A} = \begin{pmatrix} 0 & 1 & 0 & 0 \\ 0 & 0 & 1 & 0 \\ 0 & 0 & 0 & 1 \\ 1 & 0 & 0 & 0 \end{pmatrix}$, $\mathbf{F} = \begin{pmatrix} t^2 \\ 0 \\ 0 \\ 0 \end{pmatrix}$, and $\mathbf{Y}(0) = \begin{pmatrix} 1 \\ 1 \\ 1 \\ -1 \end{pmatrix}$.

The system

$$V_1 c_1' = h(a - c_1) + k(c_2 - c_1)$$
$$V_2 c_2' = k(c_1 - c_2)$$

arises in the biomathematical analysis of the cellular diffusion of a substance dissolved in a surrounding blood supply. In this model it is assumed that the concentration of the substance in the blood remains at the constant value a and that the substance goes through a two-stage diffusion process, first into a membrane enclosing an individual cell and then into the cell itself. The membrane and cell volumes V_1 and V_2 are constant, whereas the concentrations $c_1(t)$ and $c_2(t)$ in the two volumes vary with time. Both h and k are positive constants.

Exercises 16–17 refer to this biomathematical model. For simplicity, assume $V_1 = V_2 = 1$ throughout.

16. Determine a fundamental matrix for this differential system.

17. Find the complete solution of the homogeneous system and investigate its behavior as $t \to \infty$.

9–8   APPLICATIONS

Electrical Circuits

The current in the electrical circuit shown in Figure 9–2 can be found by applying the elements of circuit analysis discussed in Chapters 2 and 4. Essential to the analysis is **Kirchhoff's law,** which states that *the sum of the voltage drops around any closed loop is zero.* In applying Kirchhoff's law, we use the fact that the voltage across an inductance is $v_L = L(dI/dt)$ and the voltage across a resistance is $v_R = IR$.

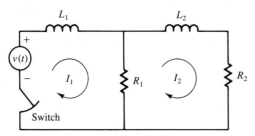

FIGURE 9–2

Let the current in the left loop of the circuit be I_1, and the current in the right loop be I_2. From the figure we conclude that the current in the resistor R_1 is $I_1 - I_2$ relative to the left loop, and $I_2 - I_1$ relative to the right loop. Applying Kirchhoff's law to the left loop, we get

$$L_1 \dot{I}_1 + R_1(I_1 - I_2) = v(t) \tag{9–21a}$$

where $\dot{I} = dI/dt$. Similarly, the sum of the voltage drops around the right loop yields

$$L_2\dot{I}_2 + R_2I_2 + R_1(I_2 - I_1) = 0 \tag{9–21b}$$

If the components of the circuit are given, the values of I_1 and I_2 can be found by solving this system of differential equations.

Example 1

Consider the circuit shown in Figure 9–2. Determine I_1 and I_2 when the switch is closed if $L_1 = L_2 = 2$ henrys, $R_1 = 3$ ohms, $R_2 = 8$ ohms, and $v(t) = 6$ volts. Assume the initial current in the circuit is zero.

SOLUTION   As previously noted, the circuit is described by the system

$$L_1\dot{I}_1 + R_1(I_1 - I_2) = v(t)$$

$$L_2\dot{I}_2 + R_2I_2 + R_1(I_2 - I_1) = 0$$

Substituting the given values yields the system

$$\left.\begin{array}{r} 2\dot{I}_1 + 3I_1 - 3I_2 = 6 \\ 2\dot{I}_2 + 11I_2 - 3I_1 = 0 \end{array}\right\} \quad I_1(0) = I_2(0) = 0$$

Writing the system in operator form, we have

$$(2D + 3)I_1 - 3I_2 = 6$$
$$-3I_1 + (2D + 11)I_2 = 0$$

Multiplying the first equation by 3 and applying the operator $2D + 3$ to the second and then adding the two equations, we obtain

$$(4D^2 + 28D + 24)I_2 = 18$$

Or dividing by 4,

$$(D^2 + 7D + 6)I_2 = \frac{9}{2}$$

The solution of this second-order nonhomogeneous linear differential equation consists of the sum of the general solution of the corresponding homogeneous equation and a particular solution of the given nonhomogeneous equation. The solution of $(D^2 + 7D + 6)I_2 = 0$ is $c_1e^{-t} + c_2e^{-6t}$. The method of undetermined coefficients can be used to show that $I_{p2} = \frac{3}{4}$ is a particular solution of the nonhomogeneous equation. Thus the expression for I_2 is

$$I_2 = c_1e^{-t} + c_2e^{-6t} + \frac{3}{4}$$

To find an expression for I_1, we substitute I_2 into the second equation of the system:

$$2(-c_1e^{-t} - 6c_2e^{-6t}) + 11\left(c_1e^{-t} + c_2e^{-6t} + \frac{3}{4}\right) - 3I_1 = 0$$

Solving this equation for I_1 yields

$$I_1 = 3c_1e^{-t} - \frac{1}{3}c_2e^{-6t} + \frac{11}{4}$$

Finally, using $I_1(0) = I_2(0) = 0$ in the expressions for I_1 and I_2, we get

$$0 = 3c_1 - \frac{1}{3}c_2 + \frac{11}{4} \qquad (I_1)$$

$$0 = c_1 + c_2 + \frac{3}{4} \qquad (I_2)$$

Solving this system, we get $c_1 = -\frac{9}{10}$ and $c_2 = \frac{3}{20}$, so the desired currents are

$$I_1 = -\frac{27}{10}e^{-t} - \frac{1}{20}e^{-6t} + \frac{11}{4}$$

$$I_2 = \frac{-9}{10}e^{-t} + \frac{3}{20}e^{-6t} + \frac{3}{4}$$ ■

Coupled Springs

Figure 9–3 shows a spring-mass diagram involving two springs and two masses. In this system the arrangement of springs makes the movements of the masses interdependent. We will derive the equations that describe the motion in Figure 9–3.

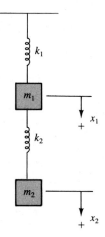

FIGURE 9–3

Consider two objects of mass m_1 and m_2, connected to springs with spring constants k_1 and k_2. The displacements from equilibrium are represented by x_1 and x_2, where x_1 and x_2 are both considered positive downward. In each case mass times acceleration is equal to the restoring force generated by the springs. Thus we may write the equations

$$m_1\ddot{x}_1 = -k_1 x_1 - k_2(x_1 - x_2)$$
$$m_2\ddot{x}_2 = -k_2(x_2 - x_1)$$

Rearranging these equations, we describe the spring-mass combination in Figure 9–3 by the system of differential equations

$$m_1\ddot{x}_1 + (k_1 + k_2)x_1 - k_2 x_2 = 0 \tag{9–22}$$
$$-k_2 x_1 + m_2\ddot{x}_2 + k_2 x_2 = 0$$

Example 2

Determine the motions of the two masses shown in Figure 9–3 if $m_1 = m_2 = 1$; $k_1 = 3$, $k_2 = 2$. Assume both masses are pulled down 1 unit from equilibrium and released with zero initial velocity.

SOLUTION   Using Equation 9–22, we have

$$\ddot{x}_1 + 5x_1 - 2x_2 = 0$$
$$-2x_1 + \ddot{x}_2 + 2x_2 = 0$$

Taking the Laplace transform of both sides of both equations yields

$$s^2 X_1 - s + 5X_1 - 2X_2 = 0$$
$$-2X_1 + s^2 X_2 - s + 2X_2 = 0$$

where $X_1 = \mathcal{L}\{x_1\}$ and $X_2 = \mathcal{L}\{x_2\}$. After simplifying, this becomes

$$X_1(s^2 + 5) - 2X_2 = s$$
$$-2X_1 + X_2(s^2 + 2) = s$$

Using Cramer's rule, X_1 is given by

$$X_1 = \frac{\begin{vmatrix} s & -2 \\ s & s^2 + 2 \end{vmatrix}}{\begin{vmatrix} s^2 + 5 & -2 \\ -2 & s^2 + 2 \end{vmatrix}} = \frac{s(s^2 + 4)}{(s^2 + 6)(s^2 + 1)}$$

which, by a partial-fractions expansion, becomes

$$X_1 = \frac{(3/5)s}{s^2 + 1} + \frac{(2/5)s}{s^2 + 6}$$

The expression for X_2 is

$$X_2 = \frac{\begin{vmatrix} s^2 + 5 & s \\ -2 & s \end{vmatrix}}{\begin{vmatrix} s^2 + 5 & -2 \\ -2 & s^2 + 2 \end{vmatrix}} = \frac{s(s^2 + 7)}{(s^2 + 1)(s^2 + 6)} = \frac{(6/5)s}{s^2 + 1} - \frac{(1/5)s}{s^2 + 6}$$

Taking the inverse transforms, we have the solution

$$x_1 = \frac{3}{5} \cos t + \frac{2}{5} \cos \sqrt{6}t$$

$$x_2 = \frac{6}{5} \cos t - \frac{1}{5} \cos \sqrt{6}t$$

Note that 2π is the period of $\cos t$ and $2\pi/\sqrt{6}$ is the period of $\cos \sqrt{6}t$. Thus the sum of these two functions is not periodic.   ■

Example 3

Determine expressions for the displacements of the two objects in the coupled spring-mass system shown in Figure 9–4.

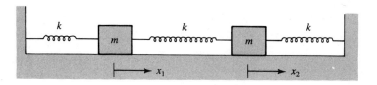

FIGURE 9–4

SOLUTION   Using Newton's second law, we find that

$$m\ddot{x}_1 = -kx_1 + k(x_2 - x_1)$$

for the object on the left and

$$m\ddot{x}_2 = -k(x_2 - x_1) - kx_2$$

for the one on the right. Writing this system in operator notation, we have

$$(mD^2 + 2k)x_1 - kx_2 = 0$$
$$-kx_1 + (mD^2 + 2k)x_2 = 0$$

Applying the operator $\frac{1}{k}(mD^2 + 2k)$ to the second equation and then adding the two equations of the system, we get

$$\left[\frac{1}{k}(mD^2 + 2k)^2 - k\right]x_2 = 0$$

Multiplying by k, this equation can be written as

$$(mD^2 + k)(mD^2 + 3k)x_2 = 0$$

The auxiliary equation for this homogeneous differential equation, which has the same form as the operator, has roots of $\pm\sqrt{k/m}i$ and $\pm\sqrt{3k/m}i$. The expression for x_2 is then

$$x_2 = c_1 \cos \omega t + c_2 \sin \omega t + c_3 \cos \sqrt{3}\,\omega t + c_4 \sin \sqrt{3}\,\omega t$$

where $\omega = \sqrt{k/m}$. Substituting x_2 into the first equation of the original system gives

$$x_1 = c_1 \cos \omega t + c_2 \sin \omega t - c_3 \cos \sqrt{3}\,\omega t - c_4 \sin \sqrt{3}\,\omega t$$

Thus we have two modes of vibration: $x_1 = x_2$ (the first two terms) and $x_1 = -x_2$ (the last two terms), each at different frequencies. ∎

Mixture Problems

Consider the two tanks shown in Figure 9–5, in which a salt solution of concentration c_i lb/gal is pumped into tank 1 at a rate g_i gal/min. A feedback loop interconnects both tanks so that solution is pumped from tank 1 to tank 2 at a rate of g_1 gal/min and from tank 2 to tank 1 at a rate of g_2 gal/min. Simultaneously, the solution in tank 2 is draining out at a rate of g_0 gal/min. The problem is to find the amount of salt in each tank.

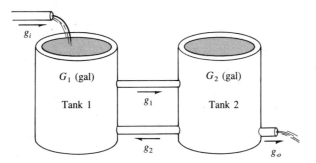

FIGURE 9–5

Let y_1 and y_2 represent the amount of salt (in pounds) in tanks 1 and 2 at any time and G_1 and G_2 represent the amount of liquid in the tank at any time. Then the concentration in each tank is given by

$$c_1 = \frac{y_1}{G_1} = \text{concentration of salt in tank 1}$$

$$c_2 = \frac{y_2}{G_2} = \text{concentration of salt in tank 2}$$

The rate at which the salt is changing in tank 1 is

$$\frac{dy_1}{dt} = c_i g_i - c_1 g_1 + c_2 g_2$$

$$= c_i g_i - \frac{y_1}{G_1} g_1 + \frac{y_2}{G_2} g_2 \tag{9-23}$$

Also, in tank 2 we have

$$\frac{dy_2}{dt} = \frac{y_1}{G_1} g_1 - \frac{y_2}{G_2} g_2 - \frac{y_2}{G_2} g_0 \tag{9-24}$$

Equations 9–23 and 9–24 form a system in y_1 and y_2. If G_{10} and G_{20} are the initial volumes of the two tanks, then

$$G_1 = G_{10} + g_i t - g_1 t + g_2 t \qquad G_2 = G_{20} - g_0 t - g_2 t + g_1 t$$

COMMENT: In many mixture problems the rates are assumed to be balanced; that is, $g_i = g_0$, $g_i + g_2 = g_1$, and $g_0 + g_2 = g_1$. In this problem we assume the volumes are constant; that is, $G_1 = G_{10}$ and $G_2 = G_{20}$.

Example 4

Assume both tanks in Figure 9–5 are filled with 100 gal of salt solutions of concentrations c_1 and c_2, respectively. Pure water is pumped into tank 1 at the rate of 5 gal/min. The solution is thoroughly mixed and pumped into and out of tank 2 at the rate of 5 gal/min. Assume no solution is pumped from tank 2 to tank 1. The system of equations describing this process is obtained by using Equations 9–23 and 9–24 with $c_i = 0$, $g_i = g_1 = g_0 = 5$, and $g_2 = 0$. Thus

$$\frac{dy_1}{dt} = -5\frac{y_1}{100}$$

$$\frac{dy_2}{dt} = 5\frac{y_1}{100} - 5\frac{y_2}{100}$$

The specific solution requires only that we know the initial concentrations.   ∎

EXERCISES FOR SECTION 9–8

1. Derive the system of equations for the current in the circuit shown in Figure 9–6.

2. At $t = 0$ the switch in Figure 9–6 is closed. Prior to $t = 0$, the currents in the circuit are zero. Determine the current in the circuit for $t > 0$, if $L_1 = L_2 = 1$, $R = 3$, $R_1 = 2$, and $v(t) = 12$.

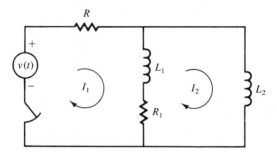

FIGURE 9–6

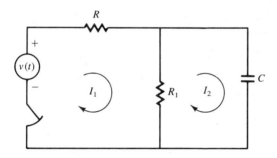

FIGURE 9–7

3. Referring to Figure 9–1, we assume that $L_1 = L_2 = 1$, $R_1 = 6$, $R = 1$, and $R_2 = 3$. Prior to $t = 0$, the generator has an output of 6 v; at $t = 0$ it is increased to 12 v. Determine the initial current and the current for $t > 0$.

4. Derive the system of differential equations for the system shown in Figure 9–7.

5. Solve the system derived in Exercise 4, given $v(t) = 6$, $R = R_1 = C = 1$. Assume the initial current in the circuit is zero.

6. Solve the system in Example 2 if $m_1 = m_2 = 1$, $k_1 = 3$, and $k_2 = 2$. Assume there is no initial displacement but that the initial velocities are each 1 ft/sec (downward).

7. Solve the system in Example 2 if $m_1 = m_2 = 1$, $k_1 = 8$, and $k_2 = 3$. Assume the initial velocities are zero but the initial displacement of m_1 is 1 ft and of m_2 is 2 ft.

8. Repeat Exercise 7, but assume the initial velocities are $v_1(0) = -2$ ft/sec and $v_2(0) = -3$ ft/sec.

9. Show that the system of coupled springs shown in Figure 9–8 is represented by

$$m_1 \ddot{x}_1 = -k_1 x_1 - k_2(x_1 - x_2)$$
$$m_2 \ddot{x}_2 = -k_3 x_2 - k_2(x_2 - x_1)$$

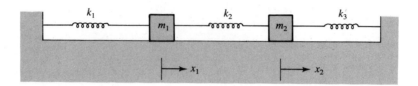

FIGURE 9–8

10. Determine the motion of the blocks in Exercise 9 if $m_1 = m_2 = 1$, $k_1 = 1$, $k_2 = 1$, $k_3 = 1$. Assume $x_1(0) = 1$, $\dot{x}_1(0) = 0$, $x_2(0) = -1$, $\dot{x}_2(0) = 0$.

Each of the mixture problems in Exercises 11–13 refers to the system of tanks shown in Figure 9–5.

11. Solve for the amount of salt at any time in two 100-gal tanks if the input is pure water and $g_i = g_0 = 5$, $g_1 = 7$, and $g_2 = 2$. Assume $y_1(0) = 10$ and $y_2(0) = 0$.

12. Solve for the amount of salt at any time in two 100-gal tanks if the input is a salt solution with a concentration of 2 lb/gal and $g_i = g_1 = g_0 = 3$ gal/min. Assume $g_2 = 0$ and $y_1(0) = y_2(0) = 0$.

13. Solve for the amount of salt at any time in two 100-gal tanks if the input is a salt solution with a concentration of 2 lb/gal, $g_i = g_0 = 5$, $g_1 = 7$, and $g_2 = 2$. Assume $y_1(0) = y_2(0) = 0$.

REVIEW EXERCISES FOR CHAPTER 9

In Exercises 1 and 2 solve the given system.

1. $y_1' = 2y_1 + y_2$
  $y_2' = y_1 - y_2$

2. $y_1' = y_1 - 3y_2$
  $y_2' = 2y_1 + 9y_2$

In Exercises 3–5 determine the number of arbitrary coefficients in the general solution of the system.

3. $y_1' - y_1 + y_2' + y_2 = x$
  $y_1' - 3y_1 + y_2' + y_2 = 1$

4. $y_1' + y_1 + y_2'' = 0$
  $y_1'' + y_2' - y_2 = x^2$

5. $y_1'' - y_1 + y_2''' = 0$
  $y_1' + y_1 + y_2'' = 0$

In Exercises 6–12 solve the given system.

6. $y_1' + y_1 + y_2' = x$
$\quad\quad y_1' + 2y_2 = 1$

7. $y_1' + 2y_2' = 2$
$\quad\quad y_1' - y_2' = x$

8. $2y_1' + y_1 - y_2' + y_2 = x$
$\quad\quad y_1' + 2y_1 - y_2' - y_2 = 0$

9. $y_1' - y_1 + y_2' = e^x$
$\quad\quad y_1' + y_2' + y_2 = x^2$

10. $y_1'' + y_1 + y_2'' + y_2 = 1$
$\quad\quad y_1'' + y_1 + y_2'' - y_2 = 0$

11. $\left. \begin{array}{l} \dot{y}_1 + \dot{y}_2 = 0 \\ 3\dot{y}_1 - 2\dot{y}_2 = 0 \end{array} \right\}$ $y_1(0) = y_2(0) = 1$

12. $\left. \begin{array}{l} \dot{y}_1 + \dot{y}_2 = t \\ 3\dot{y}_1 - 2\dot{y}_2 = 1 \end{array} \right\}$ $y_1(0) = y_2(0) = 1$

In Exercises 13–15 use the method of Laplace transforms to solve the given system.

13. $\left. \begin{array}{l} \ddot{y}_1 + 2y_1 + \ddot{y}_2 = 0 \\ y_1 - \dot{y}_2 = 0 \end{array} \right\}$ $y_1(0) = \dot{y}_1(0) = y_2(0) = \dot{y}_2(0) = 1$

14. $\left. \begin{array}{l} \ddot{y}_1 - \dot{y}_2 = u_1(t) \\ \dot{y}_1 - \dot{y}_2 = 0 \end{array} \right\}$ $y_1(0) = \dot{y}_1(0) = y_2(0) = 1$

15. $\left. \begin{array}{l} \ddot{y}_1 - 3y_1 + 2\dot{y}_2 = tu_1(t) \\ \dot{y}_1 + \dot{y}_2 = 0 \end{array} \right\}$ $y_1(0) = \dot{y}_1(0) = y_2(0) = 0$

Solve the systems in Exercises 16 and 17 by the method of Laplace transforms and determine any limitations on the initial conditions.

16. $y_1' + y_1 + y_2 = 0$
$\quad\quad y_1' + y_2' + 2y_2 = 0$

17. $y_1' + y_1 + y_2' = 0$
$\quad\quad y_1' + y_2' + 2y_2 = 0$

18. Find currents $i_1(t)$ and $i_2(t)$ in Figure 9–9 if $R_1 = 1$, $R_2 = 2$, $L_1 = L_2 = 1$ H, and $E(t) = \sin t$ v. Assume $i_1(0) = i_2(0) = 0$.

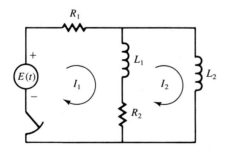

FIGURE 9–9

19. Consider the coupled spring-mass system shown in Figure 9–10.
   (a) Derive the system of equations governing the motion of the masses.

(b) Solve the system for the case $m_1 = m_2 = 1$ and $k = 4$. Assume initial conditions of $x_1(0) = 1$ and $\dot{x}_1(0) = x_2(0) = \dot{x}_2(0) = 0$.

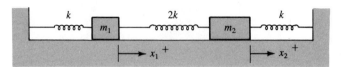

FIGURE 9–10

20. An electrical feedback circuit is designed so that the rate of change of current output of one branch is equal to the current output of the succeeding branch. See Figure 9–11. If $i_1(0) = 1$, $i_2(0) = 0$, and $i_3(0) = 0$, find all three currents at any time t.

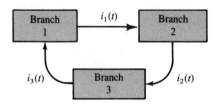

FIGURE 9–11

21. Two 100-gallon tanks initially contain 20 lb of salt in each tank. Pure water flows into the first tank at the rate of 10 gal/min, and brine is removed at the same rate from the second tank, as shown in Figure 9–12. The two tanks are interconnected so that liquid flows from tank 1 to tank 2 at 15 gal/min and from tank 2 to tank 1 at 5 gal/min. Determine the amount of salt in each of the two tanks at any time t.

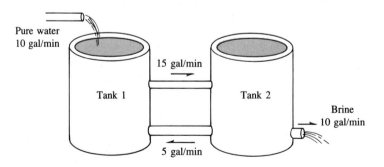

FIGURE 9–12

Appendix

Matrices are used in Section 9–4 in a procedure for solving systems of first-order linear differential equations. This appendix contains a brief review of the algebra of matrices.

DEFINITION

> A matrix is a rectangular array of numbers or functions.

We denote matrices by capital boldfaced letters and display them by enclosing their elements in parentheses. Some examples of matrices are

$$\mathbf{A} = \begin{pmatrix} 1 & -1 & 3 \\ 2 & 1 & 0 \end{pmatrix} \quad \mathbf{B} = \begin{pmatrix} t & t^2 \\ e^t & 1 \end{pmatrix} \quad \mathbf{C} = \begin{pmatrix} e^t \\ \sin t \\ 0 \end{pmatrix}$$

The dimension of a matrix refers to the number of rows and columns in the array. We specify the number of rows first, followed by the number of columns. Thus, an $m \times n$ matrix has m rows and n columns. The matrices given above are, respectively, 2×3, 2×2, and 3×1.

If a matrix has only one row it is called a **row matrix**; if it has only one column it is called a **column matrix**. Row and column matrices are often called **vectors**. If all of the entries of a matrix are zeros, the matrix is called **the zero matrix** and is denoted by **0**.

COMMENT: When distinguishing between matrices and numbers, the latter are sometimes referred to as scalars. A scalar may also be thought of as a 1 × 1 matrix.

To identify specific elements or entries within a matrix, we use a double subscript notation to indicate the row and column location of the element. The first number of the subscript refers to the row location and the second to the column location. Thus, the notation a_{ij} means the element in the ith row and the jth column of a matrix **A**. The scheme used in the following 3×4 matrix is typical.

$$\mathbf{A} = \begin{pmatrix} a_{11} & a_{12} & a_{13} & a_{14} \\ a_{21} & a_{22} & a_{23} & a_{24} \\ a_{31} & a_{32} & a_{33} & a_{34} \end{pmatrix}$$

The **transpose** of a matrix $\mathbf{A}$ is denoted by $\mathbf{A}^T$ and is the matrix obtained from $\mathbf{A}$ by interchanging the rows and columns. For example, if

$$\mathbf{A} = \begin{pmatrix} 2 & 1 & 3 \\ 3 & 0 & 1 \end{pmatrix}, \quad \text{then } \mathbf{A}^T = \begin{pmatrix} 2 & 3 \\ 1 & 0 \\ 3 & 1 \end{pmatrix}$$

Two matrices are equal if they have the same dimension and if the corresponding elements of the matrices are equal. We formalize the equality of matrices in the following definition.

EQUALITY OF MATRICES

The $m \times n$ matrices $\mathbf{A}$ and $\mathbf{B}$ are **equal** if

$$a_{ij} = b_{ij}$$

for $i = 1$ to m and $j = 1$ to n.

Example 1

(a) The following matrices are equal.

$$\begin{pmatrix} 2 & 1 \\ -1 & 7 \end{pmatrix} = \begin{pmatrix} \sqrt{4} & 1 \\ -1 & \frac{21}{3} \end{pmatrix}$$

(b) The following matrices have the same dimension but are not equal.

$$\begin{pmatrix} 3 & 5 & 0 \\ 2 & 3 & -2 \end{pmatrix} = \begin{pmatrix} 3 & 5 & 7 \\ 2 & 1 & -2 \end{pmatrix} \qquad \blacksquare$$

Two matrices can be added if they have the same dimension. The sum of two matrices is obtained by adding corresponding elements as follows.

MATRIX ADDITION

If $\mathbf{A}$ and $\mathbf{B}$ are $m \times n$ matrices, then $\mathbf{C} = \mathbf{A} + \mathbf{B}$ is given by

$$c_{ij} = a_{ij} + b_{ij}$$

for $i = 1$ to m and $j = 1$ to n. $\mathbf{C}$ is called the **sum** of $\mathbf{A}$ and $\mathbf{B}$.

Example 2

(a) $\begin{pmatrix} 2 & 5 \\ -3 & 1 \end{pmatrix} + \begin{pmatrix} -2 & 2 \\ 1 & 0 \end{pmatrix} = \begin{pmatrix} 2 + (-2) & 5 + 2 \\ -3 + 1 & 1 + 0 \end{pmatrix} = \begin{pmatrix} 0 & 7 \\ -2 & 1 \end{pmatrix}$

(b) $\begin{pmatrix} 3 & 1 \\ -2 & 4 \end{pmatrix} + \begin{pmatrix} 5 \\ 2 \end{pmatrix}$

is undefined because the two matrices do not have the same dimension. ∎

The operation of multiplying each element of a matrix $\mathbf{A}$ by a number c is called **scalar multiplication**. The next definition shows the process.

SCALAR MULTIPLICATION

If $\mathbf{A}$ is an $m \times n$ matrix and c is a real number, then the **scalar multiplication** of $\mathbf{A}$ by c is denoted by $c\mathbf{A}$ and given by

$$c\mathbf{A} = (ca_{ij})$$

for $i = 1$ to m and $j = i$ to n.

Example 3

Let $\mathbf{A} = \begin{pmatrix} 1 & 2 \\ -1 & 3 \end{pmatrix}$

then

$$2\mathbf{A} = \begin{pmatrix} 2 \cdot 1 & 2 \cdot 2 \\ 2 \cdot -1 & 2 \cdot 3 \end{pmatrix} = \begin{pmatrix} 2 & 4 \\ -2 & 6 \end{pmatrix}$$

and

$$(-1)\mathbf{A} = \begin{pmatrix} -1 \cdot 1 & -1 \cdot 2 \\ -1 \cdot -1 & -1 \cdot 3 \end{pmatrix} = \begin{pmatrix} -1 & -2 \\ 1 & -3 \end{pmatrix}$$ ∎

COMMENT: By convention $(-1)\mathbf{A}$ *is written* $-\mathbf{A}$. *In this way, subtraction of matrices is included in the definition of matrix addition.*

In addition to the operations of addition and scalar multiplication it is possible to multiply certain pairs of matrices. The process for multiplying two matrices is given in the following definition.

MATRIX MULTIPLICATION

Let A be an $m \times p$ matrix and B be a $p \times n$ matrix. Then $C = AB$ is an $m \times n$ matrix for which

$$c_{ij} = a_{i1}b_{1j} + a_{i2}b_{2j} + \cdots + a_{ip}b_{pj}$$

In words, to obtain the product $C = AB$ each entry c_{ij} in the product matrix C is obtained by multiplying each element in the ith row of A by its corresponding element in the jth column of B and adding these products. (See Figure A-1.)

FIGURE A–1

Multiplication of matrices

COMMENT: The definition of AB requires that the number of columns of A be equal to the number of rows of B. In diagram form, we have

$$A_{m \times p} \cdot B_{p \times n} = C_{m \times n}$$

must be equal

If A is $m \times p$ and B is $p \times n$, then the product C is $m \times n$.

Example 4

Let $A = \begin{pmatrix} 2 & 1 \\ 3 & 0 \end{pmatrix}$ and $B = \begin{pmatrix} 1 & 2 & 1 \\ 0 & 1 & 4 \end{pmatrix}$

Find AB and BA.

SOLUTION   Since there are 2 columns in A and 2 rows in B, we can form the product AB. The product matrix AB will be a 2×3 matrix. The six elements in the product $C = AB$ are generated as follows:

$$c_{11} = \begin{pmatrix} 2 & 1 \\ 3 & 0 \end{pmatrix}\begin{pmatrix} 1 & 2 & 1 \\ 0 & 1 & 4 \end{pmatrix} = 2 \cdot 1 + 1 \cdot 0 = 2$$

$$c_{12} = \begin{pmatrix} 2 & 1 \\ 3 & 0 \end{pmatrix}\begin{pmatrix} 1 & 2 & 1 \\ 0 & 1 & 4 \end{pmatrix} = 2 \cdot 2 + 1 \cdot 1 = 5$$

$$c_{13} = \begin{pmatrix} 2 & 1 \\ 3 & 0 \end{pmatrix} \begin{pmatrix} 1 & 2 & 1 \\ 0 & 1 & 4 \end{pmatrix} = 2 \cdot 1 + 1 \cdot 4 = 6$$

$$c_{21} = \begin{pmatrix} 2 & 1 \\ 3 & 0 \end{pmatrix} \begin{pmatrix} 1 & 2 & 1 \\ 0 & 1 & 4 \end{pmatrix} = 3 \cdot 1 + 0 \cdot 0 = 3$$

$$c_{22} = \begin{pmatrix} 2 & 1 \\ 3 & 0 \end{pmatrix} \begin{pmatrix} 1 & 2 & 1 \\ 0 & 1 & 4 \end{pmatrix} = 3 \cdot 2 + 0 \cdot 1 = 6$$

$$c_{22} = \begin{pmatrix} 2 & 1 \\ 3 & 0 \end{pmatrix} \begin{pmatrix} 1 & 2 & 1 \\ 0 & 1 & 4 \end{pmatrix} = 3 \cdot 1 + 0 \cdot 4 = 3$$

The desired matrix product, then, is

$$\mathbf{C} = \mathbf{AB} = \begin{pmatrix} 2 & 5 & 6 \\ 3 & 6 & 3 \end{pmatrix}$$

To conclude this example, we note that the product **BA** cannot be found because **B** has 3 columns and **A** has only 2 rows. ∎

Example 5

Let $\mathbf{A} = (2 \quad 1 \quad -1)$ and $\mathbf{B} = \begin{pmatrix} 1 \\ 5 \\ -2 \end{pmatrix}$. Find **AB** and **BA**.

SOLUTION **A** is a 1×3 matrix and **B** is a 3×1 matrix. Since **A** has 3 columns and **B** has 3 rows, the product **AB** can be formed. **AB** is a 1×1 matrix whose only entry is

$$\mathbf{AB} = (2 \quad 1 \quad -1) \begin{pmatrix} 1 \\ 5 \\ -2 \end{pmatrix} = (2 \cdot 1 + 1 \cdot 5 + -1 \cdot -2) = (9)$$

Note that the product **BA** will be a 3×3 matrix, based on the dimensions of **A** and **B**.

$$\mathbf{BA} = \begin{pmatrix} 1 \\ 5 \\ -2 \end{pmatrix} (2 \quad 1 \quad -1) = \begin{pmatrix} 1 \cdot 2 & 1 \cdot 1 & 1 \cdot -1 \\ 5 \cdot 2 & 5 \cdot 1 & 5 \cdot -1 \\ -2 \cdot 2 & -2 \cdot 1 & -2 \cdot -1 \end{pmatrix} = \begin{pmatrix} 2 & 1 & -1 \\ 10 & 5 & -5 \\ -4 & -2 & 2 \end{pmatrix}$$

∎

We can see from Examples 4 and 5 that the products **AB** and **BA** do not give the same results; in Example 4 the product **BA** was undefined and in Example 5 the product **AB** was not the same size as **BA**. The next example shows that even when **AB** and **BA** are both defined and the same size, they may not be equal.

Example 6

Let $\mathbf{A} = \begin{pmatrix} 1 & -1 \\ 2 & 0 \end{pmatrix}$ and $\mathbf{B} = \begin{pmatrix} 0 & 1 \\ 1 & -1 \end{pmatrix}$

Find $\mathbf{AB}$ and $\mathbf{BA}$.

SOLUTION

$$\mathbf{AB} = \begin{pmatrix} 0-1 & 1+1 \\ 0+0 & 2+0 \end{pmatrix} = \begin{pmatrix} -1 & 2 \\ 0 & 2 \end{pmatrix}$$

$$\mathbf{BA} = \begin{pmatrix} 0+2 & 0+0 \\ 1-2 & -1+0 \end{pmatrix} + \begin{pmatrix} 2 & 0 \\ -1 & -1 \end{pmatrix}$$

■

COMMENT: We conclude from Examples 4, 5, and 6 that matrix multiplication is not commutative. However, matrix multiplication is associative, $\mathbf{A(BC)} = \mathbf{(AB)C}$, *and distributive over matrix addition,* $\mathbf{A(B + C)} = \mathbf{AB} + \mathbf{AC}$.

A **square** matrix has the same number of rows and columns. This number is called the **order** of the matrix. A familiar operator on square matrices, which is of great utility in the concise statement of mathematical ideas, is the **determinant**. The **determinant** of a square matrix $\mathbf{A}$, denoted by det $\mathbf{A}$, det(a_{ij}), or

$$\begin{vmatrix} a_{11} & \cdots & a_{1n} \\ a_{21} & \cdots & a_{2n} \\ \vdots & & \vdots \\ a_{n1} & \cdots & a_{nn} \end{vmatrix}$$

is defined here in a recursive manner.

DETERMINANT OF A SQUARE MATRIX

Let $\mathbf{A} = (a_{ij})$ be a square matrix of order n. The determinant of $\mathbf{A}$ is denoted by det $\mathbf{A}$ and is defined as follows:

a.   If $n = 1$, then det $\mathbf{A} = a_{11}$.

b.   If $n = 2$, then det $\mathbf{A} = a_{11}a_{22} - a_{12}a_{21}$.

c.   For $n \geq 2$, let M_{ij} denote the $(n - 1) \times (n - 1)$ matrix obtained by deleting the ith row and jth column from $\mathbf{A}$. Furthermore, define the **cofactor** of the entry a_{ij} in $\mathbf{A}$ as the number

$$A_{ij} = (-1)^{i+j} \det (M_{ij})$$

Then

$$\det \mathbf{A} = \sum_{j=1}^{n} a_{ij}A_{ij}$$

This evaluation of det $\mathbf{A}$ is called an **expansion by elements of the ith row;** a similar expression for det $\dot{\mathbf{A}}$, obtained by an expansion of elements in the jth column is

$$\det \mathbf{A} = \sum_{i=1}^{n} a_{ij}A_{ij}$$

Example 7

Find the determinant of the following matrices.

(a) $\begin{pmatrix} 2 & 3 \\ 4 & 7 \end{pmatrix}$ (b) $\begin{pmatrix} 2 & 3 & 1 \\ -4 & 2 & 3 \\ 1 & 0 & -7 \end{pmatrix}$

SOLUTION   (a) $\begin{vmatrix} 2 & 3 \\ 4 & 7 \end{vmatrix} = (2)(7) - (3)(4) = 14 - 12 = 2$

(b) Using the elements of the first row

$$\begin{vmatrix} 2 & 3 & 1 \\ -4 & 2 & 3 \\ 1 & 0 & -7 \end{vmatrix} = (-1)^{1+1}(2)\begin{vmatrix} 2 & 3 \\ 0 & -7 \end{vmatrix} + (-1)^{1+2}(3)\begin{vmatrix} -4 & 3 \\ 1 & -7 \end{vmatrix}$$

$$+ (-1)^{1+3}(1)\begin{vmatrix} -4 & 2 \\ 1 & 0 \end{vmatrix}$$

$$= (2)(-14) - 3(25) + (1)(-2) = -105$$

It would be easier to use the elements of the third row because of the zero entry:

$$\begin{vmatrix} 2 & 3 & 1 \\ -4 & 2 & 3 \\ 1 & 0 & -7 \end{vmatrix} = (+1)\begin{vmatrix} 3 & 1 \\ 2 & 3 \end{vmatrix} - 7\begin{vmatrix} 2 & 3 \\ -4 & 2 \end{vmatrix} = 7 - 7(16) = -105 \quad \blacksquare$$

IDENTITY MATRIX

The $n \times n$ square matrix $\mathbf{I}$ with 1s on the main diagonal and 0s elsewhere is called the **identity matrix of order n.**

The identity matrix of order 2 is

$$I = \begin{pmatrix} 1 & 0 \\ 0 & 1 \end{pmatrix}$$

and the identity matrix of order 3 is

$$I = \begin{pmatrix} 1 & 0 & 0 \\ 0 & 1 & 0 \\ 0 & 0 & 1 \end{pmatrix}$$

The result obtained by multiplying an $n \times n$ matrix A by the identity matrix I of order n is to leave the matrix A unchanged. For example,

$$\begin{pmatrix} 2 & -3 \\ 1 & 4 \end{pmatrix} \begin{pmatrix} 1 & 0 \\ 0 & 1 \end{pmatrix} = \begin{pmatrix} 2 & -3 \\ 1 & 4 \end{pmatrix}$$

IDENTITY PROPERTY

Let A be a square matrix of order n and I be the identity matrix of order n. Then

$$AI = IA = A$$

A square matrix A that has the property that a square matrix B can be found so that

$$AB = BA = I$$

is said to be **invertible** and the matrix B is called the **inverse** matrix of A.

INVERSE MATRIX

Let A be a square matrix of order n. The **inverse matrix** of A is donated by A^{-1} and has the property that

$$AA^{-1} = A^{-1}A = I$$

COMMENT: Because $AA^{-1} = A^{-1}A = I$, *to show that a given matrix* B *is an inverse of* A, *it is sufficient to show that* $AB = I$.

Example 8

Let $\mathbf{A} = \begin{pmatrix} 2 & 1 \\ 3 & 2 \end{pmatrix}$. Show that the inverse of $\mathbf{A}$ is $\mathbf{B} = \begin{pmatrix} 2 & -1 \\ -3 & 2 \end{pmatrix}$.

SOLUTION   To show that $\mathbf{B}$ is the inverse of $\mathbf{A}$, we show that $\mathbf{AB} = \mathbf{I}$.

$$\mathbf{AB} = \begin{pmatrix} 2 & 1 \\ 3 & 2 \end{pmatrix}\begin{pmatrix} 2 & -1 \\ -3 & 2 \end{pmatrix} = \begin{pmatrix} 2 \cdot 2 + 1 \cdot -3 & 2 \cdot -1 + 1 \cdot 2 \\ 3 \cdot 2 + 2 \cdot -3 & 3 \cdot -1 + 2 \cdot 2 \end{pmatrix} = \begin{pmatrix} 1 & 0 \\ 0 & 1 \end{pmatrix} = \mathbf{I}$$

Thus, we have shown that $\mathbf{B}$ is the inverse of $\mathbf{A}$ and we have also shown that $\mathbf{A}$ is the inverse of $\mathbf{B}$.   ∎

The following theorem gives a convenient check to determine whether a square matrix has multiplicative inverse.

THEOREM A–1

An $n \times n$ matrix $\mathbf{A}$ has a multiplicative inverse $\mathbf{A}^{-1}$ if, and only if, $\det \mathbf{A} \neq 0$.

Example 9

The matrix $\mathbf{A} = \begin{pmatrix} 4 & 6 \\ 2 & 3 \end{pmatrix}$ does not have an inverse because $\det \mathbf{A} = 4 \cdot 3 - 2 \cdot 6 = 0$.   ∎

A square matrix that does not have an inverse is said to be **singular**; otherwise, it is said to be **nonsingular**. The next theorem describes a procedure that can be used to find the multiplicative inverse of a nonsingular matrix.

THEOREM A–2

Let $\mathbf{A}$ be an $n \times n$ nonsingular matrix and let $\mathbf{A}_{ij} = (-1)^{i+j} |M_{ij}|$ for all i and j, where $|M_{ij}|$ is the $(n-1) \times (n-1)$ determinant obtained by deleting the ith row and the jth column of $\mathbf{A}$. Then $\mathbf{A}^{-1}$ is

$$\mathbf{A}^{-1} = \frac{1}{\det \mathbf{A}}(\mathbf{A}_{ij})^{\mathrm{T}}$$

Example 10

Find A^{-1} for $A = \begin{pmatrix} 3 & -7 \\ 6 & 2 \end{pmatrix}$

SOLUTION   The determinant of A is

$$\begin{vmatrix} 3 & -7 \\ 6 & 2 \end{vmatrix} = 3(2) - 6(-7) = 48$$

Since det $A \neq 0$, the given matrix is nonsingular and therefore has an inverse. To find A^{-1} we compute the cofactors A_{ij} as follows.

$$A_{11} = (-1)^{1+1} (2) = 2$$
$$A_{12} = (-1)^{1+2} (6) = -6$$
$$A_{21} = (-1)^{2+1} (-7) = 7$$
$$A_{22} = (-1)^{2+2} (3) = 3$$

It then follows from Theorem A–2 that the inverse is

$$A^{-1} = \frac{1}{48} \begin{pmatrix} 2 & -6 \\ 7 & 3 \end{pmatrix}^T = \frac{1}{48} \begin{pmatrix} 2 & 7 \\ -6 & 3 \end{pmatrix} = \begin{pmatrix} \frac{1}{24} & \frac{7}{48} \\ -\frac{1}{8} & \frac{1}{16} \end{pmatrix}$$

We check the validity of A^{-1} by showing that $AA^{-1} = I$.

$$AA^{-1} = \begin{pmatrix} 3 & -7 \\ 6 & 2 \end{pmatrix} \begin{pmatrix} \frac{1}{24} & \frac{7}{48} \\ -\frac{1}{8} & \frac{1}{16} \end{pmatrix} = \begin{pmatrix} \frac{3}{24} + \frac{7}{8} & \frac{21}{48} + \frac{7}{16} \\ \frac{6}{24} - \frac{2}{8} & \frac{42}{48} + \frac{2}{16} \end{pmatrix} = \begin{pmatrix} 1 & 0 \\ 0 & 1 \end{pmatrix}$$

∎

Example 11

Find A^{-1} for $A = \begin{pmatrix} 2 & 1 & 0 \\ 4 & 1 & -1 \\ 2 & 1 & -1 \end{pmatrix}$.

SOLUTION   Expanding by minors in the first row, the determinant of A is

$$\begin{vmatrix} 2 & 1 & 0 \\ 4 & 1 & -1 \\ 2 & 1 & -1 \end{vmatrix} = (-1)^{1+1}(2) \begin{vmatrix} 1 & -1 \\ 1 & -1 \end{vmatrix} + (-1)^{1+2}(1) \begin{vmatrix} 4 & -1 \\ 2 & -1 \end{vmatrix} + 0$$

$$= 2 (-1 + 1) - (-4 + 2) = 2$$

Since det $A = 2 \neq 0$, the matrix is nonsingular. The cofactors A_{ij} are

$$A_{11} = \begin{vmatrix} 1 & -1 \\ 1 & -1 \end{vmatrix} = 0 \qquad A_{12} = - \begin{vmatrix} 4 & -1 \\ 2 & -1 \end{vmatrix} = 2 \quad A_{13} = \begin{vmatrix} 4 & 1 \\ 2 & 1 \end{vmatrix} = 2$$

$$A_{21} = -\begin{vmatrix} 1 & 0 \\ 1 & -1 \end{vmatrix} = 1 \quad A_{22} = \begin{vmatrix} 2 & 0 \\ 2 & -1 \end{vmatrix} = -2 \quad A_{23} = -\begin{vmatrix} 2 & 1 \\ 2 & 1 \end{vmatrix} = 0$$

$$A_{31} = \begin{vmatrix} 1 & 0 \\ 1 & -1 \end{vmatrix} = -1 \quad A_{32} = -\begin{vmatrix} 2 & 0 \\ 4 & -1 \end{vmatrix} = 2 \quad A_{33} = \begin{vmatrix} 2 & 1 \\ 4 & 1 \end{vmatrix} = -2$$

Then, by Theorem A–2, $\mathbf{A}^{-1}$ is

$$\mathbf{A}^{-1} = \frac{1}{2}\begin{pmatrix} 0 & 2 & 2 \\ 1 & -2 & 0 \\ -1 & 2 & -2 \end{pmatrix}^{\mathrm{T}} = \frac{1}{2}\begin{pmatrix} 0 & 1 & -1 \\ 2 & -2 & 2 \\ 2 & 0 & -2 \end{pmatrix} = \begin{pmatrix} 0 & \frac{1}{2} & -\frac{1}{2} \\ 1 & -1 & 1 \\ 1 & 0 & -1 \end{pmatrix} \quad \blacksquare$$

DIFFERENTIATION AND INTEGRATION OF MATRICES

Consider a matrix $\mathbf{A}(t)$ of functions differentiable on a common interval $[a,b]$. The derivative of $\mathbf{A}$ is denoted $\mathbf{A}'$ and defined to be the matrix obtained by differentiating the elements of $\mathbf{A}(t)$; that is,

$$\mathbf{A}'(t) = \frac{d\mathbf{A}}{dt} = \left(\frac{d}{dt}a_{ij}\right)$$

For example, if

$$\mathbf{A}(t) = \begin{pmatrix} t^2 & e^{3t} \\ \sin 2t & 5 \end{pmatrix}$$

then

$$\mathbf{A}'(t) = \begin{pmatrix} 2t & 3e^{3t} \\ 2\cos 2t & 0 \end{pmatrix}$$

Integrals of a matrix $\mathbf{A}$ are also defined as elementwise integration; that is,

$$\int_{t_0}^{t} \mathbf{A}(x)\, dx = \left(\int_{t_0}^{t} a_{ij}(x)\, dx\right)$$

For example, if

$$\mathbf{A}(t) = \begin{pmatrix} 2t \\ e^{2t} \\ \cos \frac{1}{2}t \end{pmatrix}$$

Then

$$\int_{0}^{t} \mathbf{A}(x)\,dx = \begin{pmatrix} t^2 \\ \frac{1}{2}e^{2t} - \frac{1}{2} \\ 2\sin\frac{1}{2}t \end{pmatrix}$$

GAUSSIAN ELIMINATION

We conclude this section with a review of the Gaussian elimination method of solving systems of linear equations. Recall that an algebraic system of n linear equations

$$a_{11}x_1 + a_{12}x_2 + \cdots + a_{1n}x_n = b_1$$
$$a_{21}x_1 + a_{22}x_2 + \cdots + a_{2n}x_n = b_2$$
$$\vdots \qquad\qquad\qquad \vdots$$
$$a_{n1}x_1 + a_{n2}x_2 + \cdots + a_{nn}x_n = b_n$$

can be written in matrix form as

$$\begin{pmatrix} a_{11} & a_{12} & \cdots & a_{1n} & \vline & b_1 \\ a_{21} & a_{22} & \cdots & a_{2n} & \vline & b_2 \\ \vdots & & & \vdots & \vline & \vdots \\ a_{n1} & a_{n2} & \cdots & a_{nn} & \vline & b_n \end{pmatrix}$$

This matrix is called the **augmented** matrix of the system of equations. The $n \times n$ matrix of coefficients (a_{ij}) is called the **coefficient** matrix.

The solution of a system of n linear equations can be found by transforming its augmented matrix into an equivalent matrix in which the elements below the main diagonal of the coefficient matrix are zeros. The system is then said to be in **triangular** form. The procedure used to transform a system into triangular form is called **Gaussian elimination** and is based on the following row operations.

EQUIVALENT MATRICES

The following row operations will transform a matrix into an equivalent matrix:

1. *Multiply a row by a nonzero constant.*
2. *Interchange two rows.*
3. *Replace a row with the sum of that row and any other row.*

Example 12

Use matrix notation to solve

$$x - 6y + 3z = -2$$
$$2x - 3y + z = -2$$
$$3x + 3y - 2z = 2$$

SOLUTION   The given system is represented by the augmented matrix

$$\left(\begin{array}{ccc|c} 1 & -6 & 3 & -2 \\ 2 & -3 & 1 & -2 \\ 3 & 3 & -2 & 2 \end{array}\right)$$

The following steps yield the triangular form:

$$\left(\begin{array}{ccc|c} 1 & -6 & 3 & -2 \\ 2 & -3 & 1 & -2 \\ 3 & 3 & -2 & 2 \end{array}\right) \begin{array}{c} (-2)R1 + R2 \to R2 \\ (-3)R1 + R3 \to R3 \\ \xrightarrow{\hspace{2cm}} \end{array} \left(\begin{array}{ccc|c} 1 & -6 & 3 & -2 \\ 0 & 9 & -5 & 2 \\ 0 & 21 & -11 & 8 \end{array}\right)$$

$$\xrightarrow{7R2 \;-\; 3R3 \;\to\; R3} \left(\begin{array}{ccc|c} 1 & -6 & 3 & -2 \\ 0 & 9 & -5 & 2 \\ 0 & 0 & -2 & -10 \end{array}\right)$$

This is the matrix of the triangular form of the given system, that is,

$$x - 6y + 3z = -2$$
$$9y - 5z = 2$$
$$-2z = -10$$

Solving the last equation, we get $z = 5$. Substituting $z = 5$ into the middle equation yields $y = 3$. Finally, we find that $x = 1$ by substituting $y = 3$ and $z = 5$ into the upper equation. The solution is $x = 1$, $y = 3$, $z = 5$.   ■

COMMENT: The notation

$$aR1 + R2 \to R2$$

means "multiply row 1 by a, add it to row 2, and replace row 2 with this sum."

A system of equations with a unique solution is a **consistent** system. If the system has infinitely many solutions it is **dependent**. A system with no solutions is **inconsistent**.

Example 13

Solve the system

$$x + y - z = -1$$
$$2x - y + z = 2$$
$$x - 5y + 5z = 7$$

SOLUTION   The given system is represented by the augmented matrix

$$\left(\begin{array}{ccc|c} 1 & 1 & -1 & -1 \\ 2 & -1 & 1 & 2 \\ 1 & -5 & 5 & 7 \end{array}\right)$$

The following steps lead to the solution

$$\begin{pmatrix} 1 & 1 & -1 & | & -1 \\ 2 & -1 & 1 & | & 2 \\ 1 & -5 & 5 & | & 7 \end{pmatrix} \xrightarrow[\begin{subarray}{c} (-2)R1 + R2 \to R2 \\ (-1)R1 + R3 \to R3 \end{subarray}]{} \begin{pmatrix} 1 & 1 & -1 & | & -1 \\ 0 & -3 & 3 & | & 4 \\ 0 & -6 & 6 & | & 8 \end{pmatrix}$$

$$\xrightarrow[(-2)R2 + R3 \to R3]{} \begin{pmatrix} 1 & 1 & -1 & | & -1 \\ 0 & -3 & 3 & | & 4 \\ 0 & 0 & 0 & | & 0 \end{pmatrix}$$

The fact that the third row consists of all zeros shows that x, y, and z are related only by the equations that the first two rows represent. The system is therefore *dependent*. Any of the infinitely many solutions may be found by assigning a value to z and solving for x and y. For example, if we let $z = 1$, then $x = \frac{1}{3}$ and $y = -\frac{1}{3}$. If $z = 0$, then $x = \frac{1}{3}$ and $y = -\frac{4}{3}$. In general, if $z = t$, the second row yields $y = t - \frac{4}{3}$ and the first row yields $x = \frac{1}{3}$. ∎

Example 14

Solve the system

$$\begin{aligned} x - y + z &= 8 \\ 2x + y - z &= -2 \\ -x - 5y + 5z &= 10 \end{aligned}$$

SOLUTION The matrix of the system is

$$\begin{pmatrix} 1 & -1 & 1 & | & 8 \\ 2 & 1 & -1 & | & -2 \\ -1 & -5 & 5 & | & 10 \end{pmatrix}$$

Reduce this matrix to triangular form.

$$\begin{pmatrix} 1 & -1 & 1 & | & 8 \\ 2 & 1 & -1 & | & -2 \\ -1 & -5 & 5 & | & 10 \end{pmatrix} \xrightarrow[\begin{subarray}{c} (-2)R1 + R2 \to R2 \\ R1 + R3 \to R3 \end{subarray}]{} \begin{pmatrix} 1 & -1 & 1 & | & 8 \\ 0 & 3 & -3 & | & -18 \\ 0 & -6 & 6 & | & 18 \end{pmatrix}$$

$$\xrightarrow[2R2 + R3 \to R3]{} \begin{pmatrix} 1 & -1 & 1 & | & 8 \\ 0 & 3 & -3 & | & -18 \\ 0 & 0 & 0 & | & -18 \end{pmatrix}$$

The last row represents

$$0x + 0y + 0z = -18$$

which is impossible for any x, y, and z. Therefore, the system is *inconsistent*. ∎

Answers

SECTION 1–1

1. first order, linear   **2.** first order, nonlinear   **3.** second order, nonlinear
4. second order, linear   **5.** second order, nonlinear   **6.** third order, linear
7. first order, nonlinear   **8.** second order, linear   **9.** fifth order, linear

10. first order, nonlinear   **11.** $\dfrac{dP}{dt} = kP$ **12.** $\dfrac{di}{dt} = ki$ **13.** $\dfrac{ds}{dt} = k\sqrt{s}$ **14.** $\dfrac{dI}{dt} = -kI^2$

15. $\dfrac{dy}{dx} = 2$
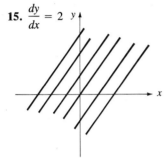

16. $y = y' + xy'$
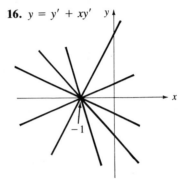

17. $y' = -\dfrac{x}{4y}$
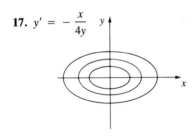

18. $y = 2xy'$
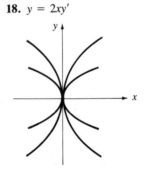

19. $x = yy'$
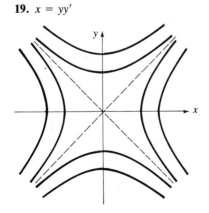

20. $3y = xy'$
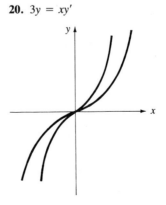

21. $y = y'$

22. $xy' = 1$

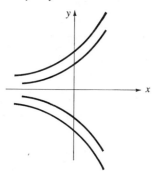

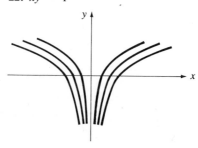

23. $1 + (y')^2 + yy'' = 0$

24. $(y')^3 = y' - xy''$

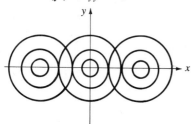

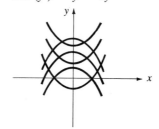

25. $xy' = 2y$ **26.** $yy' = x$

27. $y' = \dfrac{y}{x - x_m}$ where x_m is the coordinate of the man and $x_m = v_m t$

28. $\dfrac{dy}{dx} = \dfrac{y - b}{x - x_a}$, where $x_a = v_a t$

SECTION 1–2

13. $3, -\frac{1}{2}$ **14.** $-2, -1, 0$ **19.** $y = x^2 + 2x + 3$ **20.** $y = x - 2e^x$

21. $y = \frac{1}{3}e^{-x} + \frac{2}{3}e^{2x}$ **22.** $s = \sin 2t + \cos 2t$ **23.** $v = \cos 3t + t + 1 - \pi$

24. $y = 2e^{x-2} - e^{-(x-2)}$

SECTION 1–3

1. $y = \frac{2}{3}x^{3/2} + C$ **2.** $y = -x^{-1} + C$ **3.** $y = \ln(x + 2) + C$ **4.** $s = \frac{1}{2}e^{2t} + C$

5. $y = -\frac{1}{2}\cos 2x + C$ **6.** $y = 6 \ln \sec \frac{1}{2}x + C$ **7.** $y = \frac{2}{5}(x^2 + 3)^{5/2} + C$

8. $y = \frac{1}{3}(3x^2 + 2)^{1/2} + C$ **9.** $y = (1/\sqrt{3}) \text{Arctan}\, (x/\sqrt{3}) + C$ **10.** $y = \frac{1}{3}\tan 3x + C$

11. $y = \frac{1}{2}\tan^{-1}\dfrac{x}{2} + C$ **12.** $s = \frac{1}{3}\ln Ct(t^2 + 6)$ **13.** $y = (x/3) \sin 3x + \frac{1}{9}\cos 3x + C$

14. $s = -e^{-t}(t + 1) + C$ **15.** $z = t^4\left(\dfrac{\ln t}{4} - \dfrac{1}{16}\right) + C$ **16.** $y = \ln \dfrac{C(x + 3)^2}{(2x - 1)^{3/2}}$

17. $y = \ln(x+1) + 2(x+1)^{-1} - \frac{1}{2}(x+1)^{-2} + C$

18. $y = \ln\dfrac{(x-2)^4}{(x^2+4)^2} + \dfrac{1}{2}\tan^{-1}\dfrac{x}{2} + C$

19. $y = \tan^{-1}(e^x) + C$ **20.** $y = \ln(e^x + 1) + C$

21. $y = -\ln(e^x + e^{-x}) + C$ **22.** $s = \frac{1}{2}t - \frac{1}{4}\sin 2t + C$

23. $y = \frac{1}{3}\sin^3 x + C$ **24.** $y = \frac{1}{3}\cos^3 x - \cos x + C$

25. $v = \frac{1}{2}\sin x^2 + x\ln 3x - x + C$ **26.** $y = \frac{1}{27}e^{3x}(9x^2 - 6x + 2) + C$

27. $m = \frac{1}{2}\ln(t^2 + 2t + 2) - \tan^{-1}(t+1) + C$

28. $y = \frac{3}{2}\ln(x^2 + 2) + \dfrac{4}{\sqrt{2}}\tan^{-1}\dfrac{x}{\sqrt{2}} + C$

29. $y = \begin{cases} -\cos x + C, & x \le 0 \\ -1 + C, & x \ge 0 \end{cases}$

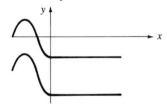

30. $y = \begin{cases} \frac{1}{2}x^2 + C, & -1 \le x \le 0 \\ \frac{1}{3}x^3 + C, & 0 \le x \le 1 \\ \frac{1}{2}x^2 - \frac{1}{6} + C, & 1 \le x \le 2 \\ \frac{1}{3}x^3 - \frac{5}{6} + C, & 2 \le x \le 3 \end{cases}$

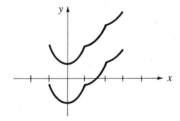

31. $y = \begin{cases} e^x + C, & x \le 0 \\ -e^x + 2 + C, & 0 \le x \le 1 \\ 2x - e + C, & x \ge 1 \end{cases}$

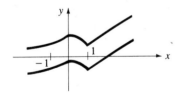

32. $y = \begin{cases} \sin x + C, & x \le 0 \\ -\cos x + 1 + C, & 0 \le x \le \pi \\ 2x + 2 - 2\pi + C, & x \ge \pi \end{cases}$

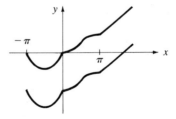

33. $y = x^3 + 1$ **34.** 20 cm   **35.** $M = -\frac{1}{2}(3 + 2x)^{-1} + \frac{61}{6}$ **36.** $q = 2(1 - e^{-2t})$

37. $y = \frac{1}{2}bkx^2 + c_1 x + c_2$

SECTION 1–4

1. theorem applies   **2.** theorem applies
3. theorem does not apply; f and f_y not continuous at $(0,0)$ **4.** theorem applies
5. theorem does not apply; f and f_y not continuous at $(1,2)$ **6.** theorem applies

7. $y = -e^{-3x} + 2e^{2x}$ **8.** $y = e^{2x} - e^{5-3x}$ **9.** $y = \cos x + \sin x$

10. $y = 2 \cos x - 3 \sin x$ **11.** no solution   **12.** no solution   **15.** $k =$ an integer

16. $k = n\pi/2$, n an integer

17. $y = -\frac{1}{2}e^{-2x} + \frac{5}{2}$

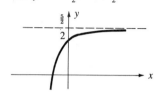

18. $y = 2\sqrt{x} + 1$

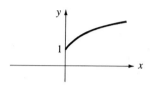

19. $y = -\pi/4 + \text{Arctan } x$

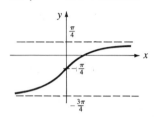

20. $y = \frac{1}{3} \sin 3x + \frac{1}{3}$

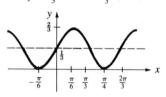

Chapter 1 Review

1. (a) second order, linear   (b) first order, nonlinear   (c) first order, linear

4. $\dfrac{-3 \pm \sqrt{29}}{2}$, $y = c_1 e^{-(3 + \sqrt{29})x/2} + c_2 e^{-(3 - \sqrt{29})x/2}$ **5.** $y = \frac{2}{9}(3x + 2)^{3/2} + c$

6. $y = \frac{1}{2}x^2 + \frac{1}{3} \ln |\sin 3x| + c$

7. $y = 5x - \frac{1}{2}e^{-2x} + c$ **8.** $y = -\frac{1}{2} \ln (1 + \cos 2x) + c$

9. $y = \frac{1}{3}t \tan 3t + \frac{1}{9} \ln |\cos 3t| + c$ **10.** $y = x \tan^{-1} 4x - \frac{1}{8} \ln(1 + 16x^2) + c$

11. $y = \ln cx^{1/2}(x^2 + 4)^{5/4}$ **12.** $y = \ln \left| c\dfrac{t - 3}{t + 2} \right|$

13. $y = \begin{cases} -2x + c, & x < 1 \\ \frac{1}{2}x^2 + 3x - \frac{11}{2} + c, & x > 1 \end{cases}$

14. $y = \frac{1}{3}x^3 + \frac{1}{3}$ **15.** Theorem does not apply since f_y is discontinuous at $(0, 0)$

16. unique solution   **18.** $y = \cos 2x$

SECTION 2–1

1. $xy = c$ **2.** $x + y = cxy$ **3.** $x^2 + y^2 = c^2$ **4.** $cy\sqrt{1 + x^2} = 1$ **5.** $1 + s^2t^2 = cs^2$
                                                $y = 0$

6. $cy(3x + 2)^{1/3} = 1$ **7.** $\sqrt{x^2 - 2} = c(1 - y)$ **8.** $x \cos y = c$

9. $2 \text{ Arctan } x + y^2 = c$ **10.** $p = 2 - ce^{-t}$ **11.** $\ln y = c + e^{-x}(x + 1)$

12. $\frac{1}{2}q^2 + \sin t - t \cos t = c$ **13.** $e^y - e^{-x} = c$ **14.** $(v - 3)^{3/4}(v + 1)^{1/4}(t + 3) = c$

15. $\dfrac{(y + 1)^{1/3}(x^2 - 1)^{1/2}}{(y + 4)^{1/3}} = c$ **16.** $\ln(1 + e^x) - \tan y = c$ **17.** $y^2 + \sec^2 x = c$

18. $y = \dfrac{ce^{2x}}{(x + 1)^2}$ **19.** $y = \dfrac{c}{x} e^{-1/x}$ **20.** $t = 2 \sec \frac{1}{2}(s + c)$ **21.** $(x - 4)(y + 2) = -4$

22. $s^2 + t^2 = 10$ **23.** $\sqrt{9 + x^2} + \ln y = 5$ **24.** $x \sec y = 6$

25. $\text{Arctan } x + \text{Arctan } y = \frac{7}{12}\pi$ **26.** $\dfrac{x - 1}{x + 1} = \dfrac{y^2}{3e^2}$

27. $y = $ constant or $y = ce^{2x}$

28. $2x + y^{-2} = c$ **29.** $y = \sqrt{2} \tan(\sqrt{2}x + c)$ **30.** $y^2 = 2x + c$

SECTION 2–2

1. 0.0023 amp   **2.** $T = 64.7°$ **3.** $t = 50.7$ min   **4.** $t = 6.97$ min

5. $m = 200e^{-0.25t}$ **6.** $t = 311$ yr   **7.** $m = 78.6$ g   **8.** (b) 15,000 yr

9. $v = 1000$ cm/sec   **10.** $I = 25e^{-0.223t}$ **11.** $P = 15e^{-0.000041h}$ and $P = 13.3$ psi

12. $t = 1.76$ hr

13. $y_0 = -\frac{1}{2}x + c$

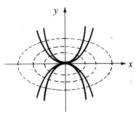

14. $y_0^2 + (x + 1)^2 = c$

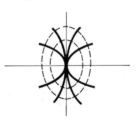

15. $y_0 = cx^4$

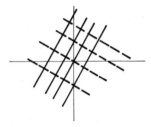

16. $y_0^2 + 2x^2 = c$

17. $xy_0 = c$

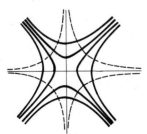

18. $3y_0^2 + x^2 = c$

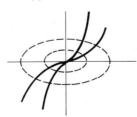

19. $y_0^2 + 2x = c$

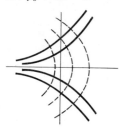

20. $y_0 + \frac{1}{2}x^2 = c$

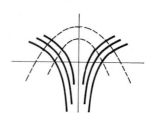

21. $x^2 + y_0^2 = c$ **22.** $x^2 + y_0^2 = 2 \ln x + c$

23. (a) $h = \left(\sqrt{h_0} - \dfrac{\sqrt{2g}B}{2A}t \right)^2$ (b) $t = 2A\sqrt{h_0}/B\sqrt{2g}$

24. (b) $h = \left[-\dfrac{10B\sqrt{2g}}{\pi}t + h_0^{5/2} \right]^{2/5}$ (c) $t = \dfrac{\pi h_0^{5/2}}{10B\sqrt{2g}}$ **25.** $y = 2x^2$

26. $y = -1/(x - 2)$ **27.** $x^2 - y^2 = 1$ **29.** 11:23 A.M.

30. $y = \dfrac{x^{-R+1}}{2(R-1)} + \dfrac{x^{R+1}}{2(R+1)} + \dfrac{R}{R^2 - 1}$ where $R = v_R/v_D$

31. $y = (T_C/2w)(e^{wx/T_C} + e^{-wx/T_C}) = \dfrac{T_C}{w} \cosh \dfrac{wx}{T_C}$

SECTION 2–3

1. $y = x \ln cx^2$ **2.** $y = x \ln cx$ **3.** $\csc\left(\dfrac{y}{x}\right) - \cot\left(\dfrac{y}{x}\right) = cx$ **4.** $cx = \sin\left(\dfrac{y}{x}\right)$

5. $\ln cy = \dfrac{x^2}{2y^2}$ **6.** $y^2 = 2x^2 \ln cx$ **7.** $-x = y \ln cx$ **8.** $y = c\sqrt{x^2 + 1}$

9. $y = \dfrac{1}{c - \sqrt{x^2 + 4}}$ **10.** $\dfrac{2}{\sqrt{3}}$ Arctan $\dfrac{2y + x}{x\sqrt{3}} = \ln cx$ **11.** $y^2 - xy + x^2 = C$

12. $\dfrac{y + x}{y + 2x} = cx$ **13.** $\ln(y - x) + \dfrac{x}{y - x} = c$ **14.** $y^2 + 2xy - x^2 = c$

15. $\dfrac{y - 4x}{y + 2x} = cx^6$ **16.** $\dfrac{y - 2x}{yx^2} = c$ **17.** Arcsin $\dfrac{y}{x} = \ln cx$

18. $-\dfrac{x^2}{2y^2} = \ln cxy$ **19.** $x(x - 2y) = 16$ **20.** $y + \sqrt{x^2 + y^2} = 1 + \sqrt{10}$

21. $x^2 - 3y^2 = 4$ **22.** $e^{y/x} = e - \ln x$

23. $XdY - (2X + Y)dX = 0$ is homogeneous in X and Y. The solution is the same as Exercise 1 with $x - 2$ and $y + 8$ replacing x and y respectively.

24. $XdY - (X + Y)dX = 0$ is homogeneous in X and Y. The solution is the same as Exercise 2 with $x + 5$ and $y - 2$ replacing x and y respectively.

25. $(X - 2Y)dY + (Y - 2X)dX = 0$ is homogeneous in X and Y. The solution is the same as Exercise 11 with $x - \frac{5}{3}$ and $y - \frac{4}{3}$ replacing x and y respectively.

26. $XdX + (Y - 2X)dY = 0$ is homogeneous in X and Y. The solution is the same as Exercise 13 with $x - 2$ and $y - 3$ replacing x and y respectively.

27. $(X + Y)dX - X\,dY = 0$ is homogeneous in X and Y. The solution is the same as Exercise 2 with $x + 2$ and $y + 1$ replacing x and y respectively.

28. $(X + Y)dY - (X - Y)dX = 0$ is homogeneous in X and Y. The solution is the same as Exercise 14 with $x + 1$ and $y + 1$ replacing x and y respectively.

29. $(y - \frac{4}{3})^2 - (x - \frac{5}{3})(y - \frac{4}{3}) + (x - \frac{5}{3})^2 = c$

30. $y + 1 = (x + 2)\ln c(x + 2)$

31. $6y - 6x + 16\ln(3x + 6y + 5) = c$ **32.** $(x - y)^2 - 2y = c$

33. $x + y + \ln(2x + y - 2) = c$ **34.** $\frac{2}{3}(x + y)^{3/2} = x + c$

35. $\ln|2x + y + 1| - (2x + y + 1)^{-1} = x + c$

SECTION 2–4

1. $x^2 + 2xy - y^2 = c$ **2.** $x^2 + 3xy + y^2 = c$ **3.** $\frac{5}{2}x^2 + xy^2 - \frac{1}{4}y^4 = c$

4. $x^3 + 4xy^2 = c$ **5.** $3x^2y + y^3 = c$ **6.** $x^2y + 3xy^2 = c$ **7.** $x^2 - 2\cos y \cos x = c$

8. $x^2e^y + x = c$ **9.** $x^3 + 3x^2y - xy^2 + y^3 = c$ **10.** $\dfrac{x}{1 + y^2} = c$ **11.** $ye^x - x^2 = c$

12. $x^2 + 4y^2 + 2y\ln x = c$ **13.** $y^2 + x^2 = c$ and $x = -y$

14. $\ln(y^2 + x^2) + 2\,\text{Arctan}\,\dfrac{y}{x} = c$ **15.** $x\cos y + 4x = c$

16. $x^4 + y^4 + 2x^2y^2 - 4xy = c$ **17.** $y = c(1 + \sin x)$

18. $x^3y^2 + \frac{1}{2}y^2 + \frac{1}{3}x^3 = c$ **19.** $y\tan x + \ln\sec x = c$ **20.** $e^{xy} + x^2y = c$

21. $3x^2y + y^3 = 32$ **22.** $x^2 + 2xy - y^2 = -16$ **23.** $x\sin y + y^2 = \pi^2$

24. $x^2e^{2y} + 2y = 4$

SECTION 2–5

1. $y = 2 + ce^{-2x}$ **2.** $y = \frac{1}{8} + ce^{-2x^2}$ **3.** $y = -\frac{3}{4} + ce^{4x}$ **4.** $i = 2 + ce^{-t}$

5. $6xy - 2x^3 - 3x^2 = c$ **6.** $y = cx^2 - \dfrac{1}{3x}$ **7.** $q = -\frac{1}{4}(2t + 1) + ce^{2t}$

8. $y = ce^{-x^2/2} - 3$ **9.** $s(t^2 + 1) = 3t + c$ **10.** $y = 1 + c(x^2 - 4)^{-1/2}$

11. $y = -2 + c\sqrt{x^2 + 1}$ **12.** $y = e^{-x} + ce^{-2x}$ **13.** $v = (r + c)\cos r$

14. $y = (x + c)\csc x$ **15.** $y = ce^x - e^{-x}$ **16.** $i = (t + c)e^{-t}$ **17.** $s = \dfrac{\ln C(t^3 + 2t)}{t^2}$

18. $y = 1 - x\cot x + c\csc x$ **19.** $y = xe^x - ex$ **20.** $y = 2 - e^{-2x}$

21. $y = 5\csc x - 10\cot x$ **22.** $r\sin\theta = \ln\sec\theta + \sqrt{3} - \ln 2$

23. $y = \pm\left(\dfrac{3x^2}{c - 2x^3}\right)^{1/2}$ **24.** $y = (cx^2 - x)^{-1}$ **25.** $y = (c\sec x - 2\tan x)^{-1}$

26. $y = \pm[(c - 2\ln(\sec x + \tan x))\cos x]^{-1/2}$ **27.** $s = (ce^{-t} - t + 1)^{-1}$

28. $y^4 = 2 + ce^{-8x^2}$ **29.** $y = \left(\dfrac{x}{2 + cx^{1/2}}\right)^{1/3}$ **30.** $s = \pm\dfrac{t^{3/2}}{(c - t)^{1/2}}$

31. $y = -\ln(-x\ln x + cx)$ **32.** $\ln y = 1 + ce^{-x}$ **33.** $y = \pm(2x/3 - 1 + c/x^2)^{1/2}$

SECTION 2–6

1. $v = 5(t - 1 + e^{-t})$ **2.** $v = te^{-0.01t}$ **3.** $v = 2000 - 400\sqrt{25 - t}$

4.

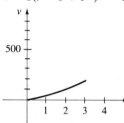

5. 420 fps   **6.** $v = \begin{cases} 100t - 500(1 - e^{-t/5}), & 0 \le t \le 2 \\ 200 - 500(e^{-(t-2)/5} - e^{-t/5}), & t > 2 \end{cases}$

7. (a) $v = 320(1 - e^{-t/10})$

(b) $v = \dfrac{16(1 + 0.9e^{120-4t})}{1 - 0.9e^{120-4t}}$ $\left(\text{NOTE: } \dfrac{16 - v(30)}{16 + v(30)} \approx -0.9\right)$

8. (a) $m\dot{v} = W \sin 30°$ (b) $v = 16t + 2$ (c) $x = 8t^2 + 2t + 5$ (d) 1.25 sec
9. (a) $v = (16 - 4\sqrt{3})t$ (b) $x = (8 - 2\sqrt{3})t^2$ (c) 113.4 ft and 45.4 ft/sec
10. (a) $v = 16(4 - \sqrt{3})(1 - e^{-t/4})$ (b) $x = 16(4 - \sqrt{3})(t - 4 + 4e^{-t/4})$
(c) 77.9 ft and 25.9 ft/sec   **11.** (a) $Q = 30 - 25e^{-t/5}$ lb   (b) 30 lb
12. $Q = 40 - 15e^{-0.1t}$ gal   **13.** (a) $Q = 4t + 100 - 22{,}500\sqrt{2}(2t + 50)^{-3/2}$ lb
(b) 51.5 lb   **14.** 898.8 lb   **15.** $Q = 300(1 - e^{-0.05t}) - 15te^{-0.05t}$ **16.** (a) 1215 lb
(b) 0.0042 lb   **17.** (b) 8.7 yr   **18.** $P = P_i(1 - e^{-Rt/v}) + P(0)e^{-Rt/v}$

19. $i = 0.6(1 - e^{-100t})$ **20.** $i = e^{-t} - e^{-2t}$ **21.** $i = \begin{cases} \frac{1}{10}(2t - 1 + e^{-2t}), & 0 \le t \le 2 \\ 0.4 - 5.4e^{-2t}, & t > 2 \end{cases}$

22.

23. $i = \begin{cases} 3[1 - (1 + t)^{-2}], & 0 \le t \le 1 \\ 3 - 0.75e^{-(t-1)}, & t > 1 \end{cases}$ **24.** $\frac{15}{4}$ amp

25. $q = VC(1 - e^{-t/RC})$ **26.** $i = \dfrac{V}{R}e^{-t/RC}$

29. (a) $i = (R \sin t - L \cos t + Le^{-Rt/L})/(R^2 + L^2)$ (b) $i = \dfrac{L}{R^2}(e^{-Rt/L} - 1) + \dfrac{t}{R}$

30. $i = \begin{cases} 1 - e^{-t}, & 0 \le t \le 1 \\ 3 - t - (e + 1)e^{-t}, & 1 \le t \le 2 \\ (e^2 - e - 1)e^{-t}, & t > 2 \end{cases}$ **31.** $i = \begin{cases} 2 - \dfrac{2}{5^{10}}(5 - t)^{10}, & 0 \le t \le 5 \\ 2, & t > 5 \end{cases}$

32. $i = 2$ for all t **33.** $i = \begin{cases} 1, & 0 \le t \le 1 \\ -1 + 2e^{-(t-1)}, & 1 \le t \le 2 \\ (2e - e^2)e^{-t}, & t > 2 \end{cases}$ **34.** $i(0) = \dfrac{1 - e}{1 + e}$ amp

35. $i(0) = \dfrac{1}{1 + e}$ amp

Chapter 2 Review

1. $y = \dfrac{c}{(x^2 + 2)^{3/2}}$ **2.** $y = e^{x^2}$ **3.** $s = 1 + e^{-2t}$ **4.** $y = 2 \sin (\frac{1}{2}x^2 + c)$
5. $y = c(x + 3)^{3/5}(x - 2)^{2/5}$ **6.** $t \sin t + \cos t = \frac{1}{2}s^2 + c$

7. $y - x - \ln|x + y + 1| = c$ **8.** $\frac{1}{2}(x + y) + \sqrt{x + y} - \ln(\sqrt{x + y} + 1) = x + c$

9. $\frac{1}{3} \ln |\sec 3y| - \frac{1}{2}xe^{-2x} - \frac{1}{4}e^{-2x} = c$ **10.** $\frac{3}{y} + x + \ln |xy| = c$

11. $\text{Arctan } \frac{2y}{x} - \ln(4y^2 + x^2) = c$ **12.** $\frac{1}{2}x^2 + xy - 2y = c$

13. same as #11 with $x - 1$ substituted for x and $y - 1$ substituted for y

14. same as #12 with $y - 1$ substituted for y **15.** $x \ln y + y + cx = 0$

16. $x^8(5y^2 + x^2) = c$ **17.** $xy^2 + x^3 = c$ **18.** $xy + y^2 - x^2 = c$ **19.** $v^3 + v \tan t = c$

20. $xy^2 + e^y = c$ **21.** $y = \dfrac{x}{c - x^2}$ **22.** $y = \dfrac{\sin x}{c - \cos x}$ **23.** $p = -2 + ce^{2t}$

24. $y = ce^{3x} - \frac{1}{2}e^x$ **25.** $y = \frac{1}{4}x^2 + cx^{-2}$ **26.** $y = c(2x + 1)^{3/2} - \frac{10}{3}$

27. $v = (10t + c) \sec t$ **28.** $y = \dfrac{\frac{3}{2} \text{Arctan } \frac{x}{2} + c}{x^2}$ **29.** $t = 39.2 \text{ yr}$

30. $t = 10.02 \text{ sec}$ **31.** $(y - 2)^2 = 4x$ **32.** $y = -4x + c$ **33.** $\dfrac{3y^2}{2} + x^2 = c$

34. $v = 20{,}000\left[1 - \left(1 - \dfrac{t}{15}\right)^{1/4}\right];$ $v(5) = 1930 \text{ ft/sec}$

35. $v = \begin{cases} 12e^{t/2} + 8e^{-2t}, & t \le 5 \\ (8 + 12e^{12.5})e^{-2t}, & t > 5 \end{cases}$

36. (a) $Q(t) = 4\left[(40 + t) - \dfrac{40^3}{(40 + t)^2}\right]$ (b) $Q(5) = 4\left[45 - \dfrac{40^3}{45^2}\right] = 53.6 \text{ lb}$

37. $i = \begin{cases} \frac{3}{2}t - \frac{3}{4}(1 - e^{-2t}), & 0 \le t \le \frac{1}{2} \\ \frac{3}{4}(1 - e)e^{-2t} + \frac{3}{4}, & \frac{1}{2} \le t \le 1 \end{cases}$

SECTION 3–1

1.   **2.**   **3.**

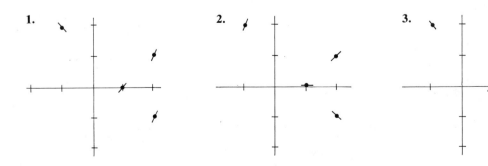

4.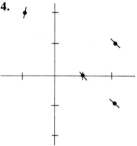

5.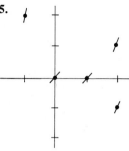

6.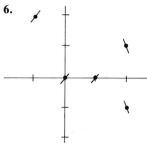

7.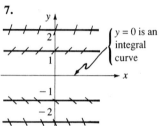

$\begin{cases} y = 0 \text{ is an} \\ \text{integral} \\ \text{curve} \end{cases}$

8.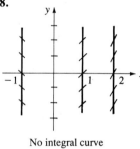

No integral curve

9.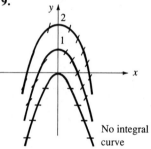

No integral curve

10.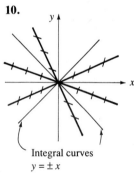

Integral curves
$y = \pm x$

11.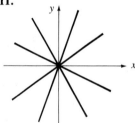

Integral curves
$y = \dfrac{-4 \pm 2\sqrt{5}}{3 \mp \sqrt{5}} x$

12.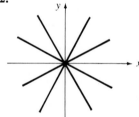

Integral curves
$y = \dfrac{2}{1 \pm \sqrt{5}} x$

13.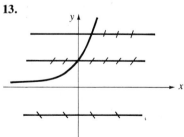

14.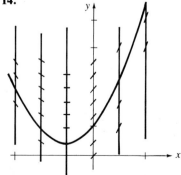

15.

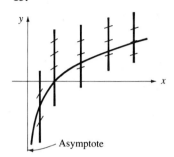

Asymptote

16.

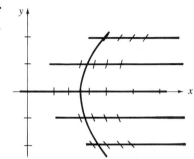

17.

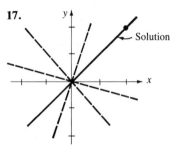

Solution

Every isocline is an integral curve

18.

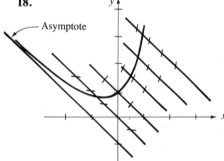

Asymptote

19.

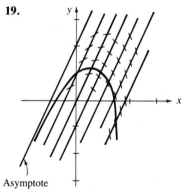

Asymptote

20.

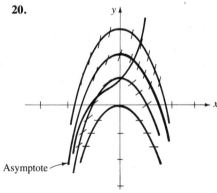

Asymptote

21.

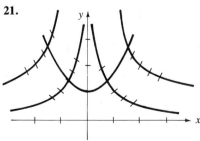

22.

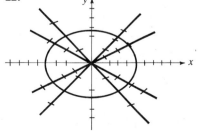

SECTION 3–2

1. $y(0.8) = 5.22$ **2.** $y(1) = 4.75$ **3.** $y(2) = 2.96$ **4.** $y(1.5) = 4.08$
5. $y(0.5) = 0.5$ **6.** $y(0.6) = -1.14$ **7.** $s(1) = 1.58$ **8.** $s(1.7) = 3.37$
9. $y(2) = 1.47$ **10.** $y(2) = 1.5$, 2% error   **11.** 5.06   **12.** 5   **13.** 4.26   **14.** 4.28
15. 0.5   **16.** -1.33 **17.** 0.188   **18.** 3.399   **19.** 1.5   **20.** negligible error
21. $s(0.4) = 0.146$ **22.** $T(2) = -9.33°$ **23.** $i(0.8) = 0.12$ amp

SECTION 3–3

1. $y(0.8) = 5.67$ **2.** $y(1) = 5.0$ **3.** $y(2) = 4.0$ **4.** $y(1.5) = 4.29$ **4.** $y(0.5) = 0.5$
6. $y(0.6) = -1.33$ **7.** $s(1) = 2.0$ **8.** $s(1.7) = 3.40$ **9.** $y(1) = 0.265$
10. $y(2) = 4.38$ **11.** 0.736 ft

SECTION 3–4

1. 1.25   **2.** 0.5   **3.** -2.359 **4.** 2.75   **5.** 3.536   **6.** 1.542
7. $y = e^{x^2}$, $y(1) = 1.284$; $h = 0.5$, percentage error $= 2.65\%$
8. $y = x^2/2$, $y(1) = 0.5$; error $= 0\%$
9. $y = e^x - x^2 - 2x - 2$, $y(1) = e - 5$; error $= 3.4\%$
10. $y = -2e^{-(x-1)} + 4$, $y(1.5) = 2.787$; $h = 0.5$, error $= 1.33\%$

SECTION 3–5

1. $y_1 = -1 - x + \dfrac{x^2}{2}$; $y_2 = -1 - x + \dfrac{x^3}{6}$; $y_3 = -1 - x + \dfrac{x^4}{24}$

2. $y_1 = \dfrac{x^2}{2}$; $y_2 = \dfrac{x^2}{2} + \dfrac{x^3}{6}$; $y_3 = \dfrac{x^2}{2} + \dfrac{x^3}{6} + \dfrac{x^4}{24}$

3. $y_1 = y_2 = y_3 = 1 + \dfrac{x^3}{3}$ **4.** $y_1 = y_2 = y_3 = \dfrac{17}{3} - 4x + \dfrac{x^3}{3}$

5. $y_1 = -2 - 2x$; $y_2 = -2 - 2x - x^2$; $y_3 = -2 - 2x - x^2 - \dfrac{x^2}{3}$

6. $y_1 = 3 - 9x + \dfrac{x^2}{2}$; $y_2 = 3 - 9x + \dfrac{55x^2}{2} - 28x^3 + \dfrac{9x^4}{4} - \dfrac{x^5}{20}$

7. The difficulty encountered is that the function of previous iteration must be substituted into the exponent. Such integration can be done only by first expanding e^y in an infinite series.

Chapter 3 Review

1.

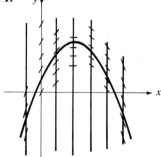

2.

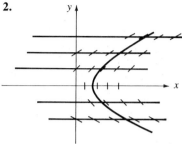

3.

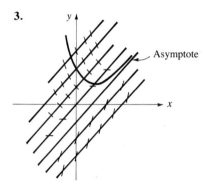

Asymptote

4.

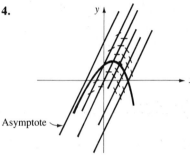

Asymptote

5. -1.47 **6.** 1.64 **7.** 1.13 **8.** 0.276 **9.** -1.52 **10.** 1.66 **11.** 1.26
12. 0.245 **13.** -1.52 **14.** 1.66 **15.** 1.25 **16.** 0.222 **17.** (a) 1.52
(b) 1.31 (c) 1.34;   D. E. solution: $y = 1.3356$ **18.** (a) 2.73 (b) 2.69
(c) 2.67;   D. E. solution $y = 2.6706$ **19.** -1.48 **20.** 1.66 **21.** 1.18 **22.** 0.25
23. $y_3 = \frac{1}{2}x^2 + x + 1$ **24.** $y_3 = 1 + x + \frac{1}{2}x^2 + \frac{1}{6}x^3$ **25.** $y_3 = -1 + x + \frac{1}{2}x^2 + \frac{1}{6}x^3$
26. $y_3 = \frac{1}{3}x^3$

SECTION 4–1

1. linear, nonhomogeneous   **2.** linear, homogeneous   **3.** linear, nonhomogeneous
4. linear, nonhomogeneous   **5.** nonlinear   **6.** linear, nonhomogeneous   **7.** nonlinear
8. nonlinear   **9.** linear, homogeneous   **10.** linear, homogeneous
11. linear, nonhomogeneous   **12.** linear, homogeneous   **13.** nonlinear   **14.** nonlinear
15. unique; $(-\infty, \infty)$ **16.** unique; $(-\infty, \infty)$ **17.** unique; $(-\infty, 1)$ **18.** unique; $(0, \infty)$
19. nonlinear; theorem does not apply   **20.** nonlinear; theorem does not apply
21. leading coefficient $= 0$ at $x = 1$; theorem does not apply
22. leading coefficient $= 0$ at $x = 0$; theorem does not apply   **23.** unique; $(-4, \infty)$
24. boundary-value problem; theorem does not apply
25. boundary-value problem; theorem does not apply   **26.** unique; $(-\infty, \infty)$

SECTION 4–2

4. (c) the theorem applies only to homogeneous differential equations, and this equation is not homogeneous

5. linearly independent   **6.** linearly independent   **7.** linearly independent

8. linearly independent   **9.** linearly independent   **10.** linearly independent

11. linearly independent   **12.** linearly dependent   **13.** linearly dependent

14. linearly dependent   **15.** linearly independent   **16.** linearly independent

17. $c_1 = -2$, $c_2 = -8$, $c_3 = 1$, $c_4 = 0$ **18.** $c_1 = 1$, $c_2 = 0$, $c_3 = -5$, $c_4 = -5$

21. y_1, y_2, and y_3 are not solutions of the same linear DE with continuous coefficients.

22. (b) $y = c_1 \sin \frac{1}{2}x + c_2 \cos \frac{1}{2}x$ **23.** (b) $y = c_1 e^{-x} + c_2 x e^{-x}$

24. (b) $y = c_1 + c_2 x + c_3 e^{2x}$

SECTION 4–3

1. $y = c_1 e^{2x} + c_2 e^{-2x}$ **2.** $y = c_1 \cosh 2x + c_2 \sinh 2x$ **3.** $y = c_1 e^{3x} + c_2 e^{-3x}$

4. $y = c_1 e^{-3x} + c_2 e^{2x}$ **5.** $y = c_1 e^{6x} + c_2 e^{x}$ **6.** $y = c_1 \sin 2x + c_2 \cos 2x$

7. $y = c_1 e^{-2x} + c_2 x e^{-2x}$ **8.** $y = c_1 \cos x + c_2 \sin x$ **9.** $y = c_1 x^2 + c_2 x^{-1}$

10. $y = c_1 + c_2 x^{-1}$ **11.** $y = c_1 \ln x + c_2$ **12.** $y = c_1 x^3 + c_2 x^{-2}$

13. $y = c_1 + c_2(2x - \sin 2x)$ **14.** $y = (c_1 + c_2 \ln x)x^2$ **15.** $y = c_1 e^{x} + c_2 x$

16. $y = c_1 + c_2 e^{x}(x - 2)$ **17.** $y = c_1 x + c_2 x \ln x$ **18.** $y = c_1 x \sin (\ln x) + c_2 x \cos (\ln x)$

19. $y = -1 + \dfrac{x}{2} \ln \left| \dfrac{x + 1}{x - 1} \right|$ **20.** $y = x^{-1/2} \cos x$

SECTION 4–4

1. $y = c_1 e^{x} + c_2 e^{2x}$ **2.** $y = c_1 e^{-2x} + c_2 e^{-3x}$ **3.** $s = c_1 + c_2 e^{-t}$ **4.** $y = c_1 e^{-x} + c_2 e^{-4x}$

5. $y = c_1 + c_2 e^{1.5x}$ **6.** $y = ce^{4x/3}$ **7.** $y = c_1 e^{2x} + c_2 e^{-2x}$ **8.** $i = c_1 e^{3t} + c_2 e^{-3t}$

9. $y = c_1 + c_2 e^{4x} + c_3 e^{-4x}$ **10.** $y = c_1 + c_2 e^{2x} + c_3 e^{-2x}$ **11.** $y = c_1 + c_2 e^{5x} + c_3 e^{-4x}$

12. $y = c_1 + c_2 e^{-x} + c_3 e^{-8x}$ **13.** $y = c_1 + c_2 e^{x/3} + c_3 e^{-2x}$ **14.** $s = c_1 + c_2 e^{-t} + c_3 e^{-3t/2}$

15. $y = c_1 e^{-x} + c_2 e^{-2x} + c_3 e^{-3x}$ **16.** $y = c_1 e^{-x} + c_2 e^{x/3} + c_2 e^{2x/3}$ **17.** $s = \frac{1}{2}(e^{2t} - e^{-2t})$

18. $y = e^{-x} - e^{3x}$ **19.** $y = e^{x}$ **20.** $y = 4 - 2e^{-3x}$ **21.** $y = 2e^{-x}$

SECTION 4–5

1. $y = c_1 e^{-4x} + c_2 x e^{-4x}$ **2.** $y = e^{x/2}(c_1 + c_2 x)$ **3.** $y = (c_1 + c_2 x)e^{-x/3}$

4. $y = c_1 e^{-x} + c_2 e^{-4x}$ **5.** $y = c_1 + c_2 x + c_3 x^2 + c_4 e^{-x} + c_5 e^{x}$

6. $s = c_1 + c_2 t + c_3 e^{t} + c_4 e^{-t}$ **7.** $y = c_1 + c_2 x + (c_3 + c_4 x)e^{-9x}$

8. $y = c_1 + c_2 x + (c_3 + c_4 x)e^{-x/3}$ **9.** $y = c_1 e^{-x} + (c_2 + c_3 x)e^{x/2}$

10. $y = c_1 e^{2x} + (c_2 + c_3 x)e^{-x}$ **11.** $y = c_1 + c_2 x + (c_3 + c_4 x)e^{-x} + c_5 e^{2x}$

12. $y = c_1 e^x + (c_2 + c_3 x + c_4 x^2)e^{2x}$ **13.** $y = c_1 \cos 3x + c_2 \sin 3x$

14. $y = c_1 + c_2 \cos 4x + c_3 \sin 4x$ **15.** $y = c_1 + c_2 \cos 5x + c_3 \sin 5x$

16. $y = e^x(c_1 \cos 2x + c_2 \sin 2x)$ **17.** $s = e^{3t}(c_1 \cos 4t + c_2 \sin 4t)$

18. $y = c_1 + e^{-x}(c_2 \cos \sqrt{3}x + c_3 \sin \sqrt{3}x)$ **19.** $y = c_1 + c_2 x + c_3 \cos x + c_4 \sin x$

20. $y = (c_1 + c_2 x)e^{-2x} + c_3 \cos 2x + c_4 \sin 2x$ **21.** $y = c_1 + (c_2 + c_3 x)e^{-9x}$

22. $y = c_1 e^{2x} + c_2 e^{-2x} + c_3 \cos 2x + c_4 \sin 2x$ **23.** $y = e^{-x/4}\left(c_1 \cos \dfrac{\sqrt{7}}{4}x + c_2 \sin \dfrac{\sqrt{7}}{4}x\right)$

24. $y = c_1 \cos x + c_2 \sin x + e^{-x/2}\left(c_3 \cos \dfrac{\sqrt{3}}{2}x + c_4 \sin \dfrac{\sqrt{3}}{2}x\right)$

25. $y = c_1 e^{-x} + c_2 \cos 2x + c_3 \sin 2x$ **26.** $y = c_1 e^{-x} + (c_2 + c_3 x)e^x$

27. $y = c_1 e^{-x} + e^{2x}(c_2 \cos 3x + c_3 \sin 3x)$

28. $y = (c_1 + c_2 x)e^{-2x} + (c_3 \cos \sqrt{3}x + c_4 \sin \sqrt{3}x)e^{2x}$

29. $y = (c_1 + c_2 x)\cos 3x + (c_3 + c_4 x)\sin 3x + c_5 e^{-x}$

30. $y = (c_1 + c_2 x)\cos x + (c_3 + c_4 x)\sin x + (c_5 + c_6 x)e^x + (c_7 + c_8 x)e^{-x}$ **31.** $y = xe^{4x}$

32. $y = (1 + x)e^x$ **33.** $y = (1 - 2x)e^{3x}$ **34.** $y = 2 + 3x + e^{-3x}$ **35.** $y = \frac{1}{2}\sin 2x$

36. $y = 3 \cos x$ **37.** $y = e^{-2x}(\cos x + 2 \sin x)$ **38.** $y = e^{3x}(2 \cos x - 5 \sin x)$

39. $y = c \sin 2x$ **40.** $y = \frac{1}{2}\sin 2x$ **41.** no solution   **42.** no solution   **43.** $y = x$

44. $\lambda = \dfrac{2n + 1}{2}; \; y = c \sin\left(\dfrac{2n + 1}{2}\right)x$

SECTION 4–6

1. (a) $x = \frac{1}{2}\sin 2t$ (b) 2   (c) $\frac{1}{2}$ **2.** 3 ft and 2π sec

3. (a) $x = -\frac{1}{4}\cos 16t, \; v = 4 \sin 16t, \; a = 64 \cos 16t$

(b) $x = 0.104$ ft, $v = 3.64$ fps, $a = -26.6$ fps^2 **4.** (a) $x = \frac{1}{3}\cos 8t + \frac{1}{4}\sin 8t$

(b) $A = \frac{5}{12}$, period $= \pi/4$, $\delta = \tan^{-1} 0.75$

(c)

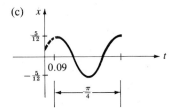

5. $x = -\frac{1}{2}\cos 4t + \frac{1}{2}\sin 4t$

6. (a) $x = -\frac{2}{3}\cos 3t$ (b) $x(0) = -\frac{2}{3}, x'(0) = 0$ (c) $\dfrac{2\pi}{3}$ **7.** $\frac{25}{6}$ lb

SECTION 4–7

1. (a) $x = \frac{1}{4}(1 + 8t)e^{-8t}$ (b)    (c)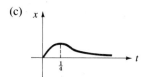

2. (a) $x = te^{-4t}$ (b) $\dfrac{1}{4e}$

3. (a) $x = -0.25e^{-4t} \sin 8t$ (b) -0.128 ft, 0.027 ft/sec, -0.006 ft/sec^2
  (c)

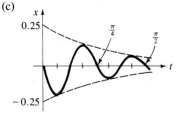

4. $x = e^{-t}(\sin 2t + 2 \cos 2t)$ **5.** (a) critically damped   (b) overdamped
   (c) critically damped   (d) damped oscillation   (e) damped oscillation

6. $f = \dfrac{1}{2\pi} \dfrac{\sqrt{c^2 - 4\,mk}}{2m}$ **7.** $t = \dfrac{\delta - \mathrm{Arctan}\left(\dfrac{1}{\omega^*}\right)}{\omega^*}$

Chapter 4 Review

1. nonlinear   **2.** nonlinear   **3.** linear, unique solution on interval that does not include -2.
4. linear, unique solution   **5.** linear, unique solution not guaranteed
6. linear, theorem does not apply: $\tan x$ is not continuous on $(-\infty, \infty)$
11. Try $c_1 = 0, c_2 = \frac{5}{2}, c_3 = -1$ **12.** Try $c_1 = c_2 = 0, c_3 = -2, c_4 = 1$ **13.** $y_2 = e^{4x}$
14. $y_2 = e^{-2x}$ **15.** $y_2 = \sqrt{x}$ **16.** $y_2 = x$ **17.** $y = c_1 + c_2e^{-x/2} + c_3e^{3x}$
18. $y = c_1 + c_2x + c_3e^{12x}$ **19.** $y = c_1 \sin \sqrt{5}x + c_2 \cos \sqrt{5}x$
20. $y = e^{-x}(c_1 \cos x + c_2 \sin x)$ **21.** $y = (c_1x + c_2)e^{-8x}$ **22.** $y = c_1e^{-\sqrt{3}x} + c_2e^{\sqrt{3}x}$
23. $y = c_1e^{(-1 + \sqrt{3})x/2} + c_2e^{(-1 - \sqrt{3})x/2}$ **24.** $y = e^{-x/2}\left(c_1 \cos \dfrac{\sqrt{3}}{2}x + c_2 \sin \dfrac{\sqrt{3}}{2}x\right)$
25. $y = c_1e^{-2x} + c_2 \cos 2x + c_3 \sin 2x$ **26.** $y = c_1e^{-x} + c_2e^{3x} + c_3e^{-x/2}$
27. $y = \cos \sqrt{2}x - \sqrt{2} \sin \sqrt{2}\,x$ **28.** $y = \frac{1}{2}e^{-2x} - \frac{11}{2}$ **29.** $y = e^{-2x}(3x + 1)$
30. $y = \frac{4}{5}e^{2x} - \frac{4}{5}e^{-x/2}$ **31.** (a) $x = -\frac{1}{3} \cos 4t - \sin 4t, v = \frac{4}{3} \sin 4t - 4 \cos 4t$
    (b) $x(1) = 0.97$ ft, $v(1) = 1.61$ ft/sec   (c) $A = \sqrt{10}/3$, period $= \pi/2$
32. (a) $x = -e^{-2t}(\cos 4t + \frac{1}{2} \sin 4t)$ (b) underdamped
33. (a) $x = 10te^{-8t}$ (b) $x_{\max} = \dfrac{5}{4e}$ ft

SECTION 5–1

1. $y = c_1 e^{-x} + c_2 e^{-2x}$ **2.** $y = c_1 e^{2x} + c_2 e^{-4x}$ **3.** $y = (c_1 x + c_2)e^{-3x}$
4. $y = (c_1 x + c_2)e^{2x}$ **5.** $y = c_1 \cos 2x + c_2 \sin 2x$ **6.** $y = c_1 \cos 4x + c_2 \sin 4x$
7. $y = c_1 e^{-x} + c_2 e^{-2x} + 2e^x$ **8.** $y = (c_1 x + c_2)e^{-3x} + x - \frac{4}{9}$
9. $y = c_1 \cos 2x + c_2 \sin 2x + \frac{1}{5} e^{-x}$ **10.** $y = c_1 \cos 4x + c_2 \sin 4x + \frac{2}{7} \sin 3x$
13. $y = c_1 e^{2x} + c_2 e^{-2x} - \frac{3}{4}$ **14.** $y = c_1 e^x + c_2 e^{-x} + \frac{1}{8} e^{3x}$ **15.** $y = c_1 e^x + c_2 e^{-x} - x$
16. $y = c_1 e^{3x} + c_2 e^{-2x} - \frac{5}{6}$ **17.** $y = e^{-2x}(c_1 + c_2 x + \frac{1}{2} x^2)$
18. $y = c_1 + c_2 e^{-x} - x + \frac{1}{2} x^2$
19. $y = c_1 + c_2 e^{-3x} + \frac{1}{4} e^x$ **20.** $y = c_1 + c_2 e^{2x} + \frac{1}{2} x e^{2x}$
21. $y = c_1 e^{\frac{1}{2}x} + c_2 e^{-2x} + \frac{25}{2}$ **22.** $y = c_1 e^{-2x} + c_2 x e^{-2x} + \frac{1}{4}$
23. $y = c_1 + c_2 e^{-4x} + \frac{1}{12} e^{2x}$ **24.** $y = c_1 + c_2 e^x + \frac{1}{2} e^{-x}$
25. $y = c_1 + c_2 e^x + \frac{1}{2} \cos x - \frac{1}{2} \sin x$ **26.** $y = c_1 + c_2 e^{-3x} - \frac{1}{13} \cos 2x + \frac{3}{26} \sin 2x$

SECTION 5–2

1. (a) $Ax + Bx^2$ (b) $Ax + Bx^2$ (c) $A \cos x + B \sin x + Cx + Dx^2$ (d) $Ax^2 e^{-x}$
(e) $(Ax^2 + Bx^3)e^{-x}$ (f) $Ax^2 e^{-x} + Be^x$ (g) Ae^x (h) $Ax + Bx^2 + (Cx^2 + Dx^3)e^{-x}$
2. $y = e^x + (c_1 + c_2 x)e^{2x}$ **3.** $y = (c_1 + c_2 x)e^{2x} + \frac{1}{4} + e^x$ **4.** $y = (c_1 + c_2 x + \frac{1}{2} x^2)e^{2x}$
5. $y = (c_1 + c_2 x)e^{2x} + \frac{4}{25} \cos x + \frac{3}{25} \sin x$ **6.** $y = (c_1 + c_2 x + \frac{1}{2} x^2 + \frac{1}{6} x^3)e^{2x}$
7. $y = (c_1 + c_2 x + \frac{1}{6} x^3)e^{2x}$ **8.** $y = c_1 \cos x + c_2 \sin x - \frac{1}{3} \sin 2x$
9. $y = c_1 \cos 2x + c_2 \sin 2x - \frac{1}{4} x \cos 2x$ **10.** $y = c_1 + c_2 e^{-4x} - \frac{1}{10} \cos 2x - \frac{1}{20} \sin 2x$
11. $y = e^x(c_1 \cos 2x + c_2 \sin 2x) + \frac{4}{17} \cos 2x + \frac{1}{17} \sin 2x$
12. $y = (c_1 \cos 2x + c_2 \sin 2x + \frac{1}{4} x \sin 2x)e^x$ **13.** $y = \frac{1}{3} \cosh 2x + c_1 e^{-x} + c_2 e^x$
14. $y = c_1 e^{-x} + c_2 e^x + \frac{1}{2} x \sinh x$ **15.** $y = c_1 \cos x + c_2 \sin x + \frac{1}{4} x(\cos x + x \sin x)$
16. $y = c_1 e^x + c_2 e^{2x} - x e^x + x e^{2x} + \frac{1}{6} e^{-x}$ **17.** $y = c_1 + c_2 e^x - 2x - x^2 - \frac{1}{3} x^3$
18. $y = c_1 + c_2 e^x - 3x$ **19.** $y = c_1 + c_2 e^{-3x} + \frac{8}{9} x + \frac{1}{6} x^2$
20. $y = \frac{3}{4} + c_1 \cos 2x + c_2 \sin 2x$ **21.** $y = \frac{3}{4} x + c_1 \cos 2x + c_2 \sin 2x$
22. $y = (c_1 + c_2 x)e^{-x} + c_3 e^{2x} + \frac{7}{100} \cos 2x - \frac{1}{100} \sin 2x$
23. $y = c_1 e^x + c_2 e^{-x} + c_3 \cos x + c_4 \sin x - \frac{1}{4} x e^{-x}$
24. $y = c_1 e^{-2x} + e^{-x}(c_2 \cos 2x + c_3 \sin 2x) - \frac{1}{24} e^x$ **25.** $y = c_1 e^{-x} + c_2 + c_3 x + \frac{1}{2} x^2$
26. $y = c_1 e^x + e^{-x}(c_2 \cos x + c_3 \sin x) - \frac{1}{2}(x^2 + 1) - \frac{1}{5}(3 \cos 2x + 4 \sin 2x)$

SECTION 5–2* (OPERATOR EXERCISES)

1. $(7D^2 + 3D + 4)y = x^2$ **2.** $(D^2 - D - 1)y = 0$ **3.** $(D^2 + 4)y = 0$
4. $(3D^3 - 7D + 4)y = 0$ **5.** $(D^3 + 6D^2 - 2D)y = \sin x$ **6.** $(2D^4 + 9D^2 - D)y = e^{2x}$
7. $D^2 + D - 6$ **8.** $D^2 - 4$ **9.** $D^3 + 2D^2 - 5D - 10$ **10.** $D^3 + 4D^2 + 3D$
11. $D^5 + 4D^3$ **12.** $D^3 + 6D^2 + 12D + 8$ **13.** $D^3 - D^2 - D + 1$
14. $D_{\cdot}^3 + 4D^2 - 5D$ **15.** $(D - 3)(D + 3)$ **16.** $(D - 3)(D + 2)$ **17.** $D(D + 3)^2$
18. $D^2(D - 10)$ **19.** $(2D - 1)(D + 2)$ **20.** $(D + 1)(D - 2)^2$
21. $(D - 2)(D + 3)(D^2 + 4)$ **22.** $(D + 2)^3$ **23.** $D - 3$ **24.** $D + 4$ **25.** $D^2 - 1$
26. $2D - 1$ **27.** $(D + 3)^2$ **28.** $(D + 3)^2(D - 1)$ **29.** D^3 **30.** $D(D - 5)^2$
31. $D^3(D - 1)$ **32.** $D^2(D - 1)^4$ **33.** $D^2 + 1$ **34.** $(D - 2)(D^2 + 25)$
35. $D^2(D^2 + 4)$ **36.** $(D^2 + 4)^2$ **37.** $(D^2 + 1)^2$ **38.** $(D - 2)^2 + 1$
39. $(D + 1)^2 + 4$ **40.** $(D - 3)^2 + 4$ **41.** $D(D^2 + 4)$ **42.** $D^2 + 4$ **43.** $D^2 + 4$
44. $D(D^2 + 4)$

SECTION 5–2*

The answers for these exercises are identical to those of Section 5–2.

SECTION 5–3

1. $y = c_1 e^{-x} + c_2 e^{2x} + \frac{1}{3}x e^{2x}$ **2.** $y = c_1 e^{-x} + c_2 x e^{-x} + x e^{-x} \ln x$

3. $y = c_1 \cos 2x + c_2 \sin 2x - \frac{1}{4}\cos 2x \ln (\sec 2x + \tan 2x)$

4. $y = c_1 \cos 2x + c_2 \sin 2x - \frac{1}{2} + \frac{1}{4}\sin 2x \ln (\sec 2x + \tan 2x)$

5. $y = c_1 \cos 2x + c_2 \sin 2x + \frac{1}{12}\sin^4 2x + \frac{1}{4}\cos^2 2x - \frac{1}{12}\cos^4 2x$

6. $y = c_1 e^x + c_2 e^{2x} - e^{2x} \cos (e^{-x})$ **7.** $y = c_1 e^{2x} + c_2 x e^{2x} + x e^{2x} \ln x$

8. $y = c_1 e^{-3x} + c_2 x e^{-3x} - \frac{1}{2}e^{-3x} \ln(x^2 + 1) + x e^{-3x} \tan^{-1} x$

9. $y = c_1 e^{-x} + c_2 x e^{-x} + \frac{1}{4}x^2 e^{-x}(2 \ln x - 3)$ **10.** $y = c_1 e^{-x} + c_2 x e^{-x} + \dfrac{e^{-x}}{2x}$

11. $y = c_1 x + c_2 x^{-1} + \frac{1}{3}x^2 \ln x - \frac{4}{9}x^2$ **12.** $y = c_1 x + c_2 x^{-1} + \frac{1}{3}x^2$

13. $y = c_1 x + c_2 x \ln x + \dfrac{1}{4x}$ **14.** $y = c_1 x + c_2 e^x - \frac{1}{2}x e^{-x} + \frac{1}{4}e^{-x}$

16. $y = c_1 x^{-1} + c_2 x^{-2}$ **17.** $y = c_1 x + c_2 x^2 - 4x \ln x + \frac{1}{10}[\sin (\ln x) + 3 \cos (\ln x)]$

18. $y = c_1 x^3 + c_2 x^{-1} - \frac{1}{9}x^2 (3 \ln x + 2)$ **19.** $y = c_1 \cos (3 \ln x) + c_2 \sin (3 \ln x)$

20. $y = c_1 x + c_2 x^2 - \frac{3}{4} - \frac{1}{2}\ln x - x \ln x - \frac{1}{2}x (\ln x)^2$

22. $y = c_1 e^{-x} + c_2 + c_3 x + \frac{1}{2}x^2$

23. $y = (c_1 + c_2 x)e^{-x} + c_3 e^{2x} + \frac{7}{100}\cos 2x - \frac{1}{100}\sin 2x$

24. $u = \displaystyle\int \dfrac{W(y_2, y_3) f(x)/b_3}{W(y_1, y_2, y_3)}\, dx; \quad v = -\int \dfrac{W(y_1, y_3) f(x)/b_3}{W(y_1, y_2, y_3)}\, dx; \quad w = \int \dfrac{W(y_1, y_2) f(x)/b_3}{W(y_1, y_2, y_3)}\, dx$

SECTION 5–4

1. $x = \sin t \sin 9t$ **2.** $x = \frac{1}{4}(\cos 8t - \sin 8t) + \frac{1}{2}\sin 4t$

3. $x = \cos t - 6 \sin t + 3 \sin 2t$ **4.** $x = 4 \cos 3t + 2t \sin 3t$ **5.** $x = t \sin 8t$

6.

7. $x = \frac{1}{8}\sin 2t - \frac{1}{4}t \cos 2t$

SECTION 5–5

1. $x_t = e^{-5t/18}\left(-\dfrac{35}{37}\cos\dfrac{\sqrt{551}}{18}t - \dfrac{625}{37\sqrt{551}}\sin\dfrac{\sqrt{551}}{18}t\right)$ **2.** $x_t = e^{-t}(\sin t + 2 \cos t)$

$x_s = \frac{35}{37}\cos t + \frac{25}{37}\sin t$ $x_s = -2 \cos t + \sin t$

3. 1.27 and 0   **4.** $\dfrac{2mf}{c\sqrt{4m^2\omega_0^2 - c^2}}$ **5.** $\dfrac{1}{2\pi f}\text{Arctan}\dfrac{2\pi fc}{k - 4\pi^2 mf^2}$

6. $x = \begin{cases} \frac{13}{8}\cos t + \frac{3}{8}, & 0 \le t < \pi \\ \frac{7}{8}\cos t - \frac{3}{8}, & \pi \le t < 2\pi \\ \frac{1}{8}\cos t + \frac{3}{8}, & 2\pi \le t < 3\pi \end{cases}$

Block stops at $x = \frac{1}{4}$ ft

7. $x = \begin{cases} 0.75 \cos 1.5t + 0.5, & 0 \le t \le \frac{2}{3}\pi \\ -0.25, & t > \frac{2}{3}\pi \end{cases}$ **8.** $x = \begin{cases} \frac{11}{9}\cos\frac{3}{2}t + \frac{7}{9}, & 0 \le t \le \frac{2}{3}\pi \\ -\frac{4}{9}, & t > \frac{2}{3}\pi \end{cases}$

SECTION 5–6

1. $q = -(0.0003 + 0.03t)e^{-100t} + 0.0003$
$i = 3te^{-100t}$

2. $q = (-5 \times 10^{-6})e^{-3000t} + (3 \times 10^{-6})e^{-5000t} + 2 \times 10^{-6}$
$i = 0.015e^{-3000t} - 0.015e^{-5000t}$

3. $q = (192 \times 10^{-6})e^{-1500t} - (216 \times 10^{-6})e^{-1333t} + 24 \times 10^{-6} - 8 \times 10^{-6}t$
$i = -0.288e^{-1500t} + 0.288e^{-1333t} - 8 \times 10^{-6}$

4. $q = (-4 \times 10^{-4})e^{-500t} + (2 \times 10^{-4})e^{-1000t} + 2 \times 10^{-4} - 4 \times 10^{-5}t$
$i = 0.2e^{-500t} - 0.2e^{-1000t} - 4 \times 10^{-5}$

5. $q = -\frac{5}{12}\sin 3t + \frac{5}{4}\sin t$, $i = -\frac{5}{4}\cos 3t + \frac{5}{4}\cos t$; no transient current

6. $q = -\frac{1}{10}\cos 3t + \frac{1}{30}\sin 3t + \frac{1}{10}e^{-t}$, $i = \frac{3}{10}\sin 3t + \frac{1}{10}\cos 3t - \frac{1}{10}e^{-t}$; transient current is $-\frac{1}{10}e^{-t}$

7. $i_t = (c_1 + c_2 t)e^{-t}$ **8.** $i_t = e^{-2t}(c_1 \cos 4t + c_2 \sin 4t)$ **9.** $i_t = e^{-t}(c_1 \cos 2t + c_2 \sin 2t)$
$i_{ss} = 2 \sin t$ $i_{ss} = 2 \cos 2t + \sin 2t$ $i_{ss} = 2 \cos t + \sin t$

10. $i_t = e^{-t}(c_1 \cos t + c_2 \sin t)$ **11.** $i_t = e^{-2t}(c_1 \cos 4t + c_2 \sin 4t)$
    $i_{ss} = \cos t + 2 \sin t$ $i_{ss} = 10 \cos 4t + 40 \sin 4t$

12. $i_t = c_1 e^{-5t} + c_2 e^{-t}$ **13.** $i_t = e^{-3t}(c_1 \cos 4t + c_2 \sin 4t)$
    $i_{ss} = 20 \cos t + 30 \sin t$ $i_{ss} = -4 \cos 5t$

Chapter 5 Review

1. $y = c_1 e^{-x} + c_2 e^{-2x} + \frac{1}{6} e^x$

2. $y = c_1 + c_2 e^{-3x} + \frac{5}{3} x$

3. $y = c_1 + c_2 e^x - \frac{1}{2} \cos x - \frac{1}{2} \sin x$ **4.** $y = c_1 \sin 2x + c_2 \cos 2x + \frac{1}{3} \sin x$

5. $y = c_1 + c_2 e^{-3x} + \frac{1}{2} e^{2x}$ **6.** $y = c_1 \cos x + c_2 \sin x + x^2 - 2$

7. $x = c_1 e^{-3t} + c_2 e^t + \frac{1}{2} t e^t$ **8.** $y = c_1 e^{2x} + c_2 e^{-3x} + 2x e^{2x} - 3 e^{3x}$ **9.** $y = 4x e^{-4x} + 2 e^{-2x}$

10. $x = \frac{9}{125} + \frac{9}{25} t - \frac{1}{10} t^2 - \frac{9}{125} e^{5t}$ **11.** $y = c_1 e^x + c_2 e^{-x} + \frac{1}{2} x e^x$

12. $y = c_1 \cos x + c_2 \sin x - \cos x \ln (\sec x + \tan x)$

13. $y = c_1 \cos x + c_2 \sin x - \frac{1}{2} \sec x + \sin x \ln (\sec x + \tan x)$

14. $y = c_1 e^{-x} + c_2 x e^{-x} + \frac{1}{2} x^2 e^{-x} \ln x - \frac{3}{4} x^2 e^{-x}$ **15.** $y = 2 \sin 2t - 4t \cos 2t$

16. $y = e^{-2t}(\frac{1}{8} \sin 4t - \frac{1}{2} \cos 4t) + \frac{1}{2} \cos 2t + \frac{1}{4} \sin 2t$

17. $i_t = 4 e^{-t} \sin t - 2 e^{-t} \cos t; i_{ss} = -\sin t + 2 \cos t$

SECTION 6–1

1. $\dfrac{6}{s^4}$, $s > 0$ **2.** $\dfrac{1}{s + 2}$, $s > -2$ **3.** $\dfrac{6s^4 + 2s^2 + 48}{s^5}$, $s > 0$

4. $\dfrac{2s - 5}{(s - 7)(s + 2)}$, $s > 7$ **5.** $\dfrac{5s + 7}{(s + 2)(s - 1)}$, $s > 1$

6. $\dfrac{s^4 + 6s^2 + 24}{s^3(s^2 + 4)}$, $s > 0$ **7.** $\dfrac{120}{(s + 3)^6}$, $s > -3$ **8.** $\dfrac{24}{(s - 2)^4}$, $s > 2$

9. $\dfrac{s(s + 3)}{(s - 3)(s^2 + 9)}$, $s > 3$ **10.** $\dfrac{2(6 - s)}{s^2 + 16}$, $s > 0$

11. $\dfrac{6s^2 - 2s^4 - 24}{s^4(s^2 - 4)}$, $s > 2$ **12.** $\dfrac{2s^3 - 2s^2 - 9s + 18}{s^2(s^2 - 9)}$, $s > 3$

13. $\dfrac{s + 3}{(s + 1)^2}$, $s > -1$ **14.** $\dfrac{4s^2 + 19s + 20}{s(s + 2)^2}$, $s > 0$

15. $\dfrac{2(4 - s)}{(s - 3)^3}$, $s > 3$ **16.** $\dfrac{4s^2 - 36s + 82}{(s - 4)^3}$, $s > 4$

17. $\dfrac{1}{s} - \dfrac{4}{2s - 1} + \dfrac{1}{s - 1}$, $s > 1$ **18.** $\dfrac{3}{2s} + \dfrac{2s}{s^2 + 4} + \dfrac{s}{2(s^2 + 16)}$, $s > 0$

SECTION 6–2

10. $\frac{1}{s}(1 + e^{-2s})$ **11.** $\frac{1}{s^2}(1 - e^{-4s})$ **12.** $\frac{2e^{-s}}{s} - \frac{e^{-s}}{s^2} + \frac{1}{s^2} + \frac{2}{s}$

13. $\frac{2}{s^3} - \frac{2e^{-2s}}{s^3} - \frac{4e^{-2s}}{s^2} + \frac{2e^{-2s}}{s}$ **14.** $-\frac{e^{-2(s-1)}}{s-1} + \frac{1}{s-1}$ **15.** $\frac{e^{-\pi s} + 1}{s^2 + 1}$

16. $\frac{1}{s}(1 + e^{-2s} - 2e^{-4s})$ **29.** (a) yes   (b) no   (c) no   (d) yes   (e) no

30. $\sin(e^t)$

SECTION 6–3

1. e^{3t} **2.** $\cos 3t$ **3.** $\frac{2}{3}\sin 3t$ **4.** $\cos t + 2\sin t$ **5.** $\frac{1}{2}t^2$ **6.** $\frac{1}{8}t^4$ **7.** $\frac{1}{12}t^4 e^{3t}$

8. $\frac{1}{2}t^2 e^{-t}$ **9.** $\cosh 4t - \frac{3}{4}\sinh 4t$ **10.** $1 + \cosh 3t$ **11.** $\frac{2}{3}e^{-5t} + \frac{1}{3}e^t$ **12.** e^{3t}

13. $\frac{3}{5}e^{-3t} + \frac{7}{5}e^{2t} - 2$ **14.** $\frac{1}{2}t^2 e^{2t}$ **15.** $\frac{1}{2}t^2 e^{-4t}$ **16.** $7e^t - 6 - 2te^t$ **17.** $2e^t - t - 2$

18. $2e^t - e^{-t} - \frac{3}{2}e^{3t/2}$ **19.** $1 + \cos\sqrt{2}\,t$ **20.** $1 + \frac{1}{\sqrt{2}}\sin\sqrt{2}\,t$ **21.** $e^t(1 + t) + \sin t$

22. $\frac{13}{56}e^{-3t} + \frac{7}{24}e^t - \frac{11}{21}\cos\sqrt{5}\,t - \frac{25}{42\sqrt{5}}\sin\sqrt{5}\,t$

SECTION 6–4

1. $\frac{2}{(s-2)^3}$ **2.** $\frac{5}{s^2 + 4s + 29}$ **3.** $\frac{1}{s^2 - 4s + 3}$ **4.** $\frac{s-1}{s^2 - 2s + 5} - \frac{15}{s^2 - 2s + 26}$

5. $\frac{a_n n!}{(s-a)^{n+1}} + \frac{a_{n-1}(n-1)!}{(s-a)^n} + \cdots + \frac{a_1}{(s-a)^2} + \frac{a_0}{(s-a)}$ **6.** $\frac{3}{s^2 - 4s + 40}$

7. $\frac{1}{2(s+1)} + \frac{s+1}{2(s^2 + 2s + 17)}$ **8.** $\frac{3}{2(s+3)} + \frac{8}{s^2 + 6s + 25} - \frac{s+3}{2(s^2 + 6s + 73)}$

9. $\frac{1}{(s-1)^2} - \frac{1}{(s-1)^2}e^{-2(s-1)}$ **10.** $\frac{1}{s-3} - \frac{1}{s-3}e^{-2(s-3)}$ **11.** $\frac{2}{s+1} - \frac{2}{s+1}e^{-5(s+1)}$

12. $\frac{1}{(s-3)^2} - \frac{1}{(s-3)^2}e^{-(s-3)} - \frac{1}{s-3}e^{-(s-3)}$ **13.** $\frac{2}{\sqrt{3}}e^{-3t/2}\sin\frac{\sqrt{3}}{2}t$

14. $\frac{2}{\sqrt{5}}e^{-3t/2}\sinh\frac{\sqrt{5}}{2}t$ **15.** $te^{-2t}(1 + \frac{3}{2}t)$ **16.** $\frac{2}{\sqrt{3}}e^{t/2}\sin\frac{\sqrt{3}}{2}t$

17. $e^{-t}(\cos 2t - \frac{1}{2}\sin 2t)$ **18.** $e^{-3t}(1 - 3t)$ **19.** $e^{3t}(\cos 2t + 2\sin 2t)$

20. $e^{-5t}(2\cos 3t - \frac{10}{3}\sin 3t)$ **21.** $\frac{1}{5} - \frac{1}{2}e^{-t} + \frac{1}{10}e^{-2t}(3\cos t + \sin t)$ **22.** $te^{-t}(1 - t + \frac{1}{6}t^2)$

SECTION 6–5

5. $1/(s^2 + 1)$ **6.** $1/s^3$ **7.** $\frac{1}{9}(1 - \cos 3t)$ **8.** $\frac{1}{2}(e^{2t} - 1)$ **9.** $\frac{1}{2}(1 - e^{-2t})$

10. $\frac{1}{3}(1 - e^{-3t})$ **11.** $\frac{1}{9}(\cosh 3t - 1)$ **12.** $1 - \cos \sqrt{2}\, t$

SECTION 6–6

1. $y = 2e^{-3t}$ **2.** $y = -e^{4t}$ **3.** $x = 3 \cos 3t$ **4.** $y = \cos 2t + \frac{1}{2} \sin 2t$

5. $y = 1 - e^{-t}$ **6.** $x = 1 - e^{2t}$ **7.** $y = 2(e^{-2t} - e^{-t})$ **8.** $x = \frac{1}{5}(e^{3t} - e^{-2t})$

9. $x = 2(1 - e^{-2t})$ **10.** $y = e^t + e^{2t}$ **11.** $y = \frac{1}{4}(1 - \cos 2t)$

12. $y = 10 \cosh t - 2 \sinh t - 10$ **13.** $y = 1 + 2t - e^{2t}$ **14.** $x = \frac{1}{2}t - \frac{5}{8}(1 - e^{-4t})$

15. $x = \dfrac{28}{13}e^{3t} - \dfrac{2}{13}\cos 2t - \dfrac{3}{13}\sin 2t$ **16.** $y = \frac{4}{25}\cos 3t + \frac{3}{25}\sin 3t - \frac{4}{25}e^{-4t}$

17. $x = 2 + e^t - e^{-2t}$ **18.** $x = 3(1 - e^{-t} + te^{-t})$ **19.** $y = e^{3t}(\frac{1}{6}t^3 + 3 - 7t)$

20. $y = \frac{1}{5}[6e^{-3t} - e^{-t}(\cos t - 2 \sin t)]$ **21.** $y = e^{-2t}(1 + \frac{1}{2}\sin 2t)$

22. $y = \frac{1}{4} + \frac{1}{4}e^{-2t} - \frac{1}{2}e^{-t}\cos t$ **23.** 2567 ft   **24.** $x = -\frac{1}{4}\cos 8t$

25. $x = \frac{1}{2}\sin 8t - \frac{1}{4}\cos 8t$ **26.** $A = \sqrt{5/16}, T = \pi/4$ **27.** $i = 1.2(1 - e^{-5t})$

28. $i = \frac{1}{29}e^{-5t} - \frac{1}{29}\cos 2t + \frac{5}{58}\sin 2t$ **29.** $i = 3(1 - e^{-t/5})$

Chapter 6 Review

1. $\dfrac{5}{s} + \dfrac{2}{s + 3}$ **2.** $\dfrac{5!}{s^6} - \dfrac{1}{(s - 2)^2}$ **3.** $\dfrac{12}{s^2 + 16} + \dfrac{7!}{(s + 1)^8}$ **4.** $\dfrac{15}{s^2 - 9} - \dfrac{2s}{s^2 - 4}$

5. $\dfrac{2}{(s + 4)^3}$ **6.** $\dfrac{1}{(s - 1)^2} + \dfrac{5}{s - 1}$ **7.** $\dfrac{1}{s + 2} + \dfrac{5}{s^2 + 6s + 34}$ **8.** $\dfrac{5(s + 3)}{s^2 + 6s + 13}$

9. $\dfrac{120}{4s^2 - 4s + 145}$ **10.** $\dfrac{2}{s} + \dfrac{2s}{s^2 + 8}$ **11.** $\dfrac{e^{-5s}}{s}$ **12.** $\dfrac{1 - e^{-(s+1)}}{s + 1}$ **13.** no   **14.** yes

15. yes   **16.** yes   **17.** $3 \cos \sqrt{3}\, t + \dfrac{2}{\sqrt{3}} \sin \sqrt{3}\, t$ **18.** $\dfrac{1}{\sqrt{2}} \sin \sqrt{2}\, t - \cos \sqrt{2}\, t$

19. $5e^{-3t} - 3$ **20.** $-5e^{-t} + 5e^{2t}$ **21.** $2 \cos 2t + \frac{7}{2} \sin 2t + 1$ **22.** $2e^{-3t} - 4 \sin t$

23. $\dfrac{3}{\sqrt{2}}(\sin \sqrt{2}\, t)e^{-3t}$ **24.** $e^t \sin 2t$ **25.** $2e^{9t}\cosh 8t + \frac{3}{8}e^{9t}\sinh 8t$

26. $\frac{2}{3}e^{-t}\sin 3t - e^{-t}\cos 3t$ **27.** $x = \frac{5}{2}(1 - \cos 2t)$ **28.** $y = \frac{17}{9} + \frac{1}{9}e^{3t} - \frac{1}{3}t$

29. $y = \frac{1}{2} - \frac{1}{2}e^{-2t} - 2te^{-2t}$ **30.** $y = \dfrac{t^2 e^{3t}}{2}$

SECTION 7–1

1. $\dfrac{s^2 - k^2}{(s^2 + k^2)^2}$ **2.** $\dfrac{2s(s^2 - 3k^2)}{(s^2 + k^2)^3}$ **3.** $\dfrac{2}{(s - 2)^3}$ **4.** $\dfrac{24}{(s - 3)^5}$ **5.** $\dfrac{n!}{(s - k)^{n+1}}$

6. $\dfrac{2s}{(s^2 + 4)^2}$ **7.** $\dfrac{k}{s^2 + k^2} - \dfrac{s^2 - k^2}{(s^2 + k^2)^2}$ **8.** $\dfrac{1}{2s^2} - \dfrac{s^2 - 4}{2(s^2 + 4)^2}$

9. $\dfrac{6ks^2 - 2k^3 + 2s^3 - 6k^2s}{(s^2 + k^2)^3}$ **10.** $\dfrac{2k(3s^2 + k^2)}{(s^2 - k^2)^3}$ **11.** $\dfrac{10(3s^2 - 18s + 2)}{(s^2 - 6s + 34)^3}$

12. $\dfrac{1}{2}\left[\dfrac{s^2 - 6s - 55}{(s^2 - 6s + 73)^2} + \dfrac{1}{(s - 3)^2}\right]$ **13.** $(e^t - e^{-t})/t$ **14.** $(e^t - e^{2t})/t$

15. $(1 - e^{5t})/t$ **16.** $(-e^{3t} + e^t)/t$ **17.** $2(\cos 2t - 1)/t$ **18.** $2(-\cos 2t + \cos t)/t$

19. $(e^{-4t}\sin t)/t$ **20.** $(\sin t)/t$ **21.** $(\sin 2t)/t$ **22.** $(\sin \frac{1}{3}t)/t$

SECTION 7–2

1. $1/(s^2 + 1)$ **2.** $s/(s^2 + 9)$ **3.** $\dfrac{1 - e^{-s}}{s^2(1 + e^{-s})}$ **4.** $\dfrac{1 + e^{-\pi s}}{(s^2 + 1)(1 - e^{-\pi s})} = \dfrac{\coth \pi s/2}{s^2 + 1}$

5. $\dfrac{2ke^{-\pi s/2k} + s(1 + e^{-\pi s/k})}{(s^2 + k^2)(1 - e^{-\pi s/k})}$ **6.** $\dfrac{1 - e^{-Ls/2}}{s(1 - e^{-Ls})}$

7. $\dfrac{2(e^{-3s} - e^{-2s} - e^{-s} + 1)}{s(1 - e^{-3s})} = \dfrac{2(e^{-s} - 1)^2(e^{-s} + 1)}{s(1 - e^{-3s})}$ **8.** $\dfrac{1 - e^{-(s - 1)}}{(s - 1)[1 + e^{-(s - 1)}]}$

9. $\dfrac{1}{(s - 1)[1 - e^{-(s - 1)}]}$ **10.** $\dfrac{k}{s} + \dfrac{h}{s}\tanh\dfrac{sp}{4}$ **11.** $\dfrac{2}{s^2(s + 1)(s + 3)} - \dfrac{4e^{-2s}}{s(s + 1)(s + 3)(1 - e^{-2s})}$

12. $\dfrac{1}{(s + 1)(s + 3)(s^2 + 1)(1 - e^{-\pi s})}$ **13.** $\dfrac{1 - e^{-s}}{s(s + 1)(s + 3)(1 + e^{-s})}$

14. $\dfrac{1}{s(s + 1)(s + 3)(1 - e^{-s})}$

SECTION 7–3

1. $\dfrac{2}{s^2(s^2 + 4)}$ **2.** $\dfrac{s}{(s + 1)(s^2 + 1)}$ **3.** $\dfrac{2}{s(s - 3)}$ **4.** $\dfrac{1}{s(s^2 + 1)}$ **5.** $\dfrac{6}{s^4(s^2 + 1)}$

6. $\dfrac{1}{s^2(s - 1)}$ **7.** $\frac{1}{4}(1 - \cos 2t)$ **8.** $e^t - t - 1$ **9.** $\frac{1}{6}(1 - \cos 3t)$ **10.** $\frac{1}{2}(e^{2t} - 1)$

11. $\frac{1}{9}(e^{-3t} + 3t - 1)$ **12.** $\frac{3}{2}(1 - \cos\sqrt{2}\, t)$ **13.** $\frac{2}{5}(e^{2t} - e^{-3t})$ **14.** $e^{-t} - e^{-2t}$

15. $(-t\cos t + \sin t)/2$ **16.** $(t\sin t)/2$ **17.** $y = te^{-t} * 5; y = 5(1 - e^{-t} - te^{-t})$

18. $y = (e^{-t} - e^{-2t}) * \sin t;$ **19.** $y = \frac{1}{3}t * (e^t - e^{-2t});$
$\quad y = \frac{1}{2}e^{-t} - \frac{1}{5}e^{-2t} + \frac{1}{10}\sin t - \frac{3}{10}\cos t$ $\quad y = -\frac{1}{12}e^{-2t} + \frac{1}{3}e^t - \frac{1}{4} - \frac{1}{2}t$

20. $y = \sin t * \sin t;$ **21.** $1 + t^2/2$ **22.** t **23.** $e^{-t} - e^{-2t}$ **24.** $t^2 + t^4/12$
$\quad y = \frac{1}{2}(\sin t - t\cos t)$

25. $\frac{1}{6}e^{2t} - \frac{1}{6}e^{-t}\left(\cos\sqrt{3}t - \sqrt{3}\sin\sqrt{3}t\right)$ **26.** $t^2 - t^4/3$ **27.** $t + t^2/2$

28. $2 + t - 2e^t + 2te^t$

SECTION 7–4

1.

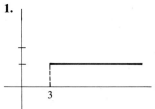

2.

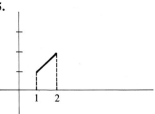

3.

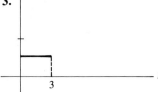

4.

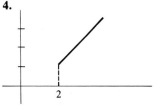

5.

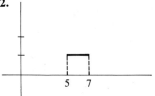

6.

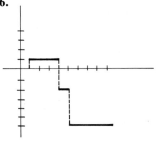

7.

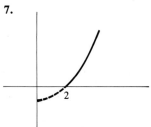

8.

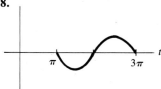

9.

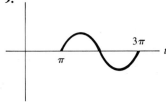

10.

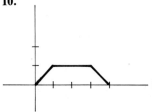

11.

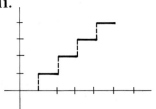

12.

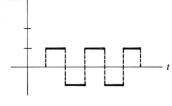

13. $f(t) = 1 - (1 - t)u_2(t)$ **14.** $f(t) = 5 + 2(t - 3)u_3(t)$ **15.** $f(t) = t^2 - t^2u_4(t)$
16. $f(t) = e^{-t} - e^{-t}u_2(t)$ **17.** $f(t) = 4[u_2(t) - u_5(t)]$ **18.** $f(t) = \cos t - \cos t\, u_{\pi/2}(t)$
19. $f(t) = t[u_1(t) - u_2(t)]$ **20.** $f(t) = 4 - 2u_1(t) - 2u_2(t)$ **21.** $f(t) = 1 - u_2(t)$
22. $f(t) = t + 2u_3(t)$ **23.** $3u_1(t) - u_2(t)$ **24.** $u_2(t) + 3u_5(t)$

SECTION 7–5

1. $\dfrac{2}{s} + \dfrac{e^{-s}}{s^2} - \dfrac{e^{-s}}{s}$ **2.** $\dfrac{3}{s} + \dfrac{e^{-2s}}{s^2}$ **3.** $\dfrac{1 - e^{-3(s+1)}}{s + 1}$ **4.** $\dfrac{3(1 + e^{-\pi s})}{s^2 + 9}$

5. $\dfrac{2}{s^3} - \dfrac{2e^{-3s}}{s^3} - \dfrac{6e^{-3s}}{s^2} - \dfrac{9e^{-5s}}{s}$ **6.** $\dfrac{1}{s^2} - \dfrac{e^{-s}}{s^2} - \dfrac{e^{-3s}}{s^2} + \dfrac{e^{-4s}}{s^2}$ **7.** $\dfrac{1}{s^2} - \dfrac{2e^{-2s}}{s^2} + \dfrac{e^{-4s}}{s^2}$

8. $\dfrac{1}{s^2} - \dfrac{e^{-s}}{s(1 - e^{-s})}$ **9.** $\dfrac{1}{s^2} - \dfrac{2e^{-s}}{s^2(1 + e^{-2s})}$ **10.** $\dfrac{1}{s^2} - \dfrac{2e^{-s}}{s^2} + \dfrac{e^{-3s}}{s^2} + \dfrac{e^{-4s}}{s}$

11. $(t - 3)u_3(t)$ **12.** $\frac{1}{2}(t - 4)^2u_4(t) - \frac{1}{2}(t - 1)^2u_1(t)$ **13.** $\frac{1}{2}\sin 2(t - 1)u_1(t)$

14. $[e^{2(t-1)} - e^{t-1}]\,u_1(t)$ **15.** $\cos 2(t - 2)u_2(t)$
16. $-e^{-t} + e^{-(t-1)}u_1(t) - e^{-(t-2)}u_2(t) + e^{-(t-3)}u_3(t) + \cdots$
    $+ u_0(t) - u_1(t) + u_2(t) - u_3(t) + \cdots$
17. $t - (t - 1)u_1(t) + (t - 2)u_2(t) - (t - 3)u_3(t) + \cdots$
18. $1 + u_1(t) + u_2(t) + u_3(t) + \cdots$
19. $e^{-at} - e^{-a(t-k)}u_k(t) + e^{-a(t-2k)}u_{2k}(t) - e^{-a(t-3k)}u_{3k}(t) + \cdots$

20. $te^{-t} + (t - 1)e^{-(t-1)}u_1(t) + (t - 2)e^{-(t-2)}u_2(t) + \cdots$ **21.** $e^{-2s}\left[\dfrac{1}{s^2} + \dfrac{2}{s}\right]$

22. $e^{-3s}\left[\dfrac{2}{s^3} + \dfrac{6}{s^2} + \dfrac{9}{s}\right]$ **23.** $-\dfrac{se^{-\pi s}}{s^2 + 1}$ **24.** $e^{-2s}\left(\dfrac{2}{s^3} - \dfrac{2}{s^2} + \dfrac{1}{s}\right)$ **25.** $\dfrac{e^{-3(s-1)}}{s - 1}$

SECTION 7–6

1. $y = 5e^{-2t} + [\frac{5}{2} - \frac{5}{2}e^{-2(t-1)}]u_1(t)$ **2.** $y = [1 - e^{-(t-1)}]u_1(t) - [1 - e^{-(t-2)}]u_2(t)$
3. $x = [2 - 2(t - 4) + (t - 4)^2 - 2e^{-(t-4)}]u_4(t)$
4. $y = [-\frac{4}{9} + \frac{4}{3}(t - 2) + \frac{4}{9}e^{-3(t-2)}]u_2(t)$
5. $y = [\frac{1}{4} - \frac{1}{4}e^{-2(t-1)} - \frac{1}{2}(t - 1)e^{-2(t-1)}]u_1(t)$ **6.** $y = u_0(t)$
7. $x = [(t - 2) - \sin(t - 2)]u_2(t)$
8. $x = [-\frac{12}{25} - \frac{3}{5}(t - 1) - \frac{1}{50}e^{-5(t-1)} + \frac{1}{2}e^{(t-1)}]u_1(t)$
9. $y(t) = t - \sin t - [(t - 2) - \sin(t - 2)]u_2(t)$ **10.** $x = [-(t - 4) + \sinh(t - 4)]u_4(t)$
11. $x(t) = 5(1 - \cos 4t) - 5[1 - \cos 4(t - 1)]u_1(t)$ **12.** 10 ft
13. $x(t) = [(t - 2) - \sin(t - 2)]u_2(t)$ **14.** $x(t) = [(t - 4) + te^{-(t-2)}]u_2(t)$
15. $x(t) = \frac{1}{3} - \frac{1}{3}e^{-3t} - te^{-3t} + \frac{1}{3}[1 - e^{-3(t-1)} - 3(t - 1)e^{-3(t-1)}]u_1(t)$

16. $x(1) = \frac{1}{3}(1 - 4e^{-3})$ ft   **17.** $i(t) = \frac{1}{3} - \frac{1}{3}e^{-9t} - [\frac{1}{3} - \frac{1}{3}e^{-9(t-2)}]u_2(t)$
    $x(2) = \frac{1}{3}(2 - 7e^{-6} - 4e^{-3})$ ft

18. $i(t) = [-\frac{1}{81} + \frac{1}{9}(t - 1) + \frac{1}{81}e^{-9(t-1)}]u_1(t)$

19. $q = CE(1 - e^{-t/RC}) - CE(1 - e^{-(t-T)/RC})u_T(t)$ **20.** $q = \dfrac{h}{R}[e^{-(t-a)/RC}u_a(t) - e^{-(t-b)/RC}u_b(t)]$

Chapter 7 Review

1. $\dfrac{s^2 + 9}{(s^2 - 9)^2}$ **2.** $\dfrac{3}{2s^2} + \dfrac{3}{2}\dfrac{s^2 - 100}{(s^2 + 100)^2}$ **3.** $\dfrac{2 - e^{-2s} - e^{-4s}}{s(1 - e^{-4s})}$ **4.** $\dfrac{1 - e^{-s} - se^{-2s}}{s^2(1 - e^{-2s})}$

5. $\dfrac{2}{(s - 3)^2} - \dfrac{4e^{-2(s-3)}}{(s - 3)(1 - e^{-2(s-3)})}$ **6.** $\dfrac{1}{(1 - e^{-\pi(s+2)})(s^2 + 4s + 5)}$ **7.** $\dfrac{2}{s^2(s^2 + 4)}$

8. $\dfrac{2}{s^3(s - 3)}$ **9.** $\dfrac{1}{s^2}(1 - e^{-s} - se^{-2s})$ **10.** $\dfrac{2}{2s - 1}(1 - e^{-2s+1})$

11. $\dfrac{e^{-2s}}{s^3}(2 + 7s + 10s^2)$ **12.** $\dfrac{e^{-(s-2)}}{s - 2}$ **13.** $\dfrac{2 \sinh 3t}{t}$ **14.** $\dfrac{\sin \frac{1}{2}t}{t}$ **15.** $\frac{7}{9}(e^{8t} - e^{-t})$

16. $\frac{1}{12}(1 - \cos 2t)$ **17.** $2u_1(t) - \frac{1}{2}(t - 4)^2 u_4(t)$ **18.** $\frac{1}{4} \sin 4(t - 2)u_2(t)$

19. $\sin 3(t - 1)u_1(t) + \frac{2}{3} \sin 3(t - 2)u_2(t)$ **20.** $\frac{8}{3}(1 + u_2(t) + u_4(t) + u_6(t) + \cdots)$

21. $\frac{1}{4}[1 - \cos 2(t - 2)]u_2(t) - \frac{1}{4}[1 - \cos 2(t - 3)]u_3(t)$

22. $[1 + (t - 2) + \frac{1}{2}(t - 2)^2 - e^{t-2}]u_2(t) - (1 + t + \frac{1}{2}t^2 - e^t)$

23. $y = e^{2t} * \sinh t;\ y = -\frac{1}{2}e^t + \frac{1}{6}e^{-t} + \frac{1}{3}e^{2t}$

SECTION 8–1

1. $y = 1 + x + x^2 + \dfrac{x^3}{3} + \dfrac{x^4}{12} + \cdots$ **2.** $y = 1 + x + \dfrac{x^2}{2} + \dfrac{x^3}{6} + \dfrac{x^4}{24} + \cdots$

3. $y = 1 + x + \dfrac{x^2}{2!} + \dfrac{x^3}{3!} + \cdots$ **4.** $y = 1 - 2x + 2x^2 - x^3 + \frac{1}{2}x^4 - \frac{1}{5}x^5 + \cdots$

5. $y = 1 + x - \frac{7}{2}x^2 + \frac{25}{6}x^3 - \frac{79}{24}x^4 + \cdots$ **6.** $y = 2 + 4x + 8x^2 + 16x^3 + 32x^4 + \cdots$

7. $y = 2 - 2(x - 1) + (x - 1)^2 - \dfrac{(x - 1)^4}{12} + \dfrac{1}{20}(x - 1)^5 - \cdots$

8. $y = 1 - (x - 1) + \frac{1}{2}(x - 1)^2 - \frac{1}{2}(x - 1)^3 + \frac{3}{8}(x - 1)^4 - \cdots$

9. $y = 1 + (x - 1) + \frac{1}{6}(x - 1)^3 - \frac{1}{24}(x - 1)^4 + \frac{1}{40}(x - 1)^5 - \cdots$

10. $y = 1 + x + \dfrac{e}{2}x^2 + \dfrac{e}{6}x^3 + \dfrac{e^2 + e}{24}x^4 \cdots$ **11.** $a_n = \dfrac{2}{n!},\ n \geq 2;\ y = 2e^x - x - 1$

12. $a_n = \dfrac{1}{n!};\ y = e^x$ **13.** $a_n = \dfrac{1}{n!};\ y = e^x$

14. $a_n = \dfrac{3(-2)^n}{4n!},\ n \geq 3;\ y = \frac{3}{4}e^{-2x} + \frac{1}{2}x^2 - \frac{1}{2}x + \frac{1}{4}$

15. $a_n = \dfrac{2(-1)^n - (-3)^n}{n!};\ y = 2e^{-x} - e^{-3x}$ **16.** $a_n = 2^{n+1};\ y = \dfrac{2}{1 - 2x}$

17. $y = 1 + x^2 - x^3 + \frac{1}{2}x^4 - \frac{1}{6}x^5 + \frac{1}{24}x^6 - \cdots + (-1)^n \dfrac{x^n}{(n - 2)!} + \cdots$

18. $y = (1 + 2x^2 + 3x^4 + 4x^6 + \cdots + (n + 1)x^{2n} + \cdots)$

$\quad + \left(x + \frac{5}{3}x^3 + \frac{7}{3}x^5 + \cdots + \dfrac{2n + 3}{3}x^{2n+1} + \cdots\right)$

19. $y = 1 + x + 3x^2 + \frac{2}{3}x^3 + \frac{3}{8}x^4$

20. $y = x - \frac{1}{3}x^3 + \left(1 - \frac{3}{2}x^2 + \frac{1}{8}x^4 + \frac{1}{240}x^6 + \cdots + \dfrac{3}{2^n n!(2n-1)(2n-3)}x^{2n} + \cdots\right)$

21. $y = \left(1 + \frac{3}{8}x^2 + \left(\dfrac{3x^2}{8}\right)^2\left(\dfrac{1}{2!}\right) + \cdots + \dfrac{1}{n!}\left(\dfrac{3x^2}{8}\right)^n + \cdots\right)$

$\qquad + \left(x + \frac{1}{4}x^3 + \cdots + \dfrac{n!}{(2n+1)!}\left(\dfrac{3}{2}\right)^n x^{2n+1} + \cdots\right)$

23. $y = 1 - t^2 - \frac{1}{48}t^3$

SECTION 8–2

1. $(-1, 1)$ **2.** $(1, 9)$ **3.** $(4 - \sqrt{17}, 4 + \sqrt{17})$ **4.** $(-1, 1)$ **5.** $(0, 2)$
6. no power series solution is predictable   **7.** no power series solution is predictable
8. $(-1, 1)$ **9.** $(\sqrt{2}, 3 - \sqrt{2})$ **10.** $(-1, 1)$ **11.** $x = 1$ is a singular point

SECTION 8–3

1. $a_n = -\dfrac{1}{n}a_{n-1}$, a_0 arbitrary;

$\qquad y = a_0\left(1 - x + \frac{1}{2}x^2 - \frac{1}{3!}x^3 + \frac{1}{4!}x^4 \cdots\right)$

2. $a_n = \dfrac{1}{n!} - \dfrac{1}{n}a_{n-1}$; a_0 arbitrary;

$\qquad y = a_0\left(1 - x + \frac{x^2}{2!} - \frac{1}{3!}x^3 + \frac{1}{4!}x^4 \cdots\right) + x + \frac{1}{3!}x^3 + \frac{1}{5!}x^5 \cdots$

3. $a_n = \dfrac{(-1)^{n-1}}{n!} - \dfrac{1}{n}a_{n-1}$; a_0 arbitrary;

$\qquad y = a_0\left(1 - x + \frac{1}{2}x^2 - \frac{1}{6}x^3 + \frac{1}{4!}x^4 \cdots\right) + x - x^2 + \frac{1}{2}x^3 \cdots$

4. $a_n = -\dfrac{1}{n(n-1)}a_{n-2}$; a_0 and a_1 arbitrary;

$\qquad y = a_0\left(1 - \frac{1}{2}x^2 + \frac{1}{4!}x^4 \cdots\right) + a_1\left(x - \frac{1}{3!}x^3 + \frac{1}{5!}x^5 \cdots\right)$

5. $a_n = -\dfrac{1}{n(n-1)}a_{n-2}$; $n \neq 4$, a_0 and a_1 arbitrary,

$\qquad a_4 = \dfrac{1}{12} + \dfrac{1}{4!}a_0$;

$\qquad y = a_0\left(1 - \frac{1}{2!}x^2 + \frac{1}{4!}x^4 \cdots\right) + a_1\left(x - \frac{1}{3!}x^3 + \frac{1}{5!}x^5 \cdots\right)$

$\qquad + 2\left(\frac{1}{4!}x^4 - \frac{1}{6!}x^6 + \frac{1}{8!}x^8 \cdots\right)$

6. $a_n = \dfrac{(-1)^n}{n!} - \dfrac{1}{n(n-1)} a_{n-2}$; a_0 and a_1 arbitrary;

$$y = a_0\left(1 - \frac{1}{2!}x^2 + \frac{1}{4!}x^4 \cdots\right) + a_1\left(x - \frac{1}{3!}x^3 + \frac{1}{5!}x^5 \cdots\right) + \frac{1}{2}x^2 - \frac{1}{3!}x^3 \cdots$$

7. $a_n = -\dfrac{1}{n}a_{n-1}$, $n \neq 3$, a_0 and a_1 arbitrary,

$a_3 = \frac{1}{6}(1 + a_1)$;

$$y = a_0 + a_1\left(x - \frac{1}{2!}x^2 + \frac{1}{3!}x^3 \cdots\right) + \frac{1}{3!}x^3 - \frac{1}{4!}x^4 + \frac{1}{5!}x^5 \cdots$$

8. $a_n = -\dfrac{1}{n}a_{n-3}$, a_0 arbitrary, $a_1 = a_2 = 0$;

$$y = a_0\left(1 - \frac{1}{3}x^3 + \frac{1}{3\cdot 6}x^6 - \frac{1}{3\cdot 6\cdot 9}x^9 \cdots\right)$$

9. $a_n = -\dfrac{1}{n}a_{n-2}$, a_0 arbitrary, $a_1 = 1$;

$$y = a_0\left(1 - \frac{1}{2}x^2 + \frac{1}{8}x^4 - \frac{1}{48}x^6 \cdots\right) + x - \frac{1}{3}x^3 + \frac{1}{15}x^5 \cdots$$

10. $a_n = -\dfrac{1}{n(n-1)}a_{n-3}$, a_0 and a_1 arbitrary, $a_2 = 0$;

$$y = a_0\left(1 - \frac{1}{3\cdot 2}x^3 + \frac{1}{6\cdot 5\cdot 3\cdot 2}x^6 \cdots\right) + a_1\left(x - \frac{1}{4\cdot 3}x^4 + \frac{1}{7\cdot 6\cdot 4\cdot 3}x^7 \cdots\right)$$

11. $a_n = -\dfrac{1}{n(n-1)}a_{n-5}$; a_0 and a_1 arbitrary, $a_2 = a_3 = a_4 = 0$;

$$y = a_0\left(1 - \frac{1}{5\cdot 4}x^5 + \frac{1}{10\cdot 9\cdot 5\cdot 4}x^{10} - \frac{1}{15\cdot 14\cdot 10\cdot 9\cdot 5\cdot 4}x^{15} \cdots\right)$$
$$+ a_1\left(x - \frac{1}{6\cdot 5}x^6 + \frac{1}{11\cdot 10\cdot 6\cdot 5}x^{11} - \frac{1}{16\cdot 15\cdot 11\cdot 10\cdot 6\cdot 5}x^{16} \cdots\right)$$

12. $a_n = \dfrac{n-7}{n}a_{n-2}$, a_0 and a_1 arbitrary;

$$y = a_0\left(1 - \frac{5}{2}x^2 + \frac{15}{8}x^4 - \frac{5}{16}x^6 \cdots\right) + a_1\left(x - \frac{4}{3}x^3 + \frac{8}{15}x^5\right)$$

13. $a_n = \dfrac{(n-3)^2}{n(n-1)}a_{n-2}$, a_0 and a_1 arbitrary;

$$y = a_0\left(1 + \frac{1}{2}x^2 + \frac{1}{24}x^4 \cdots\right) + a_1 x$$

14. $a_n = -\dfrac{n-7}{n-1}a_{n-2}$, a_0 and a_1 arbitrary;

$$y = a_0\left(1 + 5x^2 + 5x^4 + x^6 - \frac{1}{7}x^8 + \frac{1}{21}x^{10} \cdots\right) + a_1(x + 2x^3 + x^5)$$

15. $a_n = -\dfrac{n-6}{n(n-1)}a_{n-2}$, a_0 and a_1 arbitrary;

$$y = a_0\left(1 + 2x^2 + \frac{1}{3}x^4\right) + a_1\left(x + \frac{1}{2}x^3 + \frac{1}{40}x^5 \cdots\right)$$

16. $a_n = -\dfrac{(n-3)(n+1)}{n(n-1)}a_{n-2}$, a_0 and a_1 arbitrary;

$$y = a_0\left(1 + \frac{3}{2}x^2 - \frac{5 \cdot 3}{4!}x^4 + \frac{7 \cdot 5 \cdot 3 \cdot 3}{6!}x^6 - \cdots\right) + a_1x$$

22. $a_n = \dfrac{2(n - 4)}{n(n - 1)}a_{n-2};$

$$y = a_0(1 - 2x^2) + a_1\left(x - \frac{2}{3!}x^3 - \frac{2^2}{5!}x^5 - \frac{2^3 \cdot 3}{7!}x^7 - \frac{2^4 \cdot 3 \cdot 5}{9!}x^9 \cdots\right)$$

23. $a_n = \dfrac{n - 4}{n - 1}a_{n-2};$

$$y = a_0(1 - 2x^2) + a_1(x - \tfrac{1}{2}x^3 - \tfrac{1}{8}x^5 - \tfrac{1}{16}x^7 - \tfrac{5}{128}x^9 \cdots)$$

24. $a_2 = 0,\ a_n = -\dfrac{1}{n(n - 1)}a_{n-3};$

$$y = a_0[1 - \tfrac{1}{6}(x - 2)^3 + \tfrac{1}{180}(x - 2)^6 \cdots]$$
$$+ a_1[(x - 2) - \tfrac{1}{12}(x - 2)^4 + \tfrac{1}{504}(x - 2)^7 \cdots]$$

25. $a_2 = \dfrac{1}{2}a_0;\ a_n = \dfrac{1}{n(n - 1)}(a_{n-2} - a_{n-3});$

$$y = a_0(1 + \tfrac{1}{2}(x - 1)^2 - \tfrac{1}{6}(x - 1)^3 + \tfrac{1}{24}(x - 1)^4 \cdots)$$
$$+ a_1((x - 1) + \tfrac{1}{6}(x - 1)^3 - \tfrac{1}{12}(x - 1)^4 \cdots)$$

26. $a_2 = \dfrac{1}{2}a_0;\ a_n = \dfrac{1}{n(n - 1)}[a_{n-2} - (n - 1)(n - 2)a_{n-1}];$

$$y = a_0(1 + \tfrac{1}{2}(x - 1)^2 - \tfrac{1}{6}(x - 1)^3 + \tfrac{1}{8}(x - 1)^4 \cdots)$$
$$+ a_1((x - 1) + \tfrac{1}{6}(x - 1)^3 - \tfrac{1}{12}(x - 1)^4 + \tfrac{7}{120}(x - 1)^5 \cdots)$$

27. $a_n = \dfrac{2 - n}{n(n - 1)}a_{n-2};$

$$y = a_0 + a_1((x - 1) - \tfrac{1}{3!}(x - 1)^3 + \tfrac{3}{5!}(x - 1)^5 \cdots)$$

SECTION 8–4

1. R.S.P. at 0   **2.** R.S.P. at 0, -1; I.S.P. at 1   **3.** R.S.P. at 0; I.S.P. at 1   **4.** I.S.P. at 0

5. I.S.P. at 2   **6.** $0, \frac{1}{2}; y = a_0 \displaystyle\sum_{n=0}^{\infty} \frac{(-1)^n}{(2n)!}x^n + a_0^* x^{1/2} \sum_{n=0}^{\infty} \frac{(-1)^n}{(2n + 1)!}x^n$

7. $0, 4; y = a_0(1 + \frac{2}{3}x + \frac{1}{3}x^2) + a_0^* \displaystyle\sum_{n=4}^{\infty} (n - 3)x^n$ **8.** $\frac{1}{2}, \frac{3}{2}; y = a_0 x^{1/2} + a_0^* x^{1/2} \displaystyle\sum_{n=1}^{\infty} \frac{x^n}{2^{n-1} n!}$

9. $0, 0; y = \frac{1}{2}a_0 \displaystyle\sum_{n=0}^{\infty} (-1)^n (n + 1)(n + 2)x^n$ **10.** $0, 0; y = a_0(1 + x)$

11. $1, -1; y = a_0 \displaystyle\sum_{n=0}^{\infty} \frac{(-1)^n x^{2n+1}}{2^{2n} n!(n + 1)!}$ **12.** $2, -2; y = a_0 \displaystyle\sum_{n=0}^{\infty} \frac{(-1)^n x^{2n+2}}{2^{2n-1} n!(n + 2)!}$

13. $1, -\frac{1}{2}; y = a_0 x^{-1/2} \displaystyle\sum_{n=0}^{\infty} \frac{x^n}{2^n n!} + a_0^* x +$

$$a_0^* x \sum_{n=1}^{\infty} \frac{x^n}{(2n + 3)(2n + 1)(2n - 1) \cdots (17)(15)(13)(11)(9)(7)(5)}$$

14. $0, 0; y = a_0 \sum_{n=0}^{\infty} \dfrac{(-1)^n x^{2n}}{2^{2n}(n!)^2}$ **15.** $0, 1; y = a_0 \sum_{n=0}^{\infty} \dfrac{(-1)^n x^{n+1}}{n!(n+1)!}$

16. $\frac{1}{2}, -\frac{1}{2}; y = a_0 x^{1/2} \sum_{n=0}^{\infty} \dfrac{x^n}{2^n n!}$

17. $y = \sum_{n=0}^{\infty} a_n x^n$, where a_0 is arbitrary, $a_1 = 0$, and

$$a_n = \frac{1}{n(n+2)}[a_{n-2} - 2(n-1)(n-2)a_{n-1}], n \geq 2$$

18. $y = x^{-1}\sum_{n=0}^{\infty} a_n x^n + x^{1/3}\sum_{n=0}^{\infty} a_n^* x^n$, where a_0 is arbitrary, $a_1 = a_0$, and

$$a_n = \frac{1}{n(3n-4)}[(n-2)a_{n-1} - 2a_{n-2}], n \geq 2;$$

a_0^* is arbitrary, $a_1^* = \frac{1}{21}a_0^*$, and

$$a_n^* = \frac{-1}{n(3n+4)}\left[\frac{3n-2}{3}a_{n-1}^* - 2a_{n-2}^*\right], n \geq 2$$

19. $y = \sum_{n=0}^{\infty} a_n x^n + x^{3/2}\sum_{n=0}^{\infty} a_n^* x^n$, where a_0 is arbitrary, $a_1 = a_0$,

$$a_2 = -\tfrac{1}{2}a_0, a_n = \frac{1}{n(2n-3)}[(n-3)a_{n-3} - a_{n-1}], n \geq 3;$$

and a_0^* is arbitrary, $a_1^* = -\frac{1}{5}a_0^*$, $a_2^* = \frac{1}{70}a_0^*$,

$$a_n^* = \frac{1}{n(2n+3)}\left[\frac{2n-3}{2}a_{n-3} - a_{n-1}^*\right], n \geq 3$$

20. $y = x^{-1}\sum_{n=0}^{\infty} a_n x^n + x^{1/3}\sum_{n=0}^{\infty} a_n^* x^n$, where a_0 is arbitrary, $a_1 = -a_0$,

$$a_n = \frac{1}{n(3n-4)}[a_{n-1} - a_{n-2}(n-3)(n-4)], n \geq 2;$$

and a_0^* is arbitrary, $a_1^* = \frac{1}{7}a_0^*$,

$$a_n^* = \frac{1}{n(3n+4)}[a_{n-1}^* - a_{n-2}^*(n - \tfrac{8}{3})(n - \tfrac{5}{3})], n \geq 2$$

SECTION 8–5

1. $y = y_1 \ln x - \frac{3}{2}y_1 + \frac{3}{2} - \frac{5}{2}x + \frac{7}{2}x^2 - \frac{9}{2}x^3 \cdots$ **2.** $y = y_1 \ln x$

3. $y = y_1 \ln x + 4x - 3x^2 + \frac{22}{27}x^3 \cdots$ **4.** $y = y_1 \ln x - \frac{3}{2} - \frac{3}{2}x - \frac{3}{2}x^2 - \frac{11}{6}x^3 \cdots$

5. $y = y_1 \ln x + \frac{1}{4}x^2 - \frac{3}{128}x^4 + \frac{11}{13824}x^6 \cdots$ **6.** $y = y_1 \ln x - 1 - x + \frac{5}{4}x^2 \cdots$

7. $y = y_1 \ln x + x^{-1/2}\left[2 - \frac{1}{2}x^2 - \dfrac{(1 + \frac{1}{2})}{4 \cdot 2}x^3 - \dfrac{(1 + \frac{1}{2} + \frac{1}{3})}{8 \cdot 3!}x^4 \cdots\right]$

8. $y = y_1 \ln x - x - \frac{3}{4}x^2 - \frac{11}{36}x^3 \cdots$

SECTION 8–6

1. $P_4(x) = \dfrac{35x^4 - 30x^2 + 3}{8}$

$P_5(x) = \dfrac{63x^5 - 70x^3 + 15x}{8}$

5. $2[P_1(x) - P_2(x)]$ **8.** $y = cP_4(x)$ **9.** $y = cP_2(x)$ **10.** $y = cJ_2(x)$ **11.** $y = cJ_{1/2}(x)$

Chapter 8 Review

1. $y = 1 + x - \frac{1}{6}x^3 - \frac{1}{12}x^4 \cdots$ **2.** $1 + (x - 1) - \frac{1}{2}(x - 1)^2 - \frac{1}{6}(x - 1)^3 \cdots$

3. $1 + (x - 1) + \dfrac{\sin 1}{2}(x - 1)^2 + \dfrac{\cos 1}{6}(x - 1)^3 \cdots$

4. $3 + (x - 1) + \frac{1}{6}(x - 1)^2 + \frac{2}{81}(x - 1)^3 \cdots$ **5.** $y = 1 + x - \frac{1}{6}x^3 - \frac{1}{12}x^4 \cdots$

6. recursion formula: $a_n = -\dfrac{1}{n}a_{n-2}$

$y = a_0(1 - \frac{1}{2}x^2 + \frac{1}{8}x^4 - \frac{1}{48}x^6 + \cdots) + a_1(x - \frac{1}{3}x^3 + \frac{1}{15}x^5 + \cdots)$

7. recursion formula: $a_n = -\dfrac{n - 2}{n(n - 1)}a_{n-2}$

$y = a_0 + a_1\left(x - \frac{1}{6}x^3 + \dfrac{3}{5!}x^5 - \cdots\right)$

8. recursion formula: $a_n = \dfrac{(n^2 - 6n + 7)}{n(n - 1)}a_{n-2}$

$y = a_0(1 - \frac{1}{2}x^2 + \frac{1}{24}x^4 + \cdots) + a_1(x - \frac{1}{3}x^3 - \frac{1}{30}x^5 \cdots)$

9. recursion formula: $a_n = \dfrac{n - 4}{n}a_{n-2}$

$y = a_0(1 - x^2) + a_1(x - \frac{1}{3}x^3 - \frac{1}{15}x^5 \cdots)$

10. $a_2 = -\frac{1}{2}a_0$, recursion formula: $a_n = \dfrac{1}{n}a_{n-2} - \dfrac{1}{n(n - 1)}a_{n-3}$

$y = a_0(1 - \frac{1}{2}x^2 - \frac{1}{6}x^3 - \frac{1}{8}x^4 \cdots) + a_1(x + \frac{1}{3}x^3 - \frac{1}{12}x^4 - \frac{1}{15}x^5 \cdots)$

11. $a_2 = 0$, recursion formula: $a_n = \dfrac{(n - 4)(n - 2)}{n(n - 1)}a_{n-3}$

$y = a_0(1 - \frac{1}{6}x^3 - \frac{2}{45}x^6 \cdots) + a_1 x$

12. recursion formula: $a_n = \dfrac{(3 - n)a_{n-2} + (1 - n)a_{n-1}}{n(n - 1)}$

$y = a_0(1 + \frac{1}{2}x^2 - \frac{1}{6}x^3 \cdots) + a_1(x - \frac{1}{2}x^2 + \frac{1}{6}x^3 + \cdots)$

13. $a_2 = -\frac{1}{2}a_0$, recursion formula: $a_n = \dfrac{a_{n-2}(n - 3) + a_{n-3}}{n(n - 1)}$

$y = a_0(1 - \frac{1}{2}x^2 + \frac{1}{6}x^3 - \frac{1}{24}x^4 \cdots) + a_1(x + \frac{1}{12}x^4 \cdots)$

14. $y = a_0 + a_1\left[(x - 1) - \frac{1}{2}(x - 1)^2 + \frac{1}{6}(x - 1)^3 - \frac{1}{24}(x - 1)^4 \cdots + \dfrac{(-1)^{n+1}(x - 1)^n}{n!} + \cdots\right]$

15. $y = a_0[1 + \frac{1}{6}(x - 1)^3 + \frac{1}{180}(x - 1)^6 + \frac{1}{12960}(x - 1)^9 + \cdots]$

$\qquad + a_1[(x - 1) + \frac{1}{12}(x - 1)^4 + \frac{1}{252}(x - 1)^7 + \frac{1}{22860}(x - 1)^{10} + \cdots]$

16. regular singular points at 0 and -2.   **17.** regular singular points at 0, i and $-i$.

18. $y_1 = \sum\limits_{n=0}^{\infty} a_n x^{n - \sqrt{3}}; y_2 = \sum\limits_{n=0}^{\infty} a_n^* x^{n + \sqrt{3}}; a_0$ and a_0^* arbitrary

19. $y_1 = \sum\limits_{n=0}^{\infty} a_n x^n; y_2 = \sum\limits_{n=0}^{\infty} a_n x^{n+3}; a_0$ and a_0^* arbitrary

20. $y_1 = \sum\limits_{n=0}^{\infty} a_n x^{(n + 1/3)}; y_2 = \sum\limits_{n=0}^{\infty} a_n^* x^{(n - 1/3)}; a_0$ and a_0^* arbitrary

21. $y_1 = \sum\limits_{n=0}^{\infty} a_n x^n; y_2 = y_1 \ln x + \sum\limits_{n=1}^{\infty} A_n x^n$ **22.** same answer as for Exercise 21

23. $y_1 = \sum\limits_{n=0}^{\infty} a_n x^n$, where $a_n = \frac{1}{n} a_{n-1}$ for $n > 0$; therefore, $y_1 = e^x$

$\qquad y_2 = \frac{e^{2x}(x - 1)}{x^2} + e^x \ln x + \left[\frac{x}{1!} + \frac{x^2}{2 \cdot 2!} + \frac{x^3}{3 \cdot 3!} + \cdots \right]$

24. $y_1 = x - 1; y_2 = y_1 \ln x + x - \frac{1}{2}x^2 - \frac{1}{6}x^3 - \frac{1}{12}x^4 - \frac{1}{20}x^5 \cdots$

25. $y_1 = x^3 e^{-x}; y_2 = y_1 \ln x + x + x^2 - x^3 + \frac{1}{4}x^5 \cdots$

26. $y_1 = 1 + \frac{2}{3}x + \frac{1}{6}x^2$

$\qquad y_2 = x^4 \sum\limits_{n=0}^{\infty} a_n x^n$, where $a_n = \frac{n + 1}{n(n + 4)} a_{n-1}, n \geq 1$

$\qquad = x^4(1 + \frac{2}{5}x + \frac{1}{4}x^2 + \frac{4}{21}x^3 + \cdots)$

27. $y_1 = x^{-1/2}[1 - \frac{1}{2}x^2 - \frac{1}{8}x^4 - \frac{1}{144}x^6 \cdots]$

$\qquad a_{2k} = \frac{1}{(2k)(2k - 3)} a_{2k-2}; k \geq 1$

$\qquad y_2 = x^{5/2}[1 + \frac{1}{10}x^2 + \frac{1}{280}x^4 + \frac{1}{15120}x^6 \cdots]$

$\qquad a_{2k} = \frac{1}{(2k)(2k + 3)} a_{2k-2}, k \geq 1$

SECTION 9–1

1. $y_1 = c_1 e^x - c_2 e^{-x}; y_2 = c_1 e^x + c_2 e^{-x}$

2. $y_1 = c_1 e^{2x} - \frac{1}{3} c_2 e^{-2x}; y_2 = c_1 e^{2x} + c_2 e^{-2x}$

3. $y_1 = c_1(4 + 3\sqrt{2})e^{(-1+3\sqrt{2})x} + c_2(4 - 3\sqrt{2})e^{(-1-3\sqrt{2})x}$

$\qquad y_2 = c_1 e^{(-1+3\sqrt{2})x} + c_2 e^{(-1-3\sqrt{2})x}$

4. $y_1 = c_1 e^{(2+\sqrt{7})x} + c_2 e^{(2-\sqrt{7})x}$

$\qquad y_2 = -\frac{1}{3}[c_1(2 - \sqrt{7})e^{(2+\sqrt{7})x} + c_2(2 + \sqrt{7})e^{(2-\sqrt{7})x}]$

5. $y_1 = e^{7x/2}\left[\left(\frac{3}{2}c_1 + \frac{\sqrt{7}}{2}c_2 \right) \cos \frac{\sqrt{7}}{2}x + \left(\frac{3}{2}c_2 - \frac{\sqrt{7}}{2}c_1 \right) \sin \frac{\sqrt{7}}{2}x \right]$

$\qquad y_2 = e^{7x/2}\left(c_1 \cos \frac{\sqrt{7}}{2}x + c_2 \sin \frac{\sqrt{7}}{2}x \right)$

6. $y_1 = c_1 e^{(a+b)x} - c_2 e^{(a-b)x}$
$y_2 = c_1 e^{(a+b)x} + c_2 e^{(a-b)x}$

7. $y_1 = \frac{1}{4}x^2 + \frac{3}{2}x + c_1; \, y_2 = \frac{3}{2}x - \frac{1}{4}x^2 + c_2$

8. $y_1 = 1 + c_1 e^x - c_2 e^{-x}; \, y_2 = c_1 e^x + c_2 e^{-x} - 1$

9. $y_1 = c_1 \cos \sqrt{2}x + c_2 \sin \sqrt{2}x + \frac{3}{2}x; \, y_2 = -\sqrt{2}\, c_1 \sin \sqrt{2}x + \sqrt{2}c_2 \cos \sqrt{2}\, x + \frac{3}{2} - x^2$

10. $y_1 = \frac{1}{2}x + c_1; \, y_2 = \frac{1}{2}x + c_1 + 1$

11. $y_1 = -c_1 e^x + \frac{3}{2}c_2 e^{-2x}; \, y_2 = c_1 e^x + c_2 e^{-2x}$

12. $y_1 = -c_1 e^x + \frac{3}{2}c_2 e^{-2x}; \, y_2 = c_1 e^x + c_2 e^{-2x}$

13. $y_1 = -c_1 e^x + \frac{3}{2}c_2 e^{-2x} + \frac{11}{100}e^{3x}$

$y_2 = c_1 e^x + c_2 e^{-2x} - \frac{1}{100}e^{3x}$

14. $y_1 = c_1 e^{3x} + c_2 \cos 2x + c_3 \sin 2x$
$y_2 = -1 - 5c_1 e^{3x} + c_3 \cos 2x - c_2 \sin 2x$

15. $y_1 = c_1 e^x - 3c_2 e^{-x} + (c_3 - 2c_4) \cos x + (2c_3 + c_4) \sin x + 4$
$y_2 = c_1 e^x + c_2 e^{-x} + c_3 \cos x + c_4 \sin x + 1$

16. $y_1 = 2c_1 e^{-x} - 2c_2 e^{-5x} - \dfrac{23 \sin 2x + 14 \cos 2x}{145}$

$y_2 = c_1 e^{-x} + c_2 e^{-5x} + \dfrac{26 \sin 2x - 22 \cos 2x}{145}$

17. $y_1 = \sin x + c_1 x + c_2; \, y_2 = -\frac{1}{6}c_1 x^3 - \frac{1}{2}c_2 x^2 + c_3 x + c_4$ **18.** $y_1 = \sin x, \, y_2 = c_1 x + c_2$

19. $y_1 = \sin x, \, y_2 = 0$ **20.** no solution exists   **22.** no solution

23. $y_1 = g(x), \, y_2 = \frac{1}{2}x^2 - g(x) + \int g(x)dx$, where $g(x)$ is an arbitrary function of x

24. no solution   **25.** no solution

26. $y_1 = c_1 e^x + \left(\dfrac{\sqrt{3}}{2}c_3 - \dfrac{1}{2}c_2\right)e^{-x/2} \cos \dfrac{\sqrt{3}}{2}x - \dfrac{1}{2}(c_3 + \sqrt{3}\, c_2)e^{-x/2} \sin \dfrac{\sqrt{3}}{2}x$

$y_2 = c_1 e^x - \dfrac{1}{2}(\sqrt{3}c_3 + c_2)e^{-x/2} \cos \dfrac{\sqrt{3}}{2}x - \dfrac{1}{2}(c_3 - \sqrt{3}c_2)e^{-x/2} \sin \dfrac{\sqrt{3}}{2}x$

$y_3 = c_1 e^x + c_2 e^{-x/2} \cos \dfrac{\sqrt{3}}{2}x + c_3 e^{-x/2} \sin \dfrac{\sqrt{3}}{2}x$

SECTION 9–2

Only a triangularized form is given. The solutions to the systems are the same as the corresponding exercises in Section 9–1.

1. $y_1 - Dy_2 = 0; (D^2 - 1)y_2 = 0$ **2.** $3y_1 - (D + 1)y_2 = 0; (D^2 - 4)y_2 = 0$

3. $y_1 - (D + 5)y_2 = 0; (D^2 + 2D - 17)y_2 = 0$

4. $(D - 4)y_1 - 3y_2 = 0; (D^2 - 4D - 3)y_1 = 0$

5. $y_1 - (D - 2)y_2 = 0; (D^2 - 7D + 14)y_2 = 0$

6. $by_1 - (D - a)y_2 = 0; [D^2 - 2aD + (a^2 - b^2)]y_2 = 0$ **7.** $Dy_1 + Dy_2 = 3; 2Dy_1 = 3 + x$

8. $y_1 - Dy_2 = 1; (D^2 - 1)y_2 = 1$ **9.** $Dy_1 - y_2 = x^2; (D^2 + 2)y_1 = 3x$

10. $y_1 - y_2 = -1; 2Dy_1 = 1$ **11.** $6y_1 + (5D + 1)y_2 = 0; (D^2 + D - 2)y_2 = 0$

12. $6y_1 + (5D + 1)y_2 = 0; (D^2 + D - 2)y_2 = 0$

13. $6y_1 + (5D + 1)y_2 = \frac{1}{2}e^{3x}$; $(D^2 + D - 2)y_2 = -\frac{1}{10}e^{3x}$

14. $(D^2 - D + 4)y_1 + 2y_2 = 2$; $(D - 3)(D^2 + 4)y_1 = 0$

15. $y_1 - (2D^3 - D^2)y_2 = 4$; $(D^4 - 1)y_2 = -1$

16. $-y_1 + (D + 3)y_2 = \sin 2x$; $(D^2 + 6D + 5)y_2 = 2\cos 2x + 2\sin 2x$

17. $(D^2 - 1)y_1 - D^2y_2 = -2\sin x$; $D^2y_1 = -\sin x$ **18.** $(D^2 + 1)y_1 + D^2y_2 = 0$; $y_1 = \sin x$

19. $(D^2 + 1)y_1 + D^2y_2 = 0$; $y_1 + y_2 = \sin x$ **20.** $(D + 1)y_1 + Dy_2 = 2\sin x$; $0y_1 = \cos x$

21. (i) $Dy_1 + Dy_2 = x$ (ii) $Dy_1 - Dy_2 = x$ (iii) $y_1 + y_2 = 1 - \frac{1}{2}x$

$\qquad\qquad y_2 = x - 1$ $\qquad (2D - 1)y_2 = 1 - x$ $\qquad 4y_1 = 1 - 3x$

SECTION 9–3

1. $y_1 = \sin t$; $y_2 = \cos t$ **2.** $y_1 = (\cos t + \sin t)e^t$; $y_2 = (\cos t - \sin t)e^t$

3. $y_1 = e^{-t/2} + t - 1$; $y_2 = e^{-t/2} + t - \frac{1}{2}t^2 - 1$ **4.** $y_1 = \frac{1}{2}t + \frac{1}{4}t^2 - 1$; $y_2 = -\frac{1}{2}t + \frac{1}{4}t^2$

5. $y_1 = \frac{3}{2}e^{-t} + \frac{1}{2}\sin t + \frac{1}{2}\cos t + t - 2$

$\quad y_2 = \frac{3}{2}e^{-t} - \frac{1}{2}\cos t + \frac{1}{2}\sin t + t - 1$

6. $y_1 = \frac{14}{45}e^{3t} - \frac{1}{3}e^t + \frac{2}{15}e^{-2t} - \frac{1}{3}t - \frac{1}{9}$

$\quad y_2 = 1 - \frac{2}{3}e^t - \frac{1}{3}e^{-2t}$

7. $y_1 = 0$; $y_2 = 1$ **8.** $y_1 = \cos t + t - 1$; $y_2 = \sin t + t^2$

9. $y_1 = y_2(0)\sin t + y_1(0)\cos t$

$\quad y_2 = -y_1(0)\sin t + y_2(0)\cos t$

10. $y_1 = \dfrac{y_1(0) - y_2(0)}{2}$; $y_2 = -y_1$

11. no solution unless $y_1(0) = y_2(0)$; then $y_1 = y_2$, and any differentiable function is a solution

12. $y_1 = y_2 = \frac{1}{2}e^{-t}[y_1(0) + y_2(0)]$ **13.** no nontrivial solution

SECTION 9–4

1. $\begin{pmatrix} y_1' \\ y_2' \end{pmatrix} = \begin{pmatrix} 2 & 1 \\ 1 & -1 \end{pmatrix} \begin{pmatrix} y_1 \\ y_2 \end{pmatrix}$ **2.** $\begin{pmatrix} y_1' \\ y_2' \end{pmatrix} = \begin{pmatrix} 3 & 1 \\ 1 & 0 \end{pmatrix} \begin{pmatrix} y_1 \\ y_2 \end{pmatrix}$

3. $\begin{pmatrix} y_1' \\ y_2' \end{pmatrix} = \begin{pmatrix} 1 & 1 \\ 1 & 3 \end{pmatrix} \begin{pmatrix} y_1 \\ y_2 \end{pmatrix} + \begin{pmatrix} t \\ 5 \end{pmatrix}$ **4.** $\begin{pmatrix} y_1' \\ y_2' \end{pmatrix} = \begin{pmatrix} 0 & 0 \\ 1 & -1 \end{pmatrix} \begin{pmatrix} y_1 \\ y_2 \end{pmatrix} + \begin{pmatrix} e^{-t}\sin t \\ e^{-t}\cos t \end{pmatrix}$

5. $\begin{pmatrix} y_1' \\ y_2' \end{pmatrix} = \begin{pmatrix} \frac{3}{4} \\ -\frac{1}{4} \end{pmatrix}$ **6.** $\begin{pmatrix} y_1' \\ y_2' \end{pmatrix} = \frac{1}{3}\begin{pmatrix} t + 2 \\ t - 1 \end{pmatrix}$

7. $\begin{pmatrix} y_1' \\ y_2' \end{pmatrix} = \frac{1}{3}\begin{pmatrix} 1 & -4 \\ 2 & -2 \end{pmatrix} \begin{pmatrix} y_1 \\ y_2 \end{pmatrix}$ **8.** $\begin{pmatrix} y_1' \\ y_2' \end{pmatrix} = \frac{1}{3}\begin{pmatrix} -2 & 2 \\ -2 & -1 \end{pmatrix} \begin{pmatrix} y_1 \\ y_2 \end{pmatrix}$

9. $\begin{pmatrix} y_1' \\ y_2' \\ y_3' \end{pmatrix} = \begin{pmatrix} 1 & -3 & 1 \\ 1 & -5 & 3 \\ 2 & -3 & 0 \end{pmatrix} \begin{pmatrix} y_1 \\ y_2 \\ y_3 \end{pmatrix}$ **10.** $\begin{pmatrix} y_1' \\ y_2' \\ y_3' \end{pmatrix} = \begin{pmatrix} \frac{2}{3} & 0 & \frac{2}{3} \\ \frac{1}{3} & \frac{1}{2} & -\frac{1}{6} \\ -1 & \frac{1}{2} & \frac{1}{2} \end{pmatrix} \begin{pmatrix} y_1 \\ y_2 \\ y_3 \end{pmatrix}$

11. $y_1' = y_1 + 2y_2 + t$
$y_2' = y_2 - t$

12. $y_1' = -y_1 + 2y_2 + \sin t$
$y_2' = 5y_1 + 3y_2$

13. $y_1' = y_1 + 5y_2 - y_3 + t$
$y_2' = 2y_1 + y_3 + t$
$y_3' = y_1$

14. $y_1' = 2y_1 + y_3 + \sin t$
$y_2' = y_1 + 2y_2$
$y_3' = y_2 - \sin t$

21. $Y = c_1 \begin{pmatrix} 2e^{3t} \\ e^{3t} \end{pmatrix} + c_2 \begin{pmatrix} -2e^{-t} \\ e^{-t} \end{pmatrix}$

22. $Y = c_1 \begin{pmatrix} e^{-5t} \\ e^{-5t} \end{pmatrix} + c_2 \begin{pmatrix} -e^t \\ e^t \end{pmatrix}$

23. $Y = c_1 \begin{pmatrix} -4\cos 2t \\ \cos 2t + \sin 2t \end{pmatrix} e^{4t} + c_2 \begin{pmatrix} -4\sin 2t \\ \sin 2t - \cos 2t \end{pmatrix} e^{4t}$

24. $Y = c_1 \begin{pmatrix} 0 \\ 1 \end{pmatrix} e^{2t} + c_2 \begin{pmatrix} \frac{1}{3} \\ t \end{pmatrix} e^{2t}$

25. $Y = c_1 \begin{pmatrix} 0 \\ 1 \\ 0 \end{pmatrix} + c_2 \begin{pmatrix} -\cos t \\ \frac{1}{2}(\cos t - \sin t) \\ \cos t + \sin t \end{pmatrix} + c_3 \begin{pmatrix} -\sin t \\ \frac{1}{2}(\cos t + \sin t) \\ \sin t - \cos t \end{pmatrix}$

26. $Y = c_1 e^{2t} \begin{pmatrix} 0 \\ 1 \\ -1 \end{pmatrix} + c_2 e^t \begin{pmatrix} \cos t \\ \sin t \\ \cos t \end{pmatrix} + c_3 e^t \begin{pmatrix} \sin t \\ -\cos t \\ \sin t \end{pmatrix}$

27. $Y = \begin{pmatrix} 2e^{3t} & -2e^{-t} \\ e^{3t} & e^{-t} \end{pmatrix} \begin{pmatrix} c_1 \\ c_2 \end{pmatrix}$ **28.** $Y = \begin{pmatrix} e^{-5t} & -e^t \\ e^{-5t} & e^t \end{pmatrix} \begin{pmatrix} c_1 \\ c_2 \end{pmatrix}$

29. $Y = \begin{pmatrix} -4e^{4t}\cos 2t & -4e^{4t}\sin 2t \\ e^{4t}(\cos 2t + \sin 2t) & e^{4t}(\sin 2t - \cos 2t) \end{pmatrix} \begin{pmatrix} c_1 \\ c_2 \end{pmatrix}$

30. $Y = \begin{pmatrix} 0 & \frac{1}{3}e^{2t} \\ e^{2t} & te^{2t} \end{pmatrix} \begin{pmatrix} c_1 \\ c_2 \end{pmatrix}$

31. $Y = \begin{pmatrix} 0 & -\cos t & -\sin t \\ 1 & \frac{1}{2}(\cos t - \sin t) & \frac{1}{2}(\cos t + \sin t) \\ 0 & \cos t + \sin t & \sin t - \cos t \end{pmatrix} \begin{pmatrix} c_1 \\ c_2 \\ c_3 \end{pmatrix}$

32. $Y = \begin{pmatrix} 0 & e^t \cos t & e^t \sin t \\ e^{2t} & e^t \sin t & -e^t \cos t \\ -e^{2t} & e^t \cos t & e^t \sin t \end{pmatrix} \begin{pmatrix} c_1 \\ c_2 \\ c_3 \end{pmatrix}$

33. $Y = \begin{pmatrix} \dfrac{-e^{3t}}{2} + \dfrac{3e^{-t}}{2} \\ \dfrac{-e^{3t}}{4} - \dfrac{3e^{-t}}{4} \end{pmatrix}$ **34.** $Y = \begin{pmatrix} e^t \\ -e^t \end{pmatrix}$

35. $Y = \begin{pmatrix} e^{4t}(\cos 2t - 3\sin 2t) \\ e^{4t}(-\cos 2t + \frac{1}{2}\sin 2t) \end{pmatrix}$ **36.** $Y = \begin{pmatrix} e^{2t} \\ 3te^{2t} - e^{2t} \end{pmatrix}$

37. $Y = \begin{pmatrix} 0 \\ 1 \\ 0 \end{pmatrix}$ **38.** $Y = -e^t \begin{pmatrix} \sin t \\ -\cos t \\ \sin t \end{pmatrix}$

SECTION 9–5

1. $\mathbf{Y} = c_1 \begin{pmatrix} -1 \\ 1 \end{pmatrix} e^{-t} + c_2 \begin{pmatrix} 1 \\ 1 \end{pmatrix} e^{5t}$

2. $\mathbf{Y} = c_1 \begin{pmatrix} -2 \\ 1 \end{pmatrix} e^{-t} + c_2 \begin{pmatrix} 2 \\ 1 \end{pmatrix} e^{3t}$

3. $\mathbf{Y} = c_1 \begin{pmatrix} 1 \\ (-3 + \sqrt{13})/2 \end{pmatrix} e^{(3 + \sqrt{13})t/2} + c_2 \begin{pmatrix} 1 \\ (-3 - \sqrt{13})/2 \end{pmatrix} e^{(3 - \sqrt{13})t/2}$

4. $\mathbf{Y} = c_1 \begin{pmatrix} 1 \\ (-1 + \sqrt{5})/2 \end{pmatrix} e^{(3 + \sqrt{5})t/2} + c_2 \begin{pmatrix} 1 \\ (-1 - \sqrt{5})/2 \end{pmatrix} e^{(3 - \sqrt{5})t/2}$

5. $\mathbf{Y} = c_1 e^t \begin{pmatrix} 1 \\ -1 \end{pmatrix} + c_2 e^{2t} \begin{pmatrix} 2 \\ -1 \end{pmatrix}$

6. $\mathbf{Y} = c_1 \begin{pmatrix} 1 \\ -2 \end{pmatrix} e^{2t/3} + c_2 \begin{pmatrix} 1 \\ 0 \end{pmatrix}$

7. $\mathbf{Y} = c_1 \begin{pmatrix} 1 \\ -\sqrt{3} \end{pmatrix} e^{(-1 + \sqrt{3})t/2} + c_2 \begin{pmatrix} 1 \\ \sqrt{3} \end{pmatrix} e^{(-1 - \sqrt{3})t/2}$

8. $\mathbf{Y} = c_1 e^t \begin{pmatrix} 0 \\ 1 \\ 0 \end{pmatrix} + c_2 e^{2t} \begin{pmatrix} -1 \\ 2 \\ 2 \end{pmatrix} + c_3 e^{3t} \begin{pmatrix} -1 \\ 1 \\ 1 \end{pmatrix}$

9. $\mathbf{Y} = c_1 e^t \begin{pmatrix} 1 \\ 0 \\ 0 \end{pmatrix} + c_2 e^{2t} \begin{pmatrix} 2 \\ 3 \\ 1 \end{pmatrix} + c_3 e^{-t} \begin{pmatrix} 1 \\ 0 \\ 2 \end{pmatrix}$

10. $\mathbf{Y} = c_1 \begin{pmatrix} -5 \\ 2 \\ 3 \end{pmatrix} + c_2 e^{5t} \begin{pmatrix} 0 \\ 1 \\ -1 \end{pmatrix} + c_3 e^{-3t} \begin{pmatrix} -8 \\ 5 \\ 27 \end{pmatrix}$

11. $\mathbf{Y} = \begin{pmatrix} 1 \\ 1 \end{pmatrix} e^{3t}$ **12.** $\mathbf{Y} = \frac{1}{3} \begin{pmatrix} -2 \\ 1 \end{pmatrix} e^{-4t} - \frac{1}{3} \begin{pmatrix} 1 \\ 1 \end{pmatrix} e^{5t}$

13. $\mathbf{Y} = -\frac{5}{6} \begin{pmatrix} 1 \\ 0 \\ 0 \end{pmatrix} e^t - \frac{1}{2} \begin{pmatrix} -5 \\ -2 \\ 2 \end{pmatrix} e^{-t} + \frac{1}{3} \begin{pmatrix} -2 \\ 0 \\ 3 \end{pmatrix} e^{-2t}$ **14.** $\mathbf{Y} = \frac{3}{8} \begin{pmatrix} -2 \\ -1 \\ 1 \end{pmatrix} - \frac{1}{8} \begin{pmatrix} 2 \\ -3 \\ 3 \end{pmatrix} e^{8t}$

15. $\mathbf{Y} = c_1 \begin{pmatrix} 1 \\ 1 \end{pmatrix} + c_2 \begin{pmatrix} V_2 \\ -V_1 \end{pmatrix} e^{k(1/V_1 + 1/V_2)t}$

16. $\mathbf{P} = c_1 \begin{pmatrix} 1 \\ 1 \end{pmatrix} e^{-2t} + c_2 \begin{pmatrix} 5 \\ 2 \end{pmatrix} e^t$. As $t \to \infty$, $\mathbf{P} \to c_2 \begin{pmatrix} 5 \\ 2 \end{pmatrix} e^t$

SECTION 9–6

1. $\mathbf{Y} = c_1 \begin{pmatrix} -4 \cos 2t \\ \cos 2t + \sin 2t \end{pmatrix} e^{4t} + c_2 \begin{pmatrix} -4 \sin 2t \\ \sin 2t - \cos 2t \end{pmatrix} e^{4t}$

2. $\mathbf{Y} = c_1 \begin{pmatrix} \cos 3t \\ 3 \sin 3t \end{pmatrix} e^{2t} + c_2 \begin{pmatrix} \sin 3t \\ -3 \cos 3t \end{pmatrix} e^{2t}$

3. $Y = c_1 \begin{pmatrix} 3 \\ 1 \end{pmatrix} e^{-3t} + c_2 \left[\begin{pmatrix} \frac{1}{2} \\ 0 \end{pmatrix} + t \begin{pmatrix} 3 \\ 1 \end{pmatrix} \right] e^{-3t}$

4. $Y = c_1 \begin{pmatrix} 1 \\ -2 \end{pmatrix} e^{4t} + c_2 \left[\begin{pmatrix} -1 \\ 1 \end{pmatrix} + t \begin{pmatrix} 1 \\ -2 \end{pmatrix} \right] e^{4t}$

5. $Y = c_1 \begin{pmatrix} \cos t \\ -\frac{1}{2} \sin t \end{pmatrix} e^t + c_2 \begin{pmatrix} \sin t \\ \frac{1}{2} \cos t \end{pmatrix} e^t$

6. $Y = c_1 \begin{pmatrix} 0 \\ 1 \\ 0 \end{pmatrix} e^t + c_2 \begin{pmatrix} -\cos t \\ \frac{1}{2}(\cos t - \sin t) \\ \cos t + \sin t \end{pmatrix} + c_3 \begin{pmatrix} -\sin t \\ \frac{1}{2}(\cos t + \sin t) \\ \sin t - \cos t \end{pmatrix}$

7. $Y = c_1 \begin{pmatrix} 0 \\ 1 \\ -1 \end{pmatrix} e^{2t} + c_2 e^{2t} \left[\begin{pmatrix} 5 \\ -2 \\ 5 \end{pmatrix} \cos t - \begin{pmatrix} 0 \\ -1 \\ 0 \end{pmatrix} \sin t \right] + c_3 e^{2t} \left[\begin{pmatrix} 0 \\ -1 \\ 0 \end{pmatrix} \cos t + \begin{pmatrix} 5 \\ -2 \\ 5 \end{pmatrix} \sin t \right]$

8. $Y = c_1 \begin{pmatrix} -1 \\ 2 \\ 2 \end{pmatrix} e^{-3t} + c_2 \begin{pmatrix} 2 \\ 1 \\ 0 \end{pmatrix} e^{6t} + c_3 \begin{pmatrix} 2 \\ 0 \\ 1 \end{pmatrix} e^{6t}$

9. $Y = c_1 \begin{pmatrix} 1 \\ 1 \\ -1 \end{pmatrix} + c_2 \begin{pmatrix} 0 \\ 1 \\ 0 \end{pmatrix} e^{-2t} + c_3 \left[\begin{pmatrix} 0 \\ 1 \\ 0 \end{pmatrix} t + \frac{1}{8} \begin{pmatrix} -3 \\ 0 \\ 1 \end{pmatrix} \right] e^{-2t}$

10. $Y = c_1 \begin{pmatrix} 1 \\ 1 \\ 1 \end{pmatrix} e^t + e^{-t/2} \left[c_2 \begin{pmatrix} 2\cos \frac{\sqrt{3}}{2} t \\ -\cos \frac{\sqrt{3}}{2} t - \sqrt{3} \sin \frac{\sqrt{3}}{2} t \\ -\cos \frac{\sqrt{3}}{2} t + \sqrt{3} \sin \frac{\sqrt{3}}{2} t \end{pmatrix} + c_3 \begin{pmatrix} 2 \sin \frac{\sqrt{3}}{2} t \\ \sqrt{3} \cos \frac{\sqrt{3}}{2} t - \sin \frac{\sqrt{3}}{2} t \\ -\sqrt{3} \cos \frac{\sqrt{3}}{2} t - \sin \frac{\sqrt{3}}{2} t \end{pmatrix} \right]$

11. $Y' = \begin{pmatrix} -\frac{3}{2} & -\frac{5}{2} \\ \frac{5}{2} & \frac{3}{2} \end{pmatrix} Y; \quad Y = c_1 \begin{pmatrix} -5 \cos 2t \\ 3 \cos 2t - 4 \sin 2t \end{pmatrix} + c_2 \begin{pmatrix} -5 \sin 2t \\ 4 \cos 2t + 3 \sin 2t \end{pmatrix}$

12. $Y' = \begin{pmatrix} -3 & -5 \\ 1 & 1 \end{pmatrix} Y; Y = c_1 e^{-t} \left[\begin{pmatrix} -2 \\ 1 \end{pmatrix} \cos t - \begin{pmatrix} 1 \\ 0 \end{pmatrix} \sin t \right] + c_2 e^{-t} \left[\begin{pmatrix} 1 \\ 0 \end{pmatrix} \cos t + \begin{pmatrix} -2 \\ 1 \end{pmatrix} \sin t \right]$

13. $Y' = \begin{pmatrix} -3 & -4 \\ 1 & 1 \end{pmatrix} Y; Y = c_1 \begin{pmatrix} -2 \\ 1 \end{pmatrix} e^{-t} + c_2 \left[\begin{pmatrix} -1 \\ 1 \end{pmatrix} + t \begin{pmatrix} -2 \\ 1 \end{pmatrix} \right] e^{-t}$

14. $Y' = \begin{pmatrix} 1 & -3 & 1 \\ 2 & -1 & -2 \\ 2 & -3 & 0 \end{pmatrix} Y; Y = c_1 e^{2t} \begin{pmatrix} 1 \\ 0 \\ 1 \end{pmatrix} + c_2 e^{-t} \begin{pmatrix} 1 \\ 1 \\ 1 \end{pmatrix} + c_3 e^{-t} \left[\begin{pmatrix} 1 \\ 1 \\ 1 \end{pmatrix} t + \begin{pmatrix} \frac{1}{2} \\ 0 \\ 0 \end{pmatrix} \right]$

15. $Y = \begin{pmatrix} 1 \\ 3 \\ 1 \end{pmatrix} e^t - \begin{pmatrix} 1 \\ 2 \\ 2 \end{pmatrix} te^t$

16. $Y = 2 \begin{pmatrix} 0 \\ 0 \\ 1 \end{pmatrix} e^{-t} - 2 \left[\begin{pmatrix} 1 \\ 0 \\ 1 \end{pmatrix} \cos 2t - \begin{pmatrix} 0 \\ \frac{1}{2} \\ 0 \end{pmatrix} \sin 2t \right]$

SECTION 9–7

1. $Y = c_1 \begin{pmatrix} -2 \\ 1 \end{pmatrix} e^{-t} + c_2 \begin{pmatrix} 2 \\ 1 \end{pmatrix} e^{3t} + \begin{pmatrix} 5 \\ 1 \end{pmatrix}$

2. $Y = c_1 \begin{pmatrix} -2 \\ 1 \end{pmatrix} e^{-t} + c_2 \begin{pmatrix} 2 \\ 1 \end{pmatrix} e^{3t} + \begin{pmatrix} 0 \\ -1 \end{pmatrix} e^{t}$

3. $Y = c_1 \begin{pmatrix} -2 \\ 1 \end{pmatrix} e^{-t} + c_2 \begin{pmatrix} 2 \\ 1 \end{pmatrix} e^{3t} + \begin{pmatrix} -\frac{1}{2} - t \\ \frac{1}{2}t \end{pmatrix} e^{-t}$

4. $Y = c_1 \begin{pmatrix} 0 \\ 1 \\ 0 \end{pmatrix} e^{t} + c_2 \begin{pmatrix} -\cos t \\ \frac{1}{2}(\cos t - \sin t) \\ \cos t + \sin t \end{pmatrix} + c_3 \begin{pmatrix} -\sin t \\ \frac{1}{2}(\cos t + \sin t) \\ \sin t - \cos t \end{pmatrix} + \begin{pmatrix} 2 \\ -3 \\ -4 \end{pmatrix}$

5. $Y = $ (same Y_c as Exercise 4) $+ \begin{pmatrix} 0 \\ 0 \\ -e^{-t} \end{pmatrix}$

6. $Y = $ (same Y_c as Exercise 4) $+ \begin{pmatrix} -e^{t} \\ te^{t} \\ 0 \end{pmatrix}$

7. $Y = c_1 \begin{pmatrix} 1 \\ 3 \end{pmatrix} e^{5t} + c_2 \begin{pmatrix} 1 \\ 1 \end{pmatrix} e^{3t} + \begin{pmatrix} \frac{1}{2} \\ \frac{1}{2} \end{pmatrix} e^{5t}$

8. $Y = c_1 \begin{pmatrix} 1 \\ -1 \end{pmatrix} e^{t} + c_2 \begin{pmatrix} 1 \\ 3 \end{pmatrix} e^{5t} + \begin{pmatrix} \frac{1}{2} \\ \frac{3}{2} \end{pmatrix} e^{t} + \begin{pmatrix} 1 \\ -1 \end{pmatrix} te^{t}$

9. $Y = c_1 \begin{pmatrix} 2 \sin 2t \\ \cos 2t \end{pmatrix} + c_2 \begin{pmatrix} 2 \cos 2t \\ -\sin 2t \end{pmatrix}$

$\qquad + \begin{pmatrix} 2 \sin 2t \cos^3 t - 3 \cos 2t \sin t + 2 \cos 2t \sin^3 t \\ \cos 2t \cos^3 t + \frac{3}{2} \sin t \sin 2t - \sin 2t \sin^3 t \end{pmatrix}$ NOTE: The last term reduces to $\begin{pmatrix} \sin t \\ \cos t \end{pmatrix}$.

10. $Y = c_1 \begin{pmatrix} 1 \\ -1 \end{pmatrix} + c_2 \begin{pmatrix} 1 \\ -\frac{3}{2} \end{pmatrix} e^{-t} + \begin{pmatrix} 2 \\ -2 \end{pmatrix} t^2 + \begin{pmatrix} -1 \\ 3 \end{pmatrix} t + \begin{pmatrix} 2 \\ -3 \end{pmatrix}$

11. $Y = c_1 \begin{pmatrix} -1 \\ 0 \\ 1 \end{pmatrix} + c_2 \begin{pmatrix} -1 \\ 2 \\ 0 \end{pmatrix} e^{-t} + c_3 \begin{pmatrix} -1 \\ 0 \\ 2 \end{pmatrix} e^{-t} + \begin{pmatrix} 2 \\ 0 \\ -2 \end{pmatrix} t + \begin{pmatrix} -1 \\ 0 \\ 2 \end{pmatrix} + \begin{pmatrix} -2 \\ 0 \\ 2 \end{pmatrix} e^{-t} + \begin{pmatrix} -2 \\ 0 \\ 4 \end{pmatrix} te^{-t}$

12. $Y = \begin{pmatrix} -\cos t - \sin t + 1 \\ \sin t - \cos t + 1 \end{pmatrix}$

13. $Y = \begin{pmatrix} -1 \\ -1 \end{pmatrix} e^{t} + \begin{pmatrix} 1 \\ 2 \end{pmatrix} e^{2t} + \begin{pmatrix} 0 \\ -1 \end{pmatrix} e^{3t}$

14. $Y = e^{-t} \begin{pmatrix} \frac{1}{2} \cos t - \frac{1}{2} \sin t \\ -\cos t \end{pmatrix} + \begin{pmatrix} -\frac{1}{2} \\ 1 \end{pmatrix}$

15. $Y = \begin{pmatrix} 1 \\ 1 \\ 1 \\ 1 \end{pmatrix} e^{t} + \begin{pmatrix} 0 \\ -1 \\ 0 \\ 0 \end{pmatrix} t^2 + \begin{pmatrix} 0 \\ 0 \\ -2 \\ 0 \end{pmatrix} t + \begin{pmatrix} 0 \\ 0 \\ 0 \\ -2 \end{pmatrix}$

16. Fundamental matrix: $\begin{pmatrix} (k + \alpha)e^{\alpha t} & (k + \beta)e^{\beta t} \\ ke^{\alpha t} & ke^{\beta t} \end{pmatrix}$, where $\alpha, \beta = \dfrac{-2k - h \pm \sqrt{4k^2 + h^2}}{2}$

17. $C = c_1 \begin{pmatrix} k + \alpha \\ k \end{pmatrix} e^{\alpha t} + c_2 \begin{pmatrix} k + \beta \\ k \end{pmatrix} e^{\beta t} + \dfrac{kha}{\alpha\beta}\begin{pmatrix} 1 \\ 1 \end{pmatrix}$

SECTION 9–8

1. $(L_1 D + R + R_1)i_1 + i_2 R = v(t)$
$i_1 R + (L_2 D + R)i_2 = v(t)$

2. $i_1 = \dfrac{8}{\sqrt{10}} e^{-4t} \sinh \sqrt{10}t + 4 - 4e^{-4t} \cosh \sqrt{10}t$

$i_2 = 4 - 4e^{-4t} \cosh \sqrt{10}t - \dfrac{4}{\sqrt{10}}e^{-4t} \sinh \sqrt{10}t$

3. At $t = 0$, $i_1 = 2$, $i_2 = \dfrac{4}{3}$; after $t = 0$, $i_1 = 4 - 2e^{-11t/2} \cosh \dfrac{\sqrt{13}}{2}t + \dfrac{2}{\sqrt{13}}e^{-11t/2} \sinh \dfrac{\sqrt{13}}{2}t$;

$i_2 = \dfrac{2}{3}\left(4 - 2e^{-11t/2} \cosh \dfrac{\sqrt{13}}{2}t - \dfrac{4}{\sqrt{13}}e^{-11t/2} \sinh \dfrac{\sqrt{13}}{2}t\right)$

4. $i_1(R + R_1) - i_2 R_1 = v(t)$ **5.** $i_1 = 3(1 - e^{-2t})$; $i_2 = 6e^{-2t}$
$i_1 R_1 = i_2 R_1 + \dfrac{1}{C}\displaystyle\int_0^t i_2(t^*)\, dt^*$

6. $x_1 = \dfrac{2}{5\sqrt{6}} \sin \sqrt{6}t + \dfrac{3}{5} \sin t$ **7.** $x_1 = \dfrac{3}{10} \cos \sqrt{12}t + \dfrac{7}{10} \cos \sqrt{2}t$

$x_2 = \dfrac{-1}{5\sqrt{6}} \sin \sqrt{6}t + \dfrac{6}{5} \sin t$ $x_2 = -\dfrac{1}{10} \cos \sqrt{12}t + \dfrac{21}{10} \cos \sqrt{2}t$

8. $x_1 = \dfrac{3}{10} \cos \sqrt{12}t + \dfrac{7}{10} \cos \sqrt{2}t - \dfrac{9}{10\sqrt{12}} \sin \sqrt{12}t - \dfrac{11}{10\sqrt{2}} \sin \sqrt{2}t$

$x_2 = -\dfrac{1}{10} \cos \sqrt{12}t + \dfrac{21}{10} \cos \sqrt{2}t + \dfrac{3}{10\sqrt{12}} \sin \sqrt{12}t - \dfrac{33}{10\sqrt{2}} \sin \sqrt{2}t$

10. $x_1 = \cos \sqrt{3}t$, $x_2 = -x_1$ **11.** $y_1 = 10e^{-7t/100} \cosh (\sqrt{14}/100)t$

$y_2 = \dfrac{70}{\sqrt{14}}e^{-7t/100} \sinh (\sqrt{14}/100)t$

12. $y_1 = 200(1 - e^{-3t/100})$

$y_2 = 200(1 - e^{-3t/100}) - 6te^{-3t/100}$

13. $y_1 = 200\left(1 - e^{-7t/100} \cosh \dfrac{\sqrt{14}}{100}t\right) - \dfrac{200\sqrt{14}}{7}e^{-7t/100} \sinh \dfrac{\sqrt{14}}{100}t$

$y_2 = 200\left(1 - e^{-7t/100} \cosh \dfrac{\sqrt{14}}{100}t\right) - \dfrac{200 \cdot 7}{\sqrt{14}}e^{-7t/100} \sinh \dfrac{\sqrt{14}}{100}t$

Chapter 9 Review

1. $y_1 = e^{x/2}(c_1 e^{\sqrt{13}x/2} + c_2 e^{-\sqrt{13}x/2})$

$y_2 = e^{x/2}\left(\dfrac{\sqrt{13} - 3}{2} c_1 e^{\sqrt{13}x/2} - \dfrac{\sqrt{13} + 3}{2} c_2 e^{-\sqrt{13}x/2}\right)$

2. $y_1 = c_1 e^{(5 + \sqrt{10})x} + c_2 e^{(5 - \sqrt{10})x}$

$y_2 = \dfrac{-4 - \sqrt{10}}{3} c_1 e^{(5 + \sqrt{10})x} + \dfrac{-4 + \sqrt{10}}{3} c_2 e^{(5 - \sqrt{10})x}$

3. one   **4.** four   **5.** three

6. $y_1 = c_1 e^{(1 + \sqrt{3})x} + c_2 e^{(1 - \sqrt{3})x} + \frac{1}{2}x - \frac{1}{2}$

$y_2 = (1 + \sqrt{3})c_1 e^{(1 + \sqrt{3})x} + (1 - \sqrt{3})c_2 e^{(1 - \sqrt{3})x} + \frac{1}{2}$

7. $y_1 = \frac{2}{3}x + \frac{2}{3}x^2 + c_1$

$y_2 = \frac{2}{3}x - \frac{1}{6}x^2 + c_2$

8. $y_1 = \frac{1}{3}e^{-x}(2c_1 - \sqrt{2}c_2)\sin\sqrt{2}x + \frac{1}{3}e^{-x}(2c_2 + \sqrt{2}c_1)\cos\sqrt{2}x + \frac{1}{3}x + \frac{1}{9}$

$y_2 = e^{-x}(c_1 \sin\sqrt{2}x + c_2 \cos\sqrt{2}x) + \frac{2}{3}x - \frac{1}{9}$

9. $y_1 = -2e^x + 2x, \; y_2 = e^x - 2x + x^2$ **10.** $y_1 = c_1 \cos x + c_2 \sin x + \frac{1}{2}, \; y_2 = \frac{1}{2}$

11. $y_1 = y_2 = 1$ **12.** $y_1 = 1 + \frac{1}{5}t^2 + \frac{1}{5}t; \; y_2 = 1 - \frac{1}{5}t + \frac{3}{10}t^2$ **13.** $y_1 = y_2 = \cos t + \sin t$

14. $y_1 = y_2 = (e^{t-1} - t)u_1(t) + e^t$

15. $y_1 = (-\frac{1}{9} - \frac{1}{3}(t - 1) + \frac{1}{9}e^{3(t-1)})u_1(t)$

$y_2 = -y_1$

16. $y_1 = e^{-t}[c_1 \cos t + c_2 \sin t]; \; y_2 = e^{-t}[-c_2 \cos t + c_1 \sin t]; \; c_1 = y_1(0), c_2 = -y_2(0)$

17. $y_1 = \frac{2}{3}ce^{-2t/3}; \; y_2 = \frac{c}{3}e^{-2t/3},$ where $c = y_1(0) + y_2(0)$

18. $i_1 = \dfrac{10}{17}\sin t - \dfrac{6}{17}\cos t + e^{-2t}\left(\dfrac{7}{17}\cosh\sqrt{2}t + \dfrac{15}{17\sqrt{2}}\sinh\sqrt{2}t\right)$

$i_2 = \frac{4}{17}\sin t - \frac{1}{17}\cos t + c_1 e^{-t} + c_2 e^{-4t}$

19. $x_1 = \frac{1}{2}\cos 2t + \frac{1}{2}\cos 2\sqrt{5}t$

$x_2 = \frac{1}{2}\cos 2t - \frac{1}{2}\cos 2\sqrt{5}t$

20. $i_1 = \frac{1}{3}e^t + \frac{2}{3}e^{-t/2}\cos\dfrac{\sqrt{3}}{2}t$

$i_2 = \frac{1}{3}e^t + e^{-t/2}\left(-\frac{1}{3}\cos\dfrac{\sqrt{3}}{2}t - \dfrac{1}{\sqrt{3}}\sin\dfrac{\sqrt{3}}{2}t\right)$

$i_3 = \frac{1}{3}e^t + e^{-t/2}\left(-\frac{1}{3}\cos\dfrac{\sqrt{3}}{2}t + \dfrac{1}{\sqrt{3}}\sin\dfrac{\sqrt{3}}{2}t\right)$

21. $y_1 = \dfrac{10}{\sqrt{3}}(1 + \sqrt{3})e^{(-3 + \sqrt{3})t/20} + \dfrac{10}{\sqrt{3}}(\sqrt{3} - 1)e^{(-3 - \sqrt{3})t/20}$

$y_2 = 10(1 + \sqrt{3})e^{(-3 + \sqrt{3})t/20} + 10(1 - \sqrt{3})e^{(-3 - \sqrt{3})t/20}$

Index

TRIGONOMETRY IDENTITIES

1. $\sin A = \dfrac{1}{\csc A}$

2. $\cos A = \dfrac{1}{\sec A}$

3. $\tan A = \dfrac{1}{\cot A}$

4. $\tan A = \dfrac{\sin A}{\cos A}$

5. $\cot A = \dfrac{\cos A}{\sin A}$

6. $\sin^2 A + \cos^2 A = 1$

7. $1 + \tan^2 A = \sec^2 A$

8. $1 + \cot^2 A = \csc^2 A$

9. $\sin(A + B) = \sin A \cos B + \cos A \sin B$

10. $\sin(A - B) = \sin A \cos B - \cos A \sin B$

11. $\cos (A + B) = \cos A \cos B - \sin A \sin B$

12. $\cos (A \to B) = \cos A \cos B + \sin A \sin B$

13. $\tan(A + B) = \dfrac{\tan A + \tan B}{1 - \tan A \tan B}$

14. $\tan(A - B) = \dfrac{\tan A - \tan B}{1 + \tan A \tan B}$

15. $\sin 2 A = 2 \sin A \cos A$

16. $\cos 2 A = \cos^2 A - \sin^2 A = 2 \cos^2 A - 1 = 1 - 2 \sin^2 A$

17. $\tan 2A = \dfrac{2 \tan A}{1 - \tan^2 A}$

18. $\sin \frac{1}{2} A = \pm \sqrt{(1 - \cos A)/2}$

19. $\cos \frac{1}{2} A = \pm \sqrt{(1 + \cos A)/2}$

20. $\tan \frac{1}{2} A = \dfrac{\sin A}{1 + \cos A}$

21. $\sin A + \sin B = 2 \sin \frac{1}{2}(A + B)\cos \frac{1}{2}(A - B)$

22. $\sin A - \sin B = 2 \cos \frac{1}{2}(A + B)\sin \frac{1}{2}(A - B)$

23. $\cos A + \cos B = 2 \cos \frac{1}{2}(A + B)\cos \frac{1}{2}(A - B)$

24. $\cos A - \cos B = -2 \sin \frac{1}{2}(A + B)\sin \frac{1}{2}(A - B)$

25. $\sin A \cos B = \frac{1}{2}\{\sin(A + B) + \sin(A - B)\}$

26. $\cos A \sin B = \frac{1}{2}\{\sin(A + B) - \sin(A - B)\}$

27. $\cos A \cos B = \frac{1}{2}\{\cos(A + B) + \cos(A - B)\}$

28. $\sin A \sin B = \frac{1}{2}\{\cos(A - B) - \cos(A + B)\}$